Springer Aerospace Technology

The *Springer Aerospace Technology* series is devoted to the technology of aircraft and spacecraft including design, construction, control and the science. The books present the fundamentals and applications in all fields related to aerospace engineering. The topics include aircraft, missiles, space vehicles, aircraft engines, propulsion units and related subjects.

More information about this series at http://www.springer.com/series/8613

Bernd Chudoba

Stability and Control of Conventional and Unconventional Aerospace Vehicle Configurations

A Generic Approach from Subsonic to Hypersonic Speeds

Springer

Bernd Chudoba
Department of Mechanical and Aerospace Engineering
The University of Texas at Arlington
Arlington, TX, USA

ISSN 1869-1730 ISSN 1869-1749 (electronic)
Springer Aerospace Technology
ISBN 978-3-030-16855-1 ISBN 978-3-030-16856-8 (eBook)
https://doi.org/10.1007/978-3-030-16856-8

This Springer imprint is published by the registered company Springer Nature Switzerland AG
The registered company address is: Gewerbestrasse 11, 6330 Cham, Switzerland

Don’t let your preoccupation with reality
stifle your imagination.
Robert A. Cassanova and Sharon M. Garrison

Acknowledgements

A number of people cooperated in making this work possible, and I would like to acknowledge their contributions.

The presented methodology concept was developed from 1995 to 1999 at Cranfield University, England, as part of a research contract with DaimlerChrysler Aerospace Airbus GmbH under contract number EZ Future Projects 80995517. The research contract was formally funded by the European Supersonic Commercial Transport (ESCT) project with Dr. Josef Mertens serving as technical monitor for the first two years. The European trilateral technical cooperation had been established by the ESCT project managers Detlef Reimers (DaimlerChrysler Aerospace Airbus), Phil Green (British Aerospace Airbus) and Michèle Pacull (Aérospatiale). In retrospect, the following lists some of the specialists involved: Ulf Graeber, Burkhard Kiekebusch, Dirk von Reith and Dr. Alexander Van der Velden (Synaps Inc.) from DaimlerChrysler Aerospace Airbus; Les Hyde, Dr. Clyde Warsop and Alan Perry from British Aerospace Airbus; Elie Khaski and Joseph Irvoas from Aérospatiale Aéronautique Airbus, just to mention some.

The views and conclusions contained in this book, however, are those of the author and should not be interpreted as necessarily representing the official policies or endorsements, either expressed or implied, of DaimlerChrysler Aerospace Airbus or any other company.

I am especially grateful for the joint effort of Mike Cook and Dr. Howard Smith at Cranfield University. They knew when to applaud my progress and when to demand more. Mike Cook's intimate understanding of flight mechanics, his devoted ability of being a teacher for academic and technical issues are clearly an everlasting experience. Howard Smith's knowledge of aerospace vehicle design, in particular the computational side, proved to be invaluable during the method planning phase. I am thankful for their unbiased technical, academic and personal support throughout the entire research period.

The author gratefully acknowledges the dedicated skill and expertise from the following individuals, who endured without any hesitation in intensifying the author's fascination for aerospace science. I have been fortunate to receive their

h'_x, h'_y, h'_z	Scalar components of $\vec{h}'$ in F_B
i	Aerodynamic control effector variable-incidence stabiliser angle (trimmable CE)
$\vec{i}, \vec{j}, \vec{k}$	Unit vectors
I_B	Inertia matrix
I_x, I_y, I_z	Moments of inertia
I_{xy}, I_{yz}, I_{xz}	Products of inertia
I_L^*, I_M^*, I_N^*	Inertia coupling terms
$I_L^{**}, I_M^{**}, I_N^{**}$	Inertia coupling terms
$I_L^{***}, I_M^{***}, I_N^{***}$	Inertia coupling terms
K	Generalised control system gain
K_v, K_w	Attitude feedback gains
K_p, K_q, K_r	Rate feedback gains
l	Length, moment arm
L	Lift (aircraft)
L, M, N	Scalar components of $\vec{G}$ in F_B, thrust moments
L/D	Aerodynamic efficiency
L_β	Variation of aircraft rolling moment with angle of sideslip
L_{EB}	Matrix of the direction cosines
L_p	Variation of rolling moment with roll rate
L_q	Variation of rolling moment with pitch rate
L_r	Variation of rolling moment with yaw rate
L_u	Variation of rolling moment with axial velocity
L_v	Variation of rolling moment with lateral velocity
L_w	Variation of rolling moment with normal velocity
ΔL_{CE}	Sum of rolling control moments
$L_{\delta_{LoCE}}$	Rolling control moment due to LoCE deflection
$L_{\delta_{DiCE}}$	Rolling control moment due to DiCE deflection
$L_{\delta_{LaCE}}$	Rolling control moment due to LaCE deflection
L_τ	Rolling control moment due to thrust controls
m	Mass
M	Mach number
M_α	Variation of aircraft pitching moment with angle of attack
M_c	Design cruising Mach number
M_D	Design dive Mach number
M_{DF}	Demonstrated flight diving Mach number
M_p	Variation of pitching moment with roll rate
M_q	Variation of pitching moment with pitch rate
M_r	Variation of pitching moment with yaw rate
M_u	Variation of pitching moment with axial velocity
M_v	Variation of pitching moment with lateral velocity
M_w	Variation of pitching moment with normal velocity
$M_{\dot{w}}$	Variation of pitching moment with rate of change of angle of attack

ΔM_D	Pitching moment increment due to engine failure
ΔM_{CE}	Sum of pitching control moments
$M_{\delta_{LoCE}}$	Pitching control moment due to LoCE deflection
$M_{\delta_{DiCE}}$	Pitching control moment due to DiCE deflection
$M_{\delta_{LaCE}}$	Pitching control moment due to LaCE deflection
M_τ	Pitching control moment due to thrust controls
n	Load factor
N_n	Net normal force
N_p	Variation of yawing moment with roll rate
N_q	Variation of yawing moment with pitch rate
N_r	Variation of yawing moment with yaw rate
N_u	Variation of yawing moment with axial velocity
N_v	Variation of yawing moment with lateral velocity
N_w	Variation of yawing moment with normal velocity
ΔN_D	Yawing moment increment due to engine failure
ΔN_{CE}	Sum of yawing control moments
$N_{\delta_{LoCE}}$	Yawing control moment due to LoCE deflection
$N_{\delta_{DiCE}}$	Yawing control moment due to DiCE deflection
$N_{\delta_{LaCE}}$	Yawing control moment due to LaCE deflection
N_τ	Yawing control moment due to thrust controls
p, q, r	Scalar components of $\vec{\omega}$ in F_B
$\dot{p}, \dot{q}, \dot{r}$	Scalar components of $\dot{\vec{\omega}}$ in F_B, rate of change of aircraft angular velocity components
P	Power
$\bar{q}$	Aircraft dynamic pressure
$\vec{r}_{c.g._B}$	Position vector of *dm* in the frame F_B
$\vec{r}_{c.g._E}$	Position vector of *dm* in the frame F_E
R	Turning radius
S	Area; wing reference area
t	Time
$\vec{T}$	Thrust vector; time constant
T_x, T_y, T_z	Scalar components of $\vec{T}$
$T_{1/2}$	Time to half amplitude
T_2	Time to double amplitude
u, v, w	Scalar components of $\vec{V}$ in F_B, perturbed values of *U, V and W*
U, V, W	Scalar velocity components of $\vec{V}$
$\vec{v}_E$	Inertial velocity of *dm* in the Earth frame F_E
V	Volume; control volume coefficient
$\vec{V}$	Aircraft velocity vector
$V_{\alpha_{\max_{1g}}}$	Minimum speed at high incidence
V_1	Decision speed on take-off
V_2	Take-off safety speed

V_3	Steady initial climb speed with all engines operating
$\vec{V}_B$	Airspeed vector of the aircraft mass centre in the body frame
V_C	Design cruising speed
V_D	Design diving speed
V_{DF}	Demonstrated flight diving speed
V_{LOF}	Lift-off speed
V_{MCA}	Minimum control speed, take-off climb
V_{MCA-2}	Minimum control speed, take-off climb two engines inoperative
V_{MCG}	Minimum control speed, on or near ground
V_{MCL}	Minimum control speed, approach and landing
V_{MCL-1}	Minimum control speed, one engine inoperative
V_{MCL-2}	Minimum control speed, two engines inoperative
V_{MIN}	Minimum speed
V_{MO}	Maximum operating limit speed
V_{MPC-1}	Minimum power controllability speed, one engine inoperative
V_{MPC-2}	Minimum power controllability speed, two engines inoperative
V_{REF}	Reference airspeed; landing approach speed, all engines operating
V_{REF-1}	Reference airspeed with critical engine failed; landing approach speed with critical engine failed
$V_{S_{1g}}$	One-g stall speed
V_{TMD}	Minimum demonstrated threshold speed
V_Z	Climb/descent speed
W	Weight (aircraft)
W_{rp}	Load at the rotation point
x, y, z	Coordinates
x_T, y_T, z_T	Thrust line coordinates relative to the aircraft c.g.
$\hat{x}, \hat{z}$	Coordinates of c.g. relative to rotation point (main gear axel)
$\vec{x}$	State vector
$\dot{\vec{x}}$	Derivative of the state vector
x_E, y_E, z_E	Coordinates of aircraft mass centre relative to fixed axes (inertial system F_E)
$x_{c.g.}$	Centre of gravity location as fraction of the m.a.c., measured from the LE of the m.a.c., positive aft
$x_{m.p.}$	Manoeuvre point location as fraction of the m.a.c., measured from the LE of the m.a.c., positive aft
$x_{n.p.}$	Neutral point location as fraction of the m.a.c., measured from the LE of the m.a.c., positive aft
X, Y, Z	Components of resultant force acting on the aircraft
X_p	Variation of axial force component with roll rate
X_q	Variation of axial force component with pitch rate
X_r	Variation of axial force component with yaw rate
X_u	Variation of axial force component with axial velocity
X_v	Variation of axial force component with lateral velocity
X_w	Variation of axial force component with normal velocity

ΔX_D	Axial drag increment due to engine failure
ΔX_{CE}	Sum of axial control forces
$X_{\delta_{LoCE}}$	Axial control force due to LoCE deflection
$X_{\delta_{DiCE}}$	Axial control force due to DiCE deflection
$X_{\delta_{LaCE}}$	Axial control force due to LaCE deflection
X_τ	Axial control force due to thrust controls
Y_p	Variation of lateral force component with roll rate
Y_q	Variation of lateral force component with pitch rate
Y_r	Variation of lateral force component with yaw rate
Y_u	Variation of lateral force component with axial velocity
Y_v	Variation of lateral force component with lateral velocity
Y_w	Variation of lateral force component with normal velocity
ΔY_{CE}	Sum of lateral control forces
$Y_{\delta_{LoCE}}$	Lateral control force due to LoCE deflection
$Y_{\delta_{DiCE}}$	Lateral control force due to DiCE deflection
$Y_{\delta_{LaCE}}$	Lateral control force due to LaCE deflection
Y_τ	Lateral control force due to thrust controls
Z_p	Variation of normal force component with roll rate
Z_q	Variation of normal force component with pitch rate
Z_r	Variation of normal force component with yaw rate
Z_u	Variation of normal force component with axial velocity
Z_v	Variation of normal force component with lateral velocity
Z_w	Variation of normal force component with normal velocity
$Z_{\dot{w}}$	Variation of normal force component with rate of change of angle of attack
ΔZ_{CE}	Sum of normal control forces
$Z_{\delta_{LoCE}}$	Normal control force due to LoCE deflection
$Z_{\delta_{DiCE}}$	Normal control force due to DiCE deflection
$Z_{\delta_{LaCE}}$	Normal control force due to LaCE deflection
Z_τ	Normal control force due to thrust controls

Greek Letters

α	Angle of attack
β	Angle of sideslip
γ	Flight path angle
δ	Control effector deflection angle
δ_{LoCE}	Aerodynamic longitudinal control effector deflection angle
δ_{DiCE}	Aerodynamic directional control effector deflection angle
δ_{LaCE}	Aerodynamic lateral control effector deflection angle
$\hat{\delta}$	Pilot manoeuvre command, CE deflection
Δ	Increment (perturbation) of a parameter; nonzero reference value

Γ	Circulation, vortex strength; dihedral angle
ε	Principal x-axis vertical inclination angle
ζ	Damping ratio
Λ	Sweep angle
ϕ, θ, ψ	Euler angles
μ_x, μ_y	Tire-to-runway friction coefficient
$\rho_{a/c}$	Aircraft mass density
σ	Principal x-axis horizontal inclination angle; static margin
τ	Time delay; thrust control
$\hat{\tau}$	Pilot manoeuvre command; thrust controls deflection
τ_X, τ_M, τ_N	Corrections for propulsive installation
ϕ_T	Vertical thrust line inclination angle (projection on xz-plane)
ψ_T	Horizontal thrust line inclination angle (projection on xy-plane)
Φ	Perturbation velocity potential
$\vec{\omega}$	Angular velocity vector
ω_n	Undamped natural frequency

Subscripts

A	Aerodynamic
B, b	Body frame F_B
CE	Control effector
DiCE	Directional control effector
DR	Dutch roll mode
E	Earth (inertial) frame F_E
H	Horizontal CE
i, j, k	Variable indices
LaCE	Lateral control effector
limit	Limit value of a parameter
LoCE	Longitudinal control effector
max	Maximum
min	Minimum
P	Phugoid mode
R	Roll mode
s	Stability axes
S	Spiral mode
SAS	Stability augmentation system (augmented)
sf	Sideforce
SP	Short-period mode

T	Thrust
trim	Trim value
0	Reference values in reference condition
∞	Free-stream quantity

Superscripts

(a), (b)	Case (a) and case (b)
E	Inertial system, frame F_E

List of Figures

List of Tables

Chapter 1
Introduction and Objectives

1.1 Research Project Initiation and Motivation

> The modern jet transport can be described as the largest integration of technology into a self sufficient unit. All it needs to fly is a full fuel tank, a small crew, and a long runway. Its economic success depends on performance, low maintenance costs and high passenger appeal. It is unique in that all major sections are highly technical in content, from the wing tips to the nose and the tail. Designing the individual components and fitting them together into a cohesive whole is a long process that cannot be expressed in a formula. Airplane design is a combination of industrial art and technology. Usually the process of resolving the art precedes the application of formulae.
>
> William H. Cook
>
> Retired Chief of Technical Staff in the Boeing Transport Division [1]

The author's[1] motivation for the current research investigation manifests in the venture of harmonising the sensitive balance between industrial art and technology.

1.1.1 Background

In 1996 a team from Aérospatiale Airbus, British Aerospace Airbus and Daimler-Benz Aerospace Airbus were studying the 2nd Generation Supersonic Transport aircraft, initially headed by Green et al. [2]. This European tri-lateral team assessed the viability of a future SCT (Supersonic Commercial Transport) aircraft and the potential for global collaboration. Programme background and some detail of this European SCT venture has been summarised by renown specialist P. Poisson-Quinton in an Aerospace America article [3]. At the same time, the US

[1]In this book, 'the author' always refers to the present writer. Other authors are referred to by name.

B. Chudoba, *Stability and Control of Conventional and Unconventional Aerospace Vehicle Configurations*, Springer Aerospace Technology,
https://doi.org/10.1007/978-3-030-16856-8_1

funded a large-scale 2nd generation HSCT (High-Speed Civil Transport) study, outlined by Boeing's HSCT Program Manager MacKinnon in [4].

Not having participated in the Concorde development programme or any similar SCT project before, Daimler-Benz Aerospace Airbus decided to catch-up the apparent technology deficit by introducing a new industrial aerospace vehicle design process, suitable for the design of this type of aircraft and others with the objective, to integrate the individual specialist departments with their specific knowledge into the design of an advanced 2nd generation SCT. The reasoning and some detail is described by Van der Velden [5] and von Reith [6]. Clearly, the novelty associated with the SCT aircraft fortunately secured an environment of acceptance amongst the company decision making bodies, which enabled systematic development of aircraft conceptual design tools.

As described by Van der Velden and von Reith, aircraft sizing were done using a multidisciplinary numerical, multi-point optimisation environment, which had been based primarily on physical rather than statistical roots. This approach enabled evaluation of some selected advanced SCT aircraft configurations and concepts by using a consistent set of design tools. During the early SCT studies it became evident, that the implications of stability and control on hardware sizing were substantial. This understanding immediately qualified stability and control as a primary sizing-discipline, being essential during the conceptual design phase, see author [7]. However, a representation of flight mechanics was missing in the Daimler-Benz Aerospace Airbus methodology concept, which consequently demanded incorporation of the subject matter next to the classical conceptual design disciplines. This finally initiated the present research investigation at Daimler-Benz Aerospace Airbus in Spring 1995. Originally targeted for SCT type of aircraft, the research topic evolved with catalytic influence of the SCT project to what is concisely expressed by the book title "*Stability and Control of Conventional and Unconventional Aerospace Vehicle Configurations—A Generic Approach from Subsonic to Hypersonic Speeds*".

The designer's goal when sizing and placing control effectors[2] is to provide sufficient control power to meet the requirements of prescribed manoeuvres and certification guidelines. However, explicit consideration of control power and inherent airframe stability issues in aircraft conceptual design methodologies is unusual. The highly multidisciplinary complexity of the stability and control problem, permanent lack of aerodynamic information available, and a time penalty associated with the conventional approach, are the main reasons for this situation.

Traditionally, designers use their design experience and historical data to incorporate stability and control estimation into the conceptual design cycle. This approach refers to design statistics with the obvious implication of being design-constrained to conventional, known tail-aft configuration layouts. It is important to

[2]The expression 'control effector' is preferred used throughout this book rather than 'control surface', because control can alternatively be accomplished with an aerodynamic 'surface', thrust impulse, etc.

understand that this process proved to be highly successful for Airbus Industrie, a company primarily concerned with aircraft configurations based on the initial XB-47 and B707 layout. However, any departure from this layout type naturally poses the problem of having available only a sparse statistical database. Apart from some exceptions, this situation usually results in a limitation of design freedom and consequently in restricted design tool capability. The European SCT designers were immediately confronted with this problem. The existing database was restricted to the Anglo-French Concorde and some information available about the Tupolev Tu-144, representing the only realised supersonic commercial transport aircraft so far. Additional design and technology spin-off has been gained from the North American Aviation XB-70 and other relevant military fighter, reconnaissance and bomber case studies.

1.1.2 Today's Aerospace Vehicle Design Problem

Ben Rich, vice president of Lockheed and former director of the Skunk Works (1975–1991) writes in *Skunk Works* [8]. "*In my forty years at Lockheed I worked on twenty-seven different airplanes. Today's young engineer will be lucky to build even one. The life cycle of a military airplane is far different from the development and manufacturing of anything else. Obsolescence is guaranteed because outside of a secret, high-priority project environment like the Skunk Works, it usually takes eight to ten years to get an airplane from the drawing board into production and operational. Every combat airplane that flew in Operation Desert Storm in 1991 was at least ten to fifteen years old by the time it actually proved its worth on the battlefield, and we are now entering an era in which there may be twenty- to thirty-year lapse between generations of military aircraft. ... The development costs of fighters have increased by a factor of 100 since the 1950s, and unit procurement costs have risen 11 percent every year since 1963!*"

This extractly mirrors the current military aircraft industry from the perspective of the aerospace vehicle designer. To complement this view, an analogy can be drawn to the commercial transport aircraft industry. Clearly, a detailed assessment and discussion of the dynamics of the past, present, and future civil and military aerospace industry is not the scope of the current investigation. The interested reader may be referred to Newhouse [9], Lynn [10] and Hayward [11] for further reading. However, it is instructive, to relate some of the above comments to today's aerospace vehicle design problem.

Scott [12] effectively illustrates Rich's statements by discussing the relation between an aerospace vehicle designers career length and the number of new military designs worked on. He provides an overview which spans five decades (1950–2000), a forty-year career length is assumed in Table 1.1.

In light of Rich's and Scott's views, what are the consequences for the aircraft conceptual design engineering ability and engineering excellence when faced with the above developments? The aircraft industry faces a competitive and today's

Table 1.1 Designer career length versus new military designs by decade (1950–2000)

Time span	Aircraft projects
1950s–1980s	XP-5Y, A-2D, XC-120, F-4D, F-3H, B-52, A-3D, X-3, S-2F, X-2, F-10F, F-2Y, F-100, B-57, F-102, R-3Y1, F-104, A-4D, B-66, F-11F, C-130, F-101, T-37, XFY, F-8U, F-6 M, U-2, XY-3, F-105, X-13, C-133, F-107, B-58, F-106, F-5D, X-14, C-140, T-2, F-4, A-5, T-39, T-38, AQ-1, X-15, F-5A, X-1B
1960s–1990s	A-6, SR-71, SC-4A, X-21, X-19, C-141, B-70, XC-142, F-111, A-7, OV-10, X-22, X-26B, X-5A, X-24
1970s–2000s	F-14, S-8, YA-9, A-10, F-15, F-18, YF-17, B-1A, YC-15, YC-14, AV-8B, F/A-18
1980s–2010s	F-117, F-20, X-29, T-46, T-45, B-2, V-22
1990s–2020s	YF-22, YF-23, X-31, JSF, C-17
2000s–2030s	UCAV, B-3??,...

Source Data adapted, in part, from Scott [12]

especially unforgiving environment. If we assume that this will continue to be the scenario for the short-term and longer-term future, then it is of interest to reflect some of the above implications onto the quality of the aerospace product. The following immediate effects have to be discussed: The probability has to be addressed with which today's aerospace vehicle designer may participate in the design of new aircraft projects and what effect this condition has on the aerospace product. The excess-design-potential of today's B707-type of aircraft layout needs to be prognosticated with respect to changing operational constraints, and the significance of advanced conceptual aerospace vehicle design methods has to be outlined in this context.

It is a fact that the number of all-new military and commercial transport aerospace vehicle designs in the present decade has decreased spectacularly. The chance to participate in a number of different aircraft types and projects may seem small, a statement particularly valid for the commercial transport aircraft section. On the other hand, these design activities require more people for longer periods as Anderson outlines in the stimulating *Postface* of his recent book [13]. Having in mind the military scenario as illustrated by Rich and Scott, the commercial transport aircraft companies in contrast concentrate, since the early days of the Boeing 707, primarily on aircraft derivative development, as implied by the modern family-concept strategy. Loftin [14] appreciates in Chapter *Second-Generation Transports* the contributing circumstances, along which the modern family-concept strategy evolved. 'All-new' development exceptions in the post B707-era of transport aircraft evolution have been aircraft like the Boeing 747 and the current Airbus Industrie A3XX project (in 2019, Airbus announced that A380 production would end by 2021). The above remark 'all-new' is merely supported by the fact, that exceptional and excessive technological and financial risks had been and are involved in these developments, primarily due to the large aircraft scale involved, although these aircraft still resemble the initial B707 configuration.

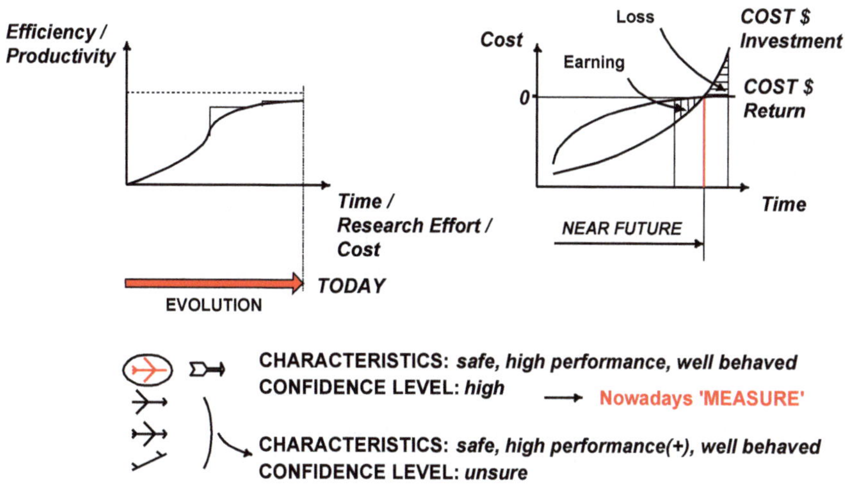

Fig. 1.1 Problem description: today's aerospace vehicle design problem

The high technology aerospace industry primarily operates to make profit. In summary, this dynamic marketplace can be characterised as an environment, where virtually no diversity exists between commercial transport aircraft realised, where development cycle periods are constantly lengthening, where development costs explode, where aerospace companies consolidate, and where the availability of creative aerospace project engineers, who have to be a cut above mere specialists in their field, is reducing. An immediate consequence of the long development cycles of today's aircraft projects is the difficulty for the aerospace vehicle designer, to educate, maintain and augment his or her creative activity and authority by involving imagination, intuition and deliberate choice. Jack van Hengst remarks that "*... in this time our doing is our learning ...*" [15].

Engineering problems have always been and will continue to be under-defined; there are many solutions, good, bad and indifferent. Bavitz from Grumman Aerospace has sketched a professional profile of the aerospace vehicle designer in an Aerospace America article [16]. He rightly characterises the individual, who has to balance industrial art and aerospace vehicle design technology towards a good solution. The aerospace industry is confronted with an ever increasing performance demand [17, 18]. It remains doubtful, if some of the present engineering circumstances are truly compatible with today's high technology demand. Figure 1.1 illustrates the present aerospace vehicle design problem with consequent effects on the aerospace product.[3]

[3]The current investigation concentrates primarily on development aspects of commercial transport aircraft.

The adopted commercial transport aircraft configuration resembles the XB-47, Dash-80, and B707 layout. This aircraft layout has evolved throughout one century of aviation history, strengthening its overall characteristics especially from the early 1950s onwards. This tail-aft configuration (TAC) type has transformed towards a dependable vehicle, which represents today's standard for air passenger transportation. It has helped to define, thus complies with the current safety standards, shows performance characteristics which justify its operation and has matured towards a well behaved man-machine synthesis. The graph on the left in Fig. 1.1 qualitatively recapitulates, how efficiency and productivity of the current aircraft arrangement strive asymptotically towards a plateau. The shallow slope of the graph visualises as a fact today's aerospace vehicle design problem. As a consequence, marginal efficiency and productivity improvements are enforced with ever-increasing research effort required, development time required, and the associated cost implications. To complement these facts, the graph on the right indicates the cost and profit issue qualitatively. The investment required to enable advancement is rising over-proportional towards infinity, whereby the return stabilises towards a finite figure. It remains the question when and how both curves intersect. It is left to the reader to decide, if the evolutionary advancement of the current commercial transport aircraft type operates in the region of relative profit or relative loss.[4]

The foregoing argumentation postulates, that the present airframe-power unit configuration is close at the limit of its development potential. Such judgement obviously requires an assessment of the physical design boundaries and the associated aerospace vehicle design potential surplus. Jones [19], Richards [20] and Gabrielli and von Kármán [21] have attempted to define aerodynamic design boundaries for aircraft. More recently, Poll and Chudoba [22] started to re-examine the power ultimately necessary for mechanical flight and the power actually expended on flight of conventional and advanced vehicles. The difficulty in assessing these boundaries becomes obvious when comparing estimates by Poll [23] and Hilbig [24] in Table 1.2. This comparison is of particular interest because one is tempted to assume consistency in the findings because both authors are well established aerodynamicists.

The above estimates of the aerodynamic (L/D) excess design-potential for the B707-type transport aircraft configuration deviate significantly from each other. By questioning the airframe aerodynamic limit it is designed to, Poll considers today's commercial transport aircraft at the limit defined by Jones. He argues, that in particular laminar flow control technology could modify this Jones's aerodynamic design limit. In contrast, Hilbig quotes a rather optimistic 36% excess design potential with reference to present technology standard. However, he does not specify how the B707-type aircraft should be shaped, whether the technology required to do so has the potential to demonstrate feasibility over the next

[4]The term *relative* is used in this context to indicate the dependency of the cost issues on company policy, financial background, etc.

Table 1.2 Foreseen excess design-potential of B707-type aircraft layout (1999 technology level assumed)

Indicator	Prof. D. I. A. Poll	Dr. R. Hilbig
L/D	Virtually no more to be gained in terms of aerodynamic efficiency; 20% improvement to fuel burn with application of laminar flow control	36% improvement to fuel burn with all measures
Propulsion	15% within the limit of the gas turbine; 40% improvement of propulsive efficiency	23% improvement to fuel burn with all measures
Structure, construction, weights	–	8% improvement to fuel burn with all measures

Source Data adapted, in part, from Poll [23] and Hilbig [24]

15–20 years. Although the latest generation of Airbus aircraft has demonstrated an impressive 30% performance advance to the initial A300 reference, it remains questionable if another 50% net performance gain over today's commercial transport can be realised by taking today's time and research effort expenditures and associated cost and profit implications into account. Without discussing this issue in more detail, it can be concluded that the question of what is attainable with respect to flight physics and aviation technology, is not yet answered conclusively.

The above outline has visualised today's struggle to satisfy the quest for reductions in Direct Operating Costs (DOC). Further problems have to be expected in the near future in the form of operational constraints. Environmental issues will offset much of the potential performance profit to be expected with an ongoing B707-type aircraft evolution. Of particular concern are restrictions on emissions, radiation and noise. It can be foreseen that these additional implications will aggravate today's time and research effort expenditures and associated cost and profit implications, further limiting the returns on investment.

Boyne [25] points out, that the *speed*, with which the Lockheed Skunk Works incorporate outstanding examples of advancing aerospace technology, is instrumental to their success. Table 1.3 compares aerospace vehicle design cycle times of selected military with civil programmes.

Table 1.3 serves the purpose of illustrating the strengths of a highly capable aircraft development environment, in particular the Skunk Works.[5] As Rich remarks [8], any company whose fortune depends as well on developing new technologies should have a Skunk Works in operation. He comments why not so many Skunk Works-style units are scattered around various industries. "... *But if Lockheed's Skunk Works has been a tremendously successful model, why haven't hundreds of other companies tried to emulate it? One reason, I think, is that most*

[5]It needs to be remarked that the presented overview serves as an illustration rather than as a technical evaluation based on consistent data.

Table 1.3 Design cycle periods of selected civil and military aircraft programmes

Aircraft	First flight	Months from go-ahead to first flight	Technological advances
Do328	1991	40	High-performance regional transport aircraft
A300	1972	42	Twin engine wide body
A320	1987	36	FBW narrow body
Concorde	1969	59	Mach 2 commercial transport; FBW
XF-104	1954	13	Mach 2 aerodynamics; sophisticated engine inlets; weapons
U-2	1955	8	Ultrahigh altitude; lightweight structure
A-12	1962	32	Mach 3 speed; advanced propulsion; titanium structure
Have blue	1977	18	Stealth
F-117A	1981	30	Stealth; weapons
YF-22	1990	46	Stealth; supersonic cruise; agility; weapons
B-2	1989	94	Stealth; flying wing configuration

Source Data adapted, in part, from Boyne [25]

other companies don't really understand the concept or its scope and limitations, while many others are loath to grant the freedom and independence from management control that really are necessary ingredients for running a successful Skunk Works enterprise. Unfortunately, the trend nowadays is towards more supervision and bureaucracy, not less. ..." Overall, the Skunk Works formula to success as expressed by Lockheed Martin [26]: "*... The key has been to identify the best individual talents in aviation, blend and equip them with every tool needed, then provide complete creative freedom so they may arrive at an optimum solution in short order ...*".

An example for implementing Skunk Works-style operation is an emulation by Boeing with Phantom Works [27]. Overall, it is of paramount importance for the quality and competitiveness of the aerospace product, to advance in-house aircraft sizing processes on a systematic and enduring basis as obligatory within a Skunk Works environment. Most industrial conceptual design environments do not have the freedom to modernise their design and analysis capability to the state-of-the-art. However, the remaining aerospace vehicle design potential of the B707-type of aircraft can only be unlocked on a profitable scale, when the available aircraft specialist design experience is efficiently blended with cutting-edge computer resources available, as to push the objective function 'cost and profit' towards short order and quality. It must be remarked that the above understanding is rarely adopted by industry decision making authorities. As a consequence, conceptual design development tool inconsistencies and inadequacy may result in local rather global design optima (systematic design errors), which are successively optimised with high-fidelity methods.

Clearly, any prospects of reducing aerospace vehicle design cycle periods and strengthening competitiveness should justify the refinement of the aircraft conceptual design process. Furthermore, commercial transport aircraft manufacturers can benefit from Skunk Works-style operation in several areas, when targeting reduced development time and research effort expenditures.

1.1.3 New Aerospace Vehicle Design Problem

At the present time, the projected rates of growth for passenger and cargo markets are such that the annual rate of consumption of aviation fuel will triple to quadruple by the year 2020 [23, 24, 28]. In view of the growing international concern about the environment, it is probable that such high levels of atmospheric pollution will be politically unacceptable and, as was the case with noise, legislation will be introduced to force change. Innovative solutions are required if, as seems likely, improvements of more than a few percent are demanded. This is because the present airframe-power unit configuration is at the limit of its profitable development potential.

Section 1.1.2 has identified the significance of advancing the current generation of aircraft conceptual design processes, an activity fundamental if there is on-going activity to probe hidden excess performance reserves of today's B707-type aircraft configuration on a profitable scale (*today's* a *Aerospace vehicle design problem*). However, even more fundamental conceptual design method shortcomings materialise, when trying to appreciate aircraft configurations different in shape than the 'conventional' B707 type. The typical conceptual design office of commercial transport aircraft manufacturers clearly has not adequate analysis and design methods at hand, to anticipate the performance potential and to evaluate the commercial feasibility of 'novel' aircraft configurations. This *new aerospace vehicle design problem*, however, provokes the question: Is it worthwhile at all to address development of a new generation of aerospace vehicle design processes and methods for advanced aircraft types, when the 'conventional' B707-type is successful in operation and has gained overall acceptance? It is the incentive of the following exerpt, to inspect the periodic commercial interest and curiosity towards advanced aircraft shapes, the fundamental vision and ingredients required to build trustworthy understanding.

Brown writes in *Wings of the Weird & Wonderful* [29]: "*The unusual and unorthodox in aerospace vehicle design or operation has always intrigued me; the superlative in performance and handling has always excited me. Occasionally fortune smiles and combines all these characteristics in one rare bird to offer the thrill of a lifetime. ... I have never been able to regard aeroplanes as inanimate objects, ... One thing they can never be, and that is dull.*" The above quote acknowledges the evolutionary and revolutionary endeavour of winged aircraft development throughout one century of aerospace vehicle design. Kroo [30] cautions the apparent stagnation of the evolutionary development of the B707-type aircraft configuration from the recent past until today by comparing the B707 and A340 commercial transports. He resumes that "*When we think about what may appear in future aerospace vehicle designs, we might look at recent history. The look may be frightening. From first appearances, anyway, nothing has happened in the last 40 years!*"

What are the reasons for this apparent stagnation in aircraft evolution? At first, it is argued that enormous economic risks are involved, consisting of investment risk and liability risk when progressing the revolutionary rather than the evolutionary track. Furthermore, any venturesome seems to be obsolete with the relative commercial success of the current B707 aircraft type. Sterk and Torenbeek [31] in contrast express rather optimisticly that—"*It is unlikely that the design trends are set merely by conservatism, for example a desire to continue a proven concept in order to avoid the large financial risks of totally new development programmes. The sharp competition always sets incentives to new and innovative concepts since new designs must be considerably improved to be competitive to (derivative of) already established and proven types.*" The author is of the opinion that the evolutionary aerospace vehicle design history has revealed considerable conservatism and idleness, being originated by insufficient design knowledge acquisition and conservation as primary reason.[6] This statement challenges rebuttal, which itself can be counter-acted by asking the question: How many truly diverse aerospace vehicle design tool-boxes can be found, coupled with an operational design knowledge-based system to take advantage of design information, experience and knowledge of past technology projects at the fingertips? As discussed above, technology awareness and true state-of-the-art capability is the particular strength of Skunk Works type environments. Bushnell [32] views the science of creativity and technology awareness as a process where you have to "*self-examine and educate yourself about everything that has been done related to the problem.*"

It has to be acknowledged, that the current aircraft development practice and consequently products have been, so far, politically acceptable, a situation obsolete when catalyst legislation will act to force change. It is then, that sharp competition will set incentives to new and innovative aircraft concepts.[7] To prepare for this scenario, it is mandatory for the *AeroScience-Triangle*—Industry and Operator, Academia and Research Institution—to rationalise the optimistic or sceptical 'opinions' in which advanced aircraft ideas are viewed, to enable comparison of a wider range of aircraft configurations on an impartial and rational basis. The technical challenge clearly lies in the design tool development activity with the goal to reduce development time, to increase design confidence, and consequently to

[6]One has to comprehend the loss of documentation belonging to several British aircraft projects and programmes [33], or the deliberate destruction of XB-35/YB-49 hardware and design documentation at Northrop [34], subsequently lost for B-2 development, just to mention some examples.

[7]An analogy may be drawn to the automobile industry, where today's strong industrial and political lobby is still promoting the powerful car with its impressive performance potential rather than the truly fuel efficient car compatible with the environment.

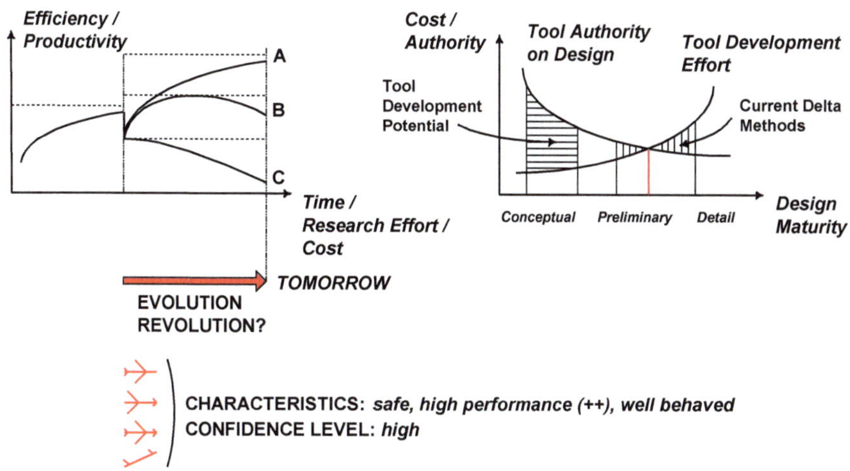

Fig. 1.2 Problem description: new aerospace vehicle design problem

increase design freedom. Systematic forming of aircraft conceptual design competence is vital, to encourage decision-making bodies to rethink their responsibility and opportunities, whilst, at the same time stimulating creativity.

Figure 1.2 illustrates the new aerospace vehicle design problem with particular emphasis on visualising design tool authority and design tool development effort as a function of project maturity.[8]

The graph on the left in Fig. 1.2 qualitatively postulates the situation when studying 'novel' aircraft configurations. The foremost who challenges advanced aircraft ideas surely has to endure an intense learning period, accompanied with initially lower efficiency and productivity. It is this phase where true commitment and an informed approach are vital. It is the undisputed responsibility of the aircraft conceptual design team to pre-select feasible aircraft arrangements with least development risk. *Curve A* represents the idealised aircraft efficiency and productivity curve. The attributes of the aircraft efficiency and productivity slope ultimately determine, if configurations under investigation represent the long awaited 'quantum leap' in aerospace vehicle design and operation. It is the objective of the present research investigation to contribute to this multidisciplinary science & engineering field or 'art' by defining the slope in advance of any serious development commitment, thereby omitting aircraft types as represented by *curves B and C.*

The graph on the right (Fig. 1.2) reiterates the relation between tool authority on design and tool development effort required relative to design maturity. It is an indisputable fact that aircraft conceptual design methods have predominant aircraft sizing authority throughout the entire development cycle. During the conceptual

[8]The current investigation concentrates primarily on development aspects of commercial transport aircraft.

period, primary design decisions are made like selection, evaluation and pre-definition of the configuration type and election of global design parameters. Subsequent design levels, the preliminary and detail design phases, check in contrast the prescribed design space with higher-order design methods but limited authority to alter global design parameters. Any modification proposed at preliminary and/or detailed design levels needs to be authorised at conceptual design level, if global design parameters are affected. Naturally, the authority of design tools on design decisions reduces when advancing the maturity of the product. However, the following phenomenon can be observed throughout the *AeroScience Triangle*. It is a prevailing practice to invest predominantly into high-fidelity design and analysis methods (e.g. CFD and FEM). As a consequence, these methods absorb the major portion of tool development resources available, whereby their design influence onto the product reduces significantly towards the end of the design cycle. This imbalance is a contributing factor to today's aerospace vehicle design problem, resulting in diminishing returns on investment.

To recapitulate, detailed (disciplinary) aerospace vehicle design investigations are today conveniently performed using all available resources like parallel supercomputers, high-level optimisation and sophisticated simulation. In contrast, the important primary aerospace vehicle design decisions (e.g. overall configuration selection) at conceptual and preliminary design level are still made using extremely simple analyses and heuristics. In this context, most existing computer-aided aerospace vehicle design methodologies refer often to statistical data without necessarily questioning and balancing the physical rationale of the solution. A further limitation might be the restriction in capability to the 'classical' aircraft arrangement only. The reason for this scenario seems to be the difficulty of not knowing how to integrate multidisciplinary simulation of both normal and radical aircraft conceptual designs in more than an ad hoc fashion.

The author firmly believes, that the multidisciplinary decision process at the conceptual design level needs to be further evolved and automated. This demands an improvement of the pertinent aerospace vehicle design disciplines involved (aerodynamics, stability and control, cost, etc.) with real-life design expertise and experience as vital ingredients. The development of an aerospace vehicle design knowledge-based system is one key element in such an undertaking. Clearly, this requires the formulation of the problem statement and the solution alternatives in the physical domain, resulting in a generic executable software solution. Summarising, a methodology concept of generic character needs to be systematically built with physical rather then statistical roots. This will enable effective exploration and comparison of normal and radical aerospace vehicle design concepts based on a consistent information database.

The recent period has been filled with exceptionally interesting developments and advances in *AeroSpace Science*. Investigations into advanced future efficient aircraft world-wide substantiate the demand for enhanced project tools, to enable rational aircraft performance-cost tradeoffs. Table 1.4 shows a selection of winged aircraft programmes from the 2001 era with promising performance-cost potential or technology demonstration function towards that aim.

Table 1.4 Recent future-efficient aircraft programmes

Civil aviation	Military aviation	Space aviation
AeroVironment HELIOS (demonstrator)	X-31 (demonstrator)	X-33 (demonstrator)
Piaggio P180JET (project)	X-32 (demonstrator)	X-34 (demonstrator)
Airbus A300-600 ST Beluga (operation)	X-35 (demonstrator)	X-37 (project)
Airbus A340 TSC (project)	X-36 (demonstrator)	X-38 (demonstrator)
Airbus A3XX-100 (project)	X-39 (project)	X-41 (project)
Airbus A3XX TSC (project)	X-40 (demonstrator)	X-42 (project)
Boeing Blended Wing Body (demonstrator)	Lockheed F117A (operation)	X-43 (project)
Lockheed Supersonic Business Jet (project)	Northrop B-2 (operation)	Lockheed Martin Venture Star (project)
Boeing Oblique Flying Wing (demonstrator)	Boeing Blended Wing Body (demonstrator)	Daimler Chrysler Hopper (project)
Airbus ESCT, Boeing HSCT (project)	Lockheed YF-22 Raptor (demonstrator)	

Note The above record outlines only a selection of aircraft technology projects and programmes considered during the 2001 era

The author perceives the above illustrated quest for advanced aircraft as an indication of how exciting the future will be! Capturing a market-share, however, requires understanding the significance of systematic technological advancement as pre-requisite. Ziegler reflects the motivation by Airbus Industrie to develop fly-by-wire technology for the A320: "*We have always been minded not to be conservative. We were not fascinated by technology, we were just being open-minded and seeing what could be build with the technology ... already available on the shelf.*" Sutter, father of the world's most profitable aircraft B747 responded at the time with "*Airbus is not going to sell a lot of airplanes by touting technology*" [10]. A company's vision, as expressed by Szodruch from Airbus Industrie, conveniently concludes the exchange of views by saying, that the "*... market cares about all this 'technical junk', and – in the end – competition will discriminate against technically and economically obsolete products!*" [35].

Bushnell [36] rightly categorises aeronautical development metrics since the early 1900s as "*... higher, faster and larger is better...*". In contrast, the metrics of today and the foreseeable future are clearly different, and hence the nature of technological improvement demanded for transition: "*These 'new' (civilian) metrics include affordability ..., productivity ..., safety and the environment ...*" Acceptance for new technology and design freedom in military and space aviation has been traditionally high for a variety of reasons. Civil aviation evolution, in contrast, has presented itself as a less driving environment. Bushnell asked in 1988: "*Is there an aerodynamic renaissance for the long haul transport?*" The question clearly showed justification when reflecting the evolution of the aerodynamic figure of merit, Mach number times lift to drag ratio with time. The B707-era (1960s until today) is characterised by an increase of M (L/D)max of around 15%; the curve being almost flat [37]. As a response, Liebeck and others started to create and evaluate alternate configurations. Consequently, one of the advanced transonic aircraft configurations under development today in the USA is known as the Blended-Wing-Body (BWB) [38], resembling a flying wing configuration for civil and particularly military application [39].

Another example of aeronautical revolution may be foreseen with the cancellation of the HSCT programme in the USA in 1999. The high-speed research team at NASA stated, that a new supersonic commercial transport aircraft "*... study will also look at alternative aircraft configurations, some of which will not be based on the four-engined layout of the present TCA* [Technology Concept Aircraft]" [40], and that the airframe element will "*... go back to basics ...*" to re-examine all aspects of the design and technology which has been used to date [41]. Davies [42] stimulates critical thinking and challenges rebuttal by judging the SCT design efforts to date: "*The cost of developing any supersonic airliner is prohibitive. The prospects of their being able to operate economically is remote. The market for such an aircraft ... is very small.*" Seebass [43] in contrast resumes, that the book he has contributed to "*... focuses much of its attention on the underlying tools for the study of such* [conventional SCT] *aircraft, as well as on unconventional configurations ...*" This translates into the prerequisite of having conceptual design forecasting tools at hand, to either prove feasibility or to identify true show-stoppers of new aircraft projects.

At last, what is the technical rationale behind future AeroSpace vehicles being especially design method demanding? Aircraft development programmes like the above mentioned Blended-Wing-Body (BWB), Supersonic Commercial Transport (SCT), Supersonic Business Jet (SSBJ) [44], Single Stage To Orbit (SSTO) conceptions [45] in particular and others, are confronted with such narrow excess design margins, that the solution-space resolution delivered with the traditional single-point optimisation design methods has not yet been able to prove or disprove technical and commercial feasibility for the above mentioned examples at all. Such vehicles may only become reality, when it is possible to take advantage of truly integrated multidisciplinary aerospace vehicle design methodologies with the ultimate objective, being able to identify vehicle performance potential with confidence in short order. One of the key technologies, obligatory when addressing multidisciplinary tasks, is currently evolving, termed Multidisciplinary Analysis (MDA or synthesis) and Multi-Disciplinary Optimisation (MDO). However, much more needs to be done.

Concluding, with today's diversity of opinion prevailing when discussing advanced aircraft arrangements, current conceptual aerospace vehicle design engineering practices demand systematic development. The author believes that only methodical background work enables construction of an acceptable confidence level into analysis and design methods for the initial design of conventional *and* advanced aerospace vehicles. With such preparation at hand, higher aircraft project risk levels can comfortably be accepted with focus on the performance/cost benefit. Ivan Shaw, Europa founder, puts it as "*If you eliminate risks, you eliminate progress*" [46].

1.2 Research Project Aims, Scope, and Objectives

It is the aim of the present research investigation, to advance aircraft conceptual design tool maturity with respect to current and future aerospace vehicle demands. The research scope and objectives have been the following:

1. Development of a generic aircraft conceptual design methodology with the primary objective to size the vehicle's stability and control surfaces, thereby reducing today's prolonged vehicle design cycle periods and to improve overall design quality.
2. Widening of the project engineer's design freedom by creation of a generic methodology concept which enables control surface sizing of subsonic to hypersonic aerospace vehicle designs of conventional and unconventional configuration layout.
3. In conventional aircraft conceptual design procedures, design for *performance* is done before design for *stability and control.* The advanced methodology shall enable evaluation of stability and control in parallel with performance during the conceptual and preliminary design phase of future efficient aircraft.
4. Transformation of flight mechanics as today's advanced analysis discipline (disconnected from design) to a generic design discipline by harmonisation of the complex balance between control power and inherent airframe stabilities (static-, dynamic-, and manoeuvre stability).
5. Integration of flight test and certification aspects relevant for the design of controls into conceptual aerospace vehicle design.
6. Assemblage, extraction, management and inclusion of appropriate aerospace vehicle design data, design information, and design knowledge, to enable an informed approach with the consequent intent "*... things should be as simple as possible, but no simpler...*".

1.3 Summary of Results

This first chapter has outlined the motivation for advancing aircraft conceptual design methods. It has been identified that the discussion of advanced conventional and advanced aircraft ideas and associated technology underlies not just technical and commercial, but as well emotional reasoning. It is this element of industrial multidisciplinary science & engineering, often dubbed 'art', which requires a balanced dialogue of pros and cons when striving for the unknown, a responsibility in particular assigned to the conceptual design environment. Only a truly informed approach to the problem appears to be competent enough to address the range of commercially challenging aerospace products in today's unforgiving but exciting environment.

References

1. Cook, W.H., "The Road to the 707 – The Inside Story of Designing the 707," *TYC Publishing Company*, Bellevue, Washington, 1991.
2. Green, P.K., Pacull, M. and Reimers, H.D., "European 2nd Generation Supersonic Commercial Transport Aircraft," ICAS 96-4.3.1, *20th Congress of the International Concil of the Aeronautical Sciences*, Sorrento, Napoli, Italy, 8-13 September 1996.
3. Poisson-Quinton, P., "Future SSTs – A European Approach," *Aerospace America*, September 1994, pp. 38–43.
4. MacKinnon, M., "Development Challenges for the HSCT," presentation, Boeing Commercial Airplane Group, March 1998.
5. Van der Velden, A., "Multi-Disciplinary SCT Design Optimisation," AIAA Paper 93-3931, *AIAA Aircraft Design, Systems, and Operations Meeting*, Monterey, CA, 11–13 August 1993.
6. von Reith, D., "Technology Assessment With Multi-Disciplinary Aerospace vehicle design Tools on the Next Generation Supersonic Commercial Transport," ICAS-96-3.5.1, 20th Congress of the International Concil of the Aeronautical Sciences, Sorrento, Napoli, Italy, 1996, pp. 1081–1086.
7. Chudoba, B., "Primary Control Surfaces on Supersonic Transport Aircraft," Dipl.-Ing. Thesis, *University of Applied Sciences (FH) Aachen*, Germany, performed at Future Projects Department, British Aerospace Airbus Limited, August 1994.
8. Rich, B.R. and Janos, L., "Skunk Works – A Personal Memoir of My Years at Lockheed," First Edition, *Little Brown and Company*, 1994.
9. Newhouse, J., "The Sporty Game," First Edition, *Alfred A. Knopf, Inc.*, 1982.
10. Lynn, M., "Birds of Prey – The War Between Boeing and Airbus," First Edition, *Mandarin*, 1995.
11. Hayward, K., "The World Aerospace Industry – Collaboration and Competition," First Edition, *Duckworth & RUSI*, 1994.
12. Scott, W.B., "Industry's Loss of Expertise Spurs Counterattack," Aerospace in Crisis, Aviation Week & Space Technology, 13 March, 2000, pp. 60–61.
13. Anderson, J.D., Jr., "Aircraft Performance and Design," First Edition, *WCB, McGraw-Hill*, 1999.
14. Loftin, L.K., Jr., "Quest for Performance – The Evolution of Modern Aircraft," NASA SP-468, *NASA Scientific and Technical Information Branch*, NASA, 1985.
15. van Hengst, J. (728JET Chief Engineer), Private Conversation, Dornier Luftfahrt GmbH, January 2000.
16. Bavitz, P.C., "So You Want to be An Aerospace Vehicle Designer?!," Aerospace America, February 1964.
17. Kingsley-Jones, M. and Lewis, P., "Airbus' A3XX Economics Target 'Unachievable', Says Boeing," Flight International, 22–28 September 1999, p. 8.
18. Anon., "Pierson Warns on A3XX Costs," Flight International, 4–10 March 1998, p. 8.
19. Jones, B.M., "The Streamline Aeroplane," *Aeronautical Journal*, 1929, 33(221), pp. 358–385.
20. Richards, E.J., "A Review of Aerodynamic Cleanness," Volume LIV, *The Journal of the Royal Aeronautical Society*, 1950, pp. 137–186.
21. Gabrielli, G. and von Kármán, T., "What Price Speed? - Specific Power Required for Propulsion of Vehicles," *Mechanical Engineering*, October 1950, pp. 775–781.
22. Poll, D.I.A. and Chudoba, B. "Prospects in Commercial Aerospace Vehicle Design – Evolution or Revolution," *FITEC'98: Farnborough International Technology Exploitation Conference*, London, 8–10 September 1998 (oral only).
23. Poll, D.I.A., "Are We on the Right Evolutionary Track for the 21st Century?," *Royal Aeronautical Society Goldstein Lecture*, Manchester University, 1997.
24. Hilbig, R., "Das Technologiekonzept 'Adaptiver Flügel'," DASA Airbus Bremen, Presentation at Technical University Munich, 17 February 2000.

25. Boyne, W.J., "Beyond the Horizons – The Lockheed Story," First Edition, *St. Martin's Press*, 1998.
26. Anon., "Company Overview – About Skunk Works Palmdale," Lockheed Martin, http://www.skunkworks.net/company_overview.html.
27. Anon., "Phantom Works: The Advanced R&D Unit of Boeing," The Boeing Company, http://www.boeing.com/news/feature/farn98/pworks.html, September 1998.
28. Jenkinson, L.R., Simpkin, P. and Rhodes, D., "Civil Jet Aerospace Vehicle Design," First Edition, Arnold, 1999.
29. Brown, E.M., "Wings of the Weird & Wonderful," First Edition, Volume 1, *Airlife Publishing Ltd.*, 1983.
30. Kroo, I., "Reinventing the Airplane: New Concepts for Flight in the 21st Century," Presentation, December 1995.
31. Sterk, F.J. and Torenbeek, E., (Editors) "Unconventional Aircraft Concepts," Papers Presented, *Symposium Organised by The Netherlands Association of Aeronautical Engineers (NVvL) and the Students Society 'Leonardo da Vinci'*, Delft University of Technology, First Edition, Delft University Press, April 1987.
32. Bushnell, D.M. (Chief Scientist NASA Langley Research Center), Private Conversation, Cranfield University, December 1998.
33. Jagger, D.H. (Former Head of Future Projects, BAe Airbus Limited, Filton), Private Conversation, BAe Airbus Limited, 1994.
34. Pape, G.R. and Campbell, J.M., "Northrop Flying Wings – A History of Jack Northrop's Visionary Aircraft," First Edition, *Schiffer Publishing Ltd.*, 1995.
35. Szodruch, J., "Aircraft Efficiency – Investment for the Future," Daimler-Benz Aerospace Airbus, 1997.
36. Bushnell, D.M., "Frontiers of the 'Responsibly Imaginable' in (Civilian) Aeronautics," AIAA Paper 98-0001, *The 1998 AIAA Dryden Lecture*, January 1997.
37. Liebeck, R.H., Page, M.A. and Rawdon, B.K., "Blended-Wing-Body Subsonic Commercial Transport," in *Future Aeronautical and Space Systems*, Edited by Noor, A.K. and Venneri, S. L., Progress in Astronautics and Aeronautics, Vol. 172, AIAA, 1997.
38. Phillips, E.H., "NASA To Fly Sub-Scale Blended Wing Body," Aviation Week & Space Technology, 7 February 2000.
39. Anon., "Global Range Aircraft," FAS – Military Analysis Network, http://www.fas.org/man/dod-101/sys/ac/gra.htm, 10 February 1998.
40. Norris, G., "US Supersonic Effort Faces Axe," Flight International, 9–15 December 1998.
41. Norris, G., "NASA Slows Down High-Speed Project," Flight International, 2–8 December 1998.
42. Davies, R.E.G., "Supersonic (Airliner) Non-Sense – A Case Study in Applied Market Research," *Paladwr Press*, 1998.
43. Sobieczky, H., (editor), "New Design Concepts For High Speed Air Transport," No. 366, *CISM Courses and Lectures*, Springer, 1997.
44. Bruner, H.S., "SSBJ: A Technological Challenge," *ICAO Journal*, August 1991.
45. Harloff, G.J. and Berkowitz, B.M., "Hypersonic Aerospace Sizing Analysis for the Preliminary Design of Aerospace Vehicles," Vol. 27, No. 2, *AIAA Journal of Aircraft*, February 1990, pp. 97–98.
46. Sarsfield, K., "Kitted for Flight – UK Kit-Maker Europa Aircraft is Revolutionising Personal Aircraft," Flight International, 11–17 January 2000, pp. 38–39.

Chapter 2
Generic Aerospace Vehicle Design—Knowledge Utilisation

2.1 Introduction

"*Progress, far from consisting in change, depends on retentiveness ... Those who cannot remember the past are condemned to fulfil it*" Santayana [1]. This quotation is particularly apt in the study of aerospace vehicle conceptual design. More than 116 years have passed since Otto Lilienthal published his breathtaking analytical study of bird flight, illuminating the foundations of modern air-transportation [2]. During this relatively brief evolutionary period, the aerospace vehicle has developed from an individual's dream to a highly sophisticated and significant force in modern society. However, it is a sobering thought that the subject is still so little understood, that Rich [3], Kroo [4] and Davies [5] could draw the conclusions noted in Chap. 1.

Lord Kelvin (1824–1907) once remarked: "*If you can measure that of which you speak, you know something of your subject; but if you cannot measure it, your knowledge is unsatisfactory.*" The trouble with many of the advanced aerospace projects is, that the range of opinion judging technical and commercial success does not comply with any accepted, thus consistent scale. It is common practice to express 'facts' in terms of *Adjectives * Adverbs*, by definition non-measurable.

This chapter outlines the quest for engineering design knowledge as a truly measurable reference, the research strategy selected, and the transformation into practical working tools.

2.2 Prelude—Design Office of Nature

Chapter 1 did illustrate the demand for an engineering scale, against which technically attainable aerospace vehicle performance and benefit optima can be evaluated and agreed, dependent on a technology level selected. However, the ability to

B. Chudoba, *Stability and Control of Conventional and Unconventional Aerospace Vehicle Configurations*, Springer Aerospace Technology,
https://doi.org/10.1007/978-3-030-16856-8_2

determine engineering feasibility-targets with such a degree of confidence appears, even at the beginning of the 21st Century, not to be in reach. The author is of the opinion that it is of particular interest to review, if the interface between human (artificial) flight and role model bird, as created by design office nature, can efficiently be utilised for today's and tomorrow's engineering tasks.

The following excerpt by renowned aeronautical specialist MacCready [6] best characterises the interdependence of artificial to biological flight. MacCready has contributed significantly to the understanding and design of ultra-high efficient aerospace vehicle designs, thus his argumentation is judged to be of special value. "*The advent of fossil fuel engines offered aeronautical engineers ten-fold to hundred-fold increases in power-to-gross-weight ratios over the ratios available for biologically powered flight creations such as birds and human powered aircraft. The tremendous achievements of engine-powered aircraft over the past eight decades have tended to obscure how numerous flight problems had already been elegantly solved by birds, many tens of millions of years ago. ... After the successes of Cayley, Lilienthal and the Wrights, and the growth of the theoretical underpinning of the field by Lancaster and Prandtl, man's aviation constructions raced far beyond those bird ancestors and the role model became virtually forgotten.*"

There is still basic research about birds and their evolution. The pool of available literature is vast, covering primarily propulsion, aerodynamics and performance aspects, as illustrated by Rayner [7, 8]. The work by Herzog [9] and Saunders [10] deserves special mention. Both scientists were striving towards an understanding of the solutions shown by catalyst nature, which enabled them to attempt technical emulation. However, the question arises whether or not there is a connection recognisable at all to today's aircraft? MacCready's answer is that "*... the connection is likely to be only the after-the-fact realization that a modern design solution could have been foretold by observing how nature has been doing it for millions of years. Using nature's designing to help us solve new aeronautical problems is rare.*" He continues that the sailplane design field is likely to appreciate evolution as a master designer of aeronautical form and function.

Today, aeronautical engineers look seriously at small-scale micro air vehicles (MAVs), which have wingspans in the category of millimetres and centimetres. These 'palm-sized' aircraft have recently attracted much attention from government agencies, private research laboratories, universities and the media for primarily military surveillance application, see Chow [11]. For this particular category of air vehicles, conventional aerospace vehicle design wisdom cannot contribute much, thus advice is best sought by master designer nature. It should be recalled, that the mystery of insect flight had been lately resolved by Ellington, a zoologist from the University of Cambridge, in 1996, see Ellington et al. [12], Alexander [13], and Winn [14].

Returning to human-scale aerospace applications, a bird has solved myriads of problems in aerodynamics, stability and control, structures, including problems scientists have not even recognised yet. Two fundamental questions arise. At first, which of the solutions provided by nature have potential for technical utilisation in advanced aerospace applications? Secondly, is it possible, to emulate in part nature's evolutionary process of species for technical aerospace product developments?

2.2.1 Technology Spin-off

The author's growing respect for and envy of nature as a designer of aeronautical species reached another local 'apogee' when trying to appreciate the diversity and quality of solutions given. MacCready categorises nature's species that have attained full flight in at least four separate routes: birds, mammals (bats), reptiles (pterosaurs) and insects. For more limited flight he includes flying fish, gliding mammals and seeds. The author believes that additional aeronautical understanding can be gained from fish, just to mention the well-known spin-off gained from sharks or the not yet investigated configuration similarities of the *Manta Birostris* ray compared to the blended-wing body (BWB) flying wing study, see Fig. 2.1.

MacCready compares biological fliers versus artificial fliers by using aeronautical terms like flight duration, flight altitude, navigation, flight manoeuvres, aerodynamic aerofoils, structure, instrumentation and stability. Without discussing detail, it can be summarised that biological and artificial flight face the same flight-physical realities of energy, momentum and viscosity. The engineer has the freedom to utilise structural strength-to-weight and propulsive power-to-weight efficiencies for aerospace applications, which are many times those available to biological devices. In contrast, the ecological niche of biological flight, where the laws of aerodynamics, biological power and biological structure properties must prevail, governs that nature finds optima which are rather similar in configuration, no matter what the starting point, see Fig. 2.2. The comparison of the wings of the *Pterosaur*, bird, and bat with the arm of man reveals that, "*... the similar bones in each form obviously bespeak their common ancestry*" [16].

When studying nature's superbly adapted solutions with view to expected technology spin-off for artificial aerospace applications, then it is of particular interest to study, what was/is perhaps nature's most remarkable aeronautical achievement. One criterion for such selection is given by the limitation of *size* for natural flight. The physical laws of energy conservation govern the minimum power required for a flying vehicle, natural or artificial, to be dependent on its shape, size, and weight. The analytical relations describing the interdependence of the above

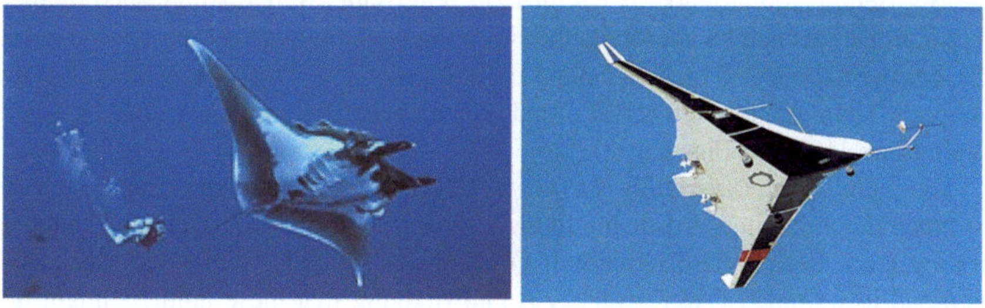

Fig. 2.1 Configuration comparison: Manta Birostris (Phillip Colla Photography) and NASA Langley Research Center/McDonnell Douglas/Stanford University Blended-Wing-Body (BWB) small scale demonstrator

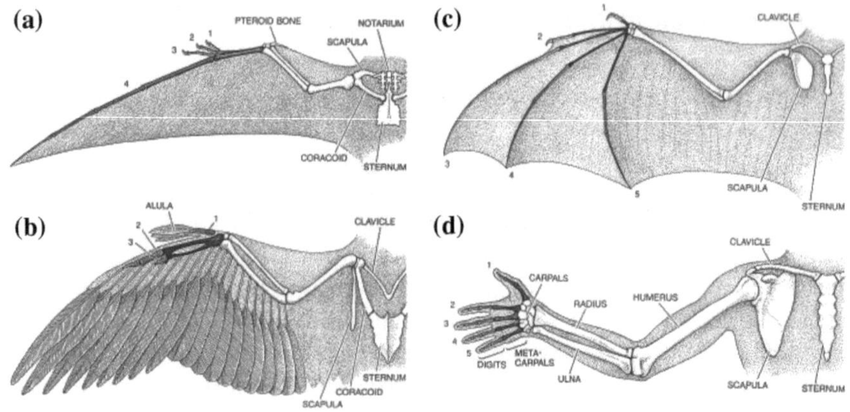

Fig. 2.2 Wings of a Pterosaur (**a**), a bird (**b**), and a bat (**c**) as evolutionary variations in comparison with the arm of man (**d**). Langston [15]

parameters are given by the scaling law, better known as 'square-cube law'. Without discussing the analysis in detail, this law means "*... that as size increases, weight increases faster than area (in fact with a square-cube law relationship between area and weight, P/W varies as* $W^{1/6}$ *or* $b^{1/2}$)" [6]. Here, *P* stands for power to fly, *W* for weight and *b* for span. MacCready continues to demonstrate that this relation holds "*... over a gigantic* 10^{12} *range of weight, from gnats to the Boeing 747 ...*" in spite of widely differing constructions [17]. This translates into the known fact, that the larger the flying vehicle, the heavier it is per unit area of the wing, the faster it must fly to stay aloft requiring more power per unit weight.

What is the limiting size of a flying animal so far experienced in nature? MacCready argues in his 1987 paper that "*... In the typical biological case, the larger the creature the less power is available per unit weight. Therefore, there is an upper limit to biological flight, when power available can just match power required; this limit is probably reached with the giant* ***pterodactyl*** *whose weight was around 100 kg.*" Fossil evidence had been discovered by D. Lawson of a giant pterodactyl with the name *Quetzalcoatlus Northropi*, a flying reptile discovered in the Big Bend National Park, Texas, in 1972. Pterosaurs were distinguished by their reptilian features and slender membranous wings, which lived during the Mesozoic era, between about 200 million and 64 million years ago. The latest estimates of the animal's span range between 11 and 12 m, according to Langston [15] and Wellnhofer [18], see Fig. 2.3.

The flight of *Quetzalcoatlus Northropi* (QN) is surprising when extrapolating the parametrics of soaring birds using today's aeronautical understanding and standard scaling laws. Before the discovery of QN, the size limits for biological flight were assumed to be much lower, even by acknowledging that these animals might not have been able to maintain active powered flight for an extended period. Consequently, all very large flying vertebrates have to minimise their fuel burn and

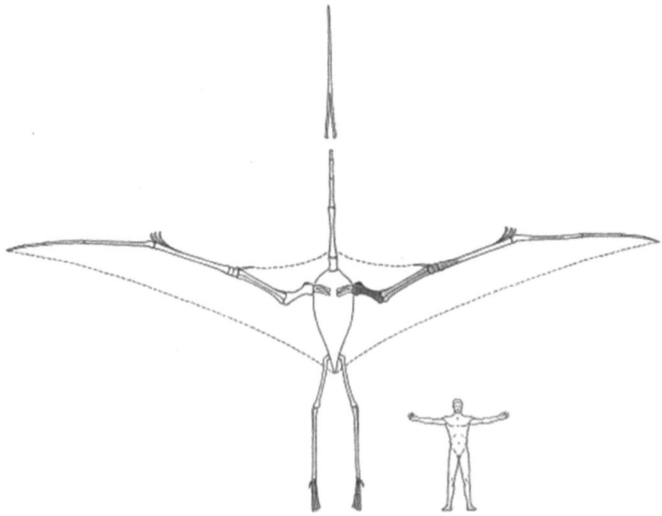

Fig. 2.3 Largest flying animal ever to inhabit the Earth is thought to have been the pterosaur **Quetzalcoatlus Northropi** [15]

extend their range by soaring. MacCready remarks that "*If the square-cube law holds, the large span requires a proportional increase in wing loading and a 2.5-fold increase in P/W, far more than could be made up for by slight aerodynamics improvements from Reynolds number*" [6]. He further elaborates that nature had three design refinements available to harmonise power available and power required, and that QN probably needed all three approaches to remain a flier, see Table 2.1.

Clearly, what is the technical substance of this remarkable nature-designed aeronautical giant *Quetzalcoatlus Northropi*, taking the design constrains into account as identified before? Today's aerospace vehicle design domain, as characterised by large size, huge power available, and subsonic to hypersonic speeds, clearly rules out any significant design directive from nature, because there has been no evolutionary pressure for these flight regimes. However, when it comes to flight phases like takeoff and landing, variable geometry, active controls, navigation, and

Table 2.1 Nature's design refinements to match power required to power available of the Pterodactyl

1	As the bird's size and/or speed increase(s), an aerodynamic scale effect comes into play and yields slightly more efficient wings (Reynolds number effect).
2	The initial selection of the creature's configuration and the ability to alter it to a certain degree during flight yields higher aerodynamic efficiency.
3	Weight can be kept below the level that would be expected if the density stayed constant as the bird grew in all three dimensions.

Source Information adapted, in part, from MacCready [16]

Table 2.2 *Quetzalcoatlus Northropi*—selected design detail

1	Inherent airframe stability.
2	Evolutionary selection of configuration layout.
3	Configuration mutation during flight (e.g. variable geometry).

overall efficiency, numerous lessons need to be learned from design office nature. Cowley recalls that these animals not only fly, "... *they expend remarkably little energy in doing so, offering the promise of a level of efficiency which far exceeds our present finesse in the practice of aerodynamics*" [19]. Nature's aeronautical achievements are of even greater brilliancy when one remembers that "... *Every design feature meets a survival purpose, some by way of aerodynamic efficiency, others by biological adaptability or sexual selection*", as noted by MacCready [17]. In short, the *Quetzalcoatlus Northropi* inherits advanced technology detail of such sophistication, man just starts to appreciate. Table 2.2 identifies some nature-evolved advanced technology detail, being of practical interest in context of the present research investigation.

Inherent Airframe Stability: The primary advantage animals in general have utilised, in contrast to artificial mechanisms, originates from the versatility of biological sensors and muscles and the feedback control loops connecting them. Clearly, birds and flying reptiles, just to concentrate on these two species, have effective active control systems. They utilise this capability favourably for adjusting 'airframe' variable stability in pitch, yaw and roll. As a fact, birds and flying reptiles are configured to enable stable, indifferent and unstable flight, dependent on mission requirement. Nature has evolved its large-scale aeronautical design-masterpiece *Quetzalcoatlus Northropi* into a statically unstable tailless 'airframe' for the same reasons.

A search for literature on bird stability and control was generally of limited success, indicating how little understanding exists for bird stability and control and associated effects on vehicle layout. MacCready has contributed significantly to the existing understanding by constructing a full-sized flying replica of *Quetzalcoatlus Northropi*, designated the QN™ replica from 1984 to 1985 [6, 16, 17, 19, 20].[1] The remarkable flight test investigations of Raven models by Hoey [21] have examined bird static stability aspects in soaring flight and the control mechanism required.[2] Nature is using active control favourably to avoid vertical stabilising and controlling surfaces for biological flight. In the longitudinal sense, indifferent or unstable flight is utilised as a performance and manoeuvrability enhancement measure.

[1]The National Air and Space Museum (NASM) had sponsored the initial studies for the Smithsonian's aviation movie, *On The Wing*, which relates manned flight to that on nature. QN™ is a registered trademark of S.C. Johnson & Son, Inc.

[2]R.G. Hoey in September 1999: "*Birds have to concern themselves not only with light-weight bones and lift, but the problem of stability—how to keep the beak-end into the wind and the pooping-end downwind.*"

Relaxed static stability enhances (a) manoeuvrability, and (b) effectively increases flight performance when taken into account during the overall vehicle design cycle. The influencing report by Ashkenas and Klyde [22] recapitulates the technical rationale of inherent airframe stability on aerospace vehicle design, and it served as catalyst for the current BWB flying wing investigations in the USA. Airbus Industrie has recently investigated the adoption of reduced and even negative manoeuvre margins for A3XX as performance enhancement measure [23]. Clearly, this research undertaking integrates the parameter *inherent stability* as a true design variable into the generic methodology concept.

<u>Configuration Selection and Configuration Mutation</u>:
Literature shows numerous attempts to classify the type of configuration arrangement flying animal resemble. The key question had been the difficulty to conclusively identify, if the tail of a bird has stabilising or controlling function at all. Nickel and Wohlfahrt offer an opinion that "*The tail of birds has virtually no stabilizing effect. ... Only to a limited extent is it used as a steering device ...*" without providing conclusive evidence [24]. Systematic investigations by Herzog [9], MacCready [16], and Hoey [21] harmonise on a different perspective. Restraining the discussion to birds and *Quetzalcoatlus Northropi*, birds utilise a tail-surface whereby *Quetzalcoatlus Northropi* is understood to be of tailless layout. The tail surface of birds serves evidently as multipurpose device, able to contribute to stability and/or control during specific flight conditions. MacCready recalls that "*The Albatross, for example, uses essentially no tail during its efficient cruising flight, but is certainly stable in pitch. Some moulting birds can fly and maneuver without a tail by moving the wings forward and back*" [16]. Birds evidently mutate their 'airframe' configuration during flight dependent on mission requirement. In technical terms, they are able to transform to either the tail-aft configuration (TAC) for low speed and manoeuvre cases, or towards the flying-wing configuration (FWC) layout for high efficiency cruise. *Quetzalcoatlus Northropi* in contrast appears to have evolved to be less a configuration transformation artist, being optimised by selection of the following parameter combination: longitudinal and lateral unstable flying wing layout[3] of low weight and large span. Latest pterodactyl research by Cunningham confirms that "*As they became more dependent on active flight control, they lost their tails and the length of their necks and heads increased*" [25].

Another configuration aspect of paramount importance for future aerospace vehicle design is the fact, that neither birds nor *Quetzalcoatlus Northropi* have evolved vertical tails to provide lateral-directional stability and control. Hoey and Gabor Miklos [26] have postulated that it may not have been possible genetically for birds and *Quetzalcoatlus Northropi* to evolve vertical tails. The configuration adaptability of these flying animals enabled them to steer very easily without having to

[3]In the large pterodactyls, there is no evidence to science of a wing attachment to the tail (feet). It is possible that the legs of QN carried a separate membrane contiguous with a short-couple tailplane put to use for certain flight manoeuvres. However, it can be expected that trim drag implications have prohibited utilisation of such short coupled auxiliary control effector to enable high efficient flight.

use a tail at all. "*Their problem was maximal lift, and they sacrificed almost everything to get this ... hollow bones, high metabolic rate ...* [thus they] *could not afford the luxury of a vertical tail*". Instead they use wing sweep to provide dihedral effect as enabled by the flexible wing arrangement. The wing tips of birds are swept aft and tilted upward, just enough to provide this stabilising characteristic. This second characteristic of the swept-wing is a weak and subtle effect, applied efficiently by flying animals to bypass the need for vertical tails and rudders that we find necessary on nearly all of artificial flying creations. Rationale is the highly destabilising effect of the available propulsion systems and their integration. Recent interest in aircraft without vertical fins is motivated by flight performance objectives and/or stealth requirements, as exemplified by the Horten flying wings, B-2 and X-36.

Chapter 1 has reviewed some of the detail why today's aircraft layout has matured to be of the 707-type tail-aft configuration (TAC). For this type of inherently stable airframe, the function of each part tends to be highly specialised. For acquisition of local efficiency optima, the separation of function is distinct. The wing provides lift efficiently (and roll control), the fuselage carries the payload (and supports the landing gear and empennage), and the tail provides pitch and yaw stability and control. As a result, the 707-type aircraft resembles a highly disintegrated configuration concept in contrast to the highly integrated flying wing configuration (FWC). The stability and control functionality of the flying wing layout is merged into the primary lifting surface. The provision of volume for fuel, payload, etc. may be blended into the lifting surface or alternatively into an adjoining fuselage.

The design process of an integrated vehicle like *Quetzalcoatlus Northropi* or the BWB flying wing is far more demanding and complex compared to the design process demanded by the conventional 707-type arrangement. As MacCready recalls "*The stability/control challenge is perhaps the paramount problem ...*" [16]. Clearly, to enable future efficient aerospace vehicle design, a truly informed approach is mandatory when addressing the complex issue of aircraft configuration selection as coupled with stability and control, certification issues, and other design disciplines. Kroo stimulates creativity by stating, that within the ecological niche of biological flight "*Nature seems not to have converged on the 707-like configuration...*" [27], a motivation worth proving for artificial flight.

2.2.2 Emulation of Nature's Evolutionary Process

Charles Darwin proposed in his famous publication *On the Origin of Species by Means of Natural Selection, or the Preservation of Favoured Races in the Struggle for Life* in 1859 [28] the theory based on the struggle for existence, descent with modification and a gradual change of the species by the forces of natural selection. Today, few people question that species change over time. Philosopher Cronin remarks that "*Darwinism is the only solution to complex design that's occurred without a designer*" [29].

Technologists are using the idea of natural selection in a modern approach to design. For artificial flight, technical solutions have to be sought fifty million times or so faster compared to natural evolution, which had a long time to perfect its flight vehicles. Two key-elements are prerequisite when venturing emulation of the design-ability of master designer nature within reasonable turn-around time spans. (a) Utilisation and advancement of existing design knowledge and statistics, and (b) an automated inverse design process that accounts for interactions of several highest-of-importance engineering disciplines.

(a) The availability of **design knowledge** is the heart of any technical design process. Nature is gaining on any design challenge in a highly diversified manner. The *evolutionary* part is originated with an immense statistical database, whereby the *revolutionary* element is enabled by allowing a multitude of design trends with the philosophy, that high risk projects enable future low-risk undertakings. It needs to be recalled that the reason for the mass extinction of the pterosaurs and other species like the dinosaurs is widely debated. It does not appear to be valid to conclude, that *Quetzalcoatlus Northropi* vanished because of being a high risk project whereby birds represent today's low risk alternatives. Nature's evolutionary process is comparable to the human learning process. The key to accelerated technical development, as contrasted by nature's long-duration evolution, is the human ability to learn, to preserve, and make readily available the steadily growing inventory of knowledge. As will be discussed in Sect. 2.3, it is exactly this element of inefficient knowledge management which obstructs man-made design creativity, resulting in constrained technology advancements.
(b) The emerging **technical processes** capable of accelerating aerospace vehicle design towards a better design compromise fall into two groups as characterised by Wood and Bauer [30]. At first, those are Computationally-Based (CB) Design such as Multi-Disciplinary Optimisation, and secondly those are Knowledged-Based (KB) Design such as Decision-Based Design. Both systems have in common to represent a design process that accounts for the effects of interactions of several engineering disciplines with their specific knowledge. CB Design assumes that a design process can be modelled in the computer alone to allow the computer to find the optimum design more efficiently than a human. Wood and Bauer characterise rightly, "*CB Design is based upon the use of explicit and critical knowledge only and does not recognize the role of tacit and intuitive knowledge and other human senses and capabilities in the design process*". In contrast, "*KB Design systems consist of tools that allow a designer to utilize skills, senses, and knowledge in pursuit of a desired outcome*". Without discussing further implications of CB- and KB Design synthesis systems in the present context, a modern design environment has to embed the computer strengths of the CB Design system, such as analysis routines and mathematical optimisation. The optimiser is a modern coupling tool for harmonising individual physical disciplines against each other, as illustrated by Gill et al. [31]. With respect to master designer nature, it is the

genetic algorithm (GA) which is a direct simulation of biological evolution. Crispin formulates "*GAs have been used in optimisation, since they mimic adaptation processes believed to play an important role in the causes of evolution ... Genetic algorithms are based on the paradigm of Darwinian evolution*" [32]. The physical rationale is the focus for the KB system, in that it must be an open, thus learning system. Without the knowledge-enrichment capability one would spend more time and resources re-creating than creating, a fact not observed in nature-made creatures.

2.3 Design Knowledge

Theodore von Kármán recalls that "*Problems never have final or universal solutions and only a constant inquisitive attitude towards science and a ceaseless and swift adaptation to new developments can maintain the security of this nation*" [33].

The author entirely shares the views by Wood and Bauer [30], that is important for the aircraft conceptual design process to recognise, that the design philosophy, not the design process, defines the design space. "*The design philosophy, which is developed by the designer, must not be constrained by known rules, constraints, or by computational tools ... The efficiency and accuracy of the conceptual design phase is directly related to the knowledge used, thus we must focus on including ever greater amounts of knowledge into the conceptual design phase.*"

2.3.1 Knowledge—A Definition

A review of the literature on knowledge shows, that confusion prevails about what data, information, and knowledge are. Such confusion has resulted in enormous expenditures on technology initiatives that rarely deliver what the initiating body really needed or thought they were getting, see for example Brézillon [34]. Davenport and Prusak [35] clarify that "*Knowledge is neither data nor information, though it is related to both, and the differences between these terms are often a matter of degree*". Clearly, data, information, and knowledge are not interchangeable concepts. The author decided to adopt the working definition of knowledge as offered by Davenport and Prusak [35] and Miles and Moore [36].

Data by itself has little relevance or purpose due to its characteristics of being a set of discrete, objective facts about events. Data represents raw material without implying any judgement or interpretation, thus it says nothing about its own importance or irrelevance. Data is important because it is essential for the creation of information.

Information can be thought of as data that makes a difference due to its impact on judgement and behaviour. Information must inform, thus it has a meaning and it is organised to some purpose. Data becomes information when its originator adds meaning and value in various ways. "*The corollary for today's managers is that having more information technology will not necessarily improve the state of information*" [35].

Knowledge represents a mixture of experience, values, contextual information, and expert insight that provides a setting for evaluating and incorporating new experiences and information. "*It originates and is applied in the minds of knowers. In organizations, it often becomes embedded not only in documents or repositories but also in organizational routines, processes, practices, and norms*" [35]. Miles and Moore [36] further classify knowledge into algorithmic knowledge and heuristics, "*... one can conceive of the algorithmic approach as utilising equations which are typically based on Newtonian physics, whereas the heuristic approach uses rules of thumb based on experience.*"

It is a characteristic of conceptual design to utilise both types of knowledge in parallel. The algorithmic approach is more complex but provides an accurate solution. "*The use of the computer assisted design process can lead to accurate algorithmic processes being made available to designers in a way that imposes little or no time penalty*" [36]. In contrast, heuristics is a rule of thumb which, at a first glance, saves time, "*... is easy to use but cannot always be guaranteed to give the correct solution.*" [36]

Derivation of knowledge[4] is clearly a human-centred activity. Knowledge can be obtained "*... from individuals or groups of knowers, or sometimes in organisational routines. It is delivered through structured media such as books and documents, and person-to-person contacts ranging from conversations to apprenticeships*" as Davenport and Prusak clarify. Knowledge develops over time and assembles via certain key components such as experience, truth, complexity, judgement, rules of thumb and intuition, values and beliefs. Clearly, it is a misconception that knowledge-building happens only via hands-on practical experience as one extreme or alternatively through scientific or academic experience on the other hand. For the aircraft conceptual design environment it is of paramount importance, to assemble the right composition and dosage of knowledge-contributing elements into the design team. Bavitz comments rightly that "*The diversified backgrounds of those ... designers are essential to the formulation of viable advanced concepts*" [37].

2.3.2 *Quest for Engineering Design Knowledge*

What has to be avoided most is that when knowledge stops evolving, it turns into opinion or dogma. It has to be re-emphasised that there is a big gap between scientific research and the engineering product, which has to be bridged by the art of the engineer as expressed by Vincenti [38]. Theodore von Kármán has

[4]"*Knowledge derives from information as information derives from data.*" [35]

characterised the difference between engineering and science as follows: "*The scientist explores what exists; the engineer creates what isn't*". Consequently, the creative, constructive knowledge of the engineer is the knowledge needed to implement that art. "*What engineers do, however, depends on what they know,...*". Thus, the main effort concentrates on producing and organising knowledge. Any shift in the knowledge of the practitioner plays a crucial role in technology development.

The aircraft conceptual design process is primarily a *design* process, itself representing the multidisciplinary sciences & engineering domain. However, the knowledge required to perform conceptual studies can not be limited to engineering *design knowledge* only. The author agrees with the classification by Vincenti in that the engineering practice is determined by the three sequential categories: (a) design, (b) construction & production, and (c) operation. The multidisciplinary aircraft conceptual design task requires all three categories to be considered per definition. The design team needs to internalise the requirements prescribed from all three categories as they determine the enabling knowledge. It is becoming clear that knowledge forms the primary focus. To complement this view, the inseparability of knowledge and its practical application is a characteristic of engineering practice. Clearly, an imbalance of knowledge and application causes some of the problems as illustrated in Chap. 1.

As outlined above, the aerospace vehicle design activity has to incorporate expertise and experience gained from the entirety of design, construction & production, and operation. Appreciating and embodiment of these effects, but concentrating on the design activity itself, *design* has been conveniently categorised by Vincenti [38] as follows.

Normal Design Its main character is evolutionary rather revolutionary. However it is a misconception that 'evolution' always proceeds via smooth continuous variation.

Radical Design It contains almost all the elements of normal design (evolution). As expressed by Vincenti, "*... the complications of novelty, however, will add the usual perplexing concerns of creative invention*." Novelty translates into largely unknown elements of the design, construction & production, and operation product development cycle and enabling knowledge.

The knowledge required for radical design is enormously diverse and complex. As a result, the activities to generate knowledge for radical design become far from normal and day-to-day. As emphasised in Chap. 1, the present research investigation focuses on the development of a *generic* stability and control methodology. The primary focus lies on control effector sizing at aircraft conceptual design level for subsonic to hypersonic aerospace vehicle designs of conventional and unconventional aircraft configuration layout. Clearly, the term 'generic' implies utilisation and advancement of normal but in particular radical aerospace vehicle design knowledge as the vital ingredient for the present research investigation.

2.3.3 Novelty and Associated Knowledge Available

As illustrated in Chap. 1, a major inconsistency can be observed in the ability to design advanced aerospace vehicles with respect to knowledge required and knowledge available. Two contributing factors have to be considered. At first, advanced and especially 'novel' vehicle design is, as a fact, characterised by permanent lack of knowledge available. As implied by novelty, design knowledge available naturally lags behind design knowledge required. The degree of this discrepancy is a measure for the design risks involved. Other factors are contributing towards the rather stagnant design technology landscape. Advanced aircraft sizing ideas have been investigated periodically throughout aviation history. This, however, has created in the past and creates today the following situation.

(a) Insufficient conservation and documentation of design knowledge gained, resulting in minimal knowledge transfer to new aerospace vehicle designer generations.
(b) Lack of development of adequate design guidelines based on then present expertise and experience.
(c) Enlarged complexity of radical design often not well understood or simply deterrent in contrast to normal design. This is reflected by only a handful of individuals in the *AeroScience Triangle*[5] capable of contributing productive and objective.
(d) Peer pressure from individuals with a position to defend.
(e) Risk-derived resistance.
(f) Lack of being familiar with existing literature prevents a systematic approach to the problem.

As discussed before, the ability to perform efficient multidisciplinary design is quickly becoming a lost skill. A wide range of technical solutions for a multitude of problems have been assessed and demonstrated in aeronautical history. Unfortunately, much of that knowledge has been either ignored for a variety of reasons or it has been simply forgotten. Some of today's conventional and unconventional design proposals could appear less risky or radical, if a true state-of-the-art aerospace vehicle design 'toolbox' would be maintained. Today, only a proportion of the historically grown design knowledge[6] is available as expertise for the aerospace vehicle design environment. As a result, a discrepancy has to be accepted between 'what can be done' to 'what could be done'. Hoey puts it as follows: "*I now understand why we can't figure out how the pyramids were built. Technology is very short-lived if it isn't used and nurtured!*" [39]

Bearing in mind some of the promising new design concepts under investigation today, the necessity of state-of-the-art technology awareness for minimising risk and cost by maximising access to naturally deficient radical design knowledge, is a

[5]The AeroScience Triangle resembles Industry and Operator, Academia, and Research Institution.
[6]Including normal and radical design knowledge.

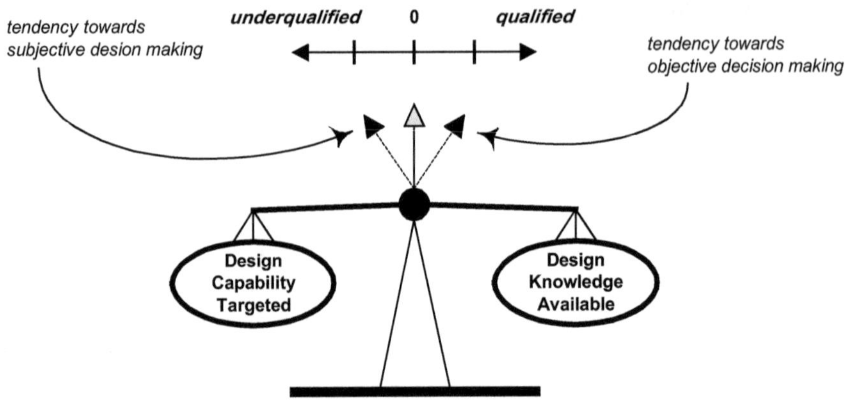

Fig. 2.4 Harmonisation of design capabilities targeted with design knowledge available

key element to advance aerospace science and technology. A mismatch between targeted novelty and associated design knowledge available results in deviation from objective decision making towards subjective decision making with the above discussed effects on development risk and cost involved. The interrelation between design capability targeted and design knowledge available is illustrated in Fig. 2.4.

The present research investigation is confronted to address the inconsistency problem of novelty and associated knowledge available. As a consequence, it is necessary to organise and advance normal and radical design knowledge until a workable balance can be provided.

2.4 Research Strategy Selected

Deficiencies in normal and radical design knowledge available to the project engineer hinder the proposal of state-of-the-art and advanced design concepts with minimum development risk and cost involved. It is the aim of the present research undertaking to build the enabling means for empowering critical evaluation of aerospace vehicle excess design potential on an impartial and rational basis, an activity which specifically translates into knowledge-demand. The aircraft conceptual design engineer has the responsibility to deliver decision-making bodies with objective and trust-worthy argumentation. The research strategy selected is defined to contribute towards that aim. Figure 2.5 sketches the research strategy conceived for the present research investigation. The modus operandi is represented via concentric spheres, where the work sequence proceeds from the outer to the inner layers. The process starts from the complete technology domain, passes in a pre-specified sequence various filtering levels until the aircraft conceptual design relevant knowledge is sufficiently assembled. The process ends when the methodology concept conceived will be evaluated against the research objectives specified in Chap. 1.

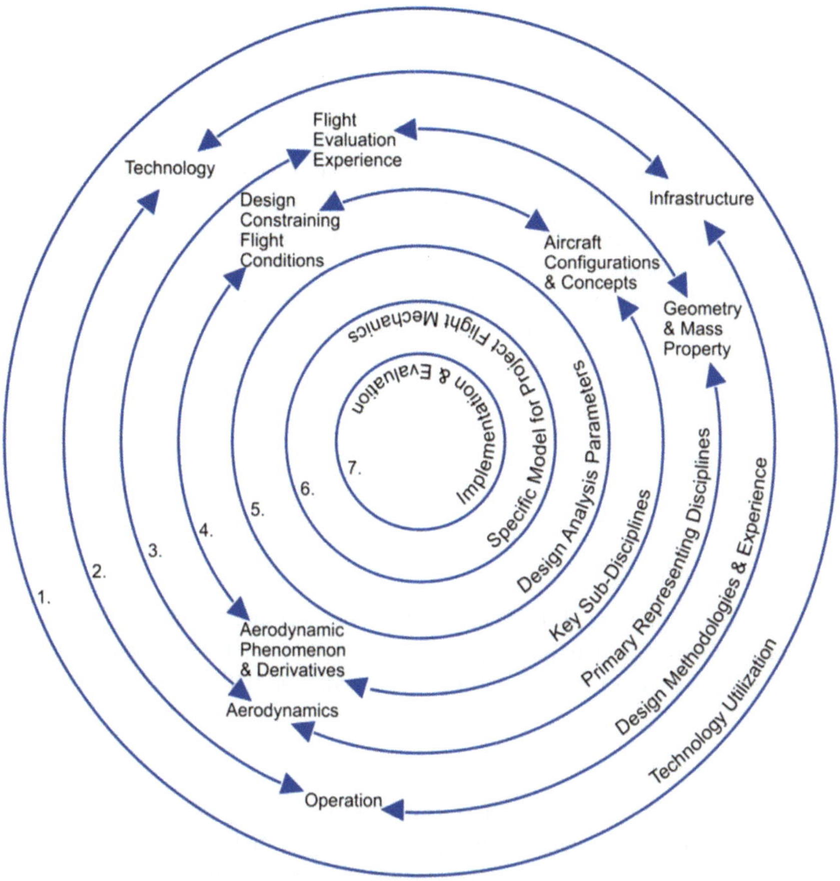

Fig. 2.5 Concentric evolution spheres represent the research strategy selected for the development of a generic stability and control methodology concept

This approach has been established respecting the aircraft project engineer's twofold qualifications required. First, the *generalist* with the capability to understand 'system aircraft' in a multidisciplinary design context, to enable authoritative decision-making (intuitive inspiration). Second, the *specialist* with the capability to understand, transform, and modify information and knowledge based on state-of-the-art disciplinary expertise (scientific distinction). Each layer of the concentric evolution sphere, as applied within the present research undertaking, is characterised briefly below.

1. Technology Utilisation

This layer represents the full technology domain, covering the complete range of engineering practice (design, construction & production, and operation). Sources for utilising normal and radical design knowledge have been: (a) public domain literature, (b) institution and company internal sources, and (c) expert advice.

2. Design Methodologies and Design Experience

This layer represents the first deduction step from the full technology domain. Here, aerospace vehicle design motivations and constraints are filtered along three categories: (1) *infrastructure* (political contents, work share, financing, development risk awareness, conservatism,…); (2) *operation* (mission specification, aircraft category, performance, safety, critical flight conditions,…); (3) *technology* (state-of-the-art technology, design philosophy, aircraft configuration,…). Clearly, this layer intends to build multidisciplinary understanding to enable informed decision-making at more disciplinary layers.

3. Primary '*Project Flight Mechanics*'[7] Representing Disciplines

Disciplinary topics interfering with flight mechanics at conceptual design level are (a) geometry and mass properties, (b) aerodynamics, and (c) flight evaluation experience.

4. Key Sub-disciplines

Key elements within the disciplines identified are (a) aircraft configurations and concepts [geometry and mass properties], (b) configuration aerodynamics [aerodynamics], and (c) design constraining flight conditions (DCFC) [flight evaluation experience].

5. Design and Analysis Parameters

The identification and extraction process of global[8] design parameters represents a key activity throughout the research period. With completion of this level, an informed methodology development sequence can be attempted.

6. Method Construction

The first five sphere layers have provided a sound understanding of the problems to be addressed. The methodology concept constructed is presented with a structogram, as detailed by the first two elements of the standard software development sequence (a) physical modelling and (b) mathematical modelling. The two successive steps (c) software engineering and (d) software execution for validation and calibration purposes are clearly beyond the scope of the present research undertaking.

7. Implementation and Evaluation

The stability and control methodology concept may be implemented into an aircraft conceptual design environment, followed by its final validation and calibration.

[7]'Project Flight Mechanics' is a synonym for the conceptual design capability targeted with *AeroMech*.

[8]Global parameters are parameters relevant for the aircraft conceptual design level.

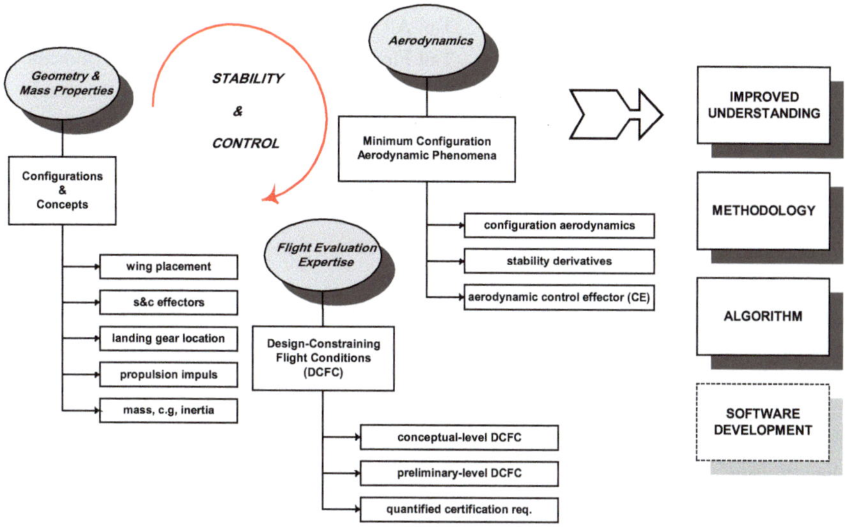

Fig. 2.6 Interdependence of subject matters to be considered for development of a generic stability and control methodology for aircraft conceptual design level

Successive evaluation needs to demonstrate its influence on aircraft conceptual design.

The research strategy outlined has been systematically followed throughout the development period, to enable an informed approach towards a generic stability and control methodology. Figure 2.6 depicts the primary disciplines interfering with stability and control at conceptual design level. The overall work sequence for the development of the generic methodology concept is shown on the right. As indicated, the final software-specific development steps have been beyond the scope of the present research undertaking. Further detail and explanation is given in the individual sections.

2.5 Design Knowledge Utilisation

A key element to enable advancements in aerospace science and technology, is effective management of the knowledge-generation and knowledge-preservation activity. As illustrated before, only improved understanding with regards to project aims and objectives will enable an informed and structured approach. An example for ineffective knowledge management is vividly illustrated by Scott [40], reviewing the conceptual design activities towards Northrop's B-2 Stealth bomber. "*In fact, none of the Northrop team that worked on the B-2 had ever been associated with the original YB-49. Managers on the B-2 even had a hard time finding a company employee who had any experience with the old flying wings. After all, that*

Table 2.3 Classification of symmetric and asymmetric aircraft configurations

TAC	**T**ail-**A**ft **C**onfiguration	Symmetric
TFC	**T**ail-**F**irst **C**onfiguration	Symmetric
TSC	**T**hree-**S**urface **C**onfiguration	Symmetric
FWC	**F**lying-**W**ing **C**onfiguration	Symmetric
OWC	**O**blique-**W**ing **C**onfiguration	Asymmetric
OFWC	**O**blique **F**lying-**W**ing **C**onfiguration	Asymmetric

effort ended about 35 years before the B-2 concept surfaced ... Even more frustrating, though, was the lack of data; engineers looked through company archives for basic flying wing information, but with no luck."

In the present context, the research challenge turned out to be the question: How far can one go with the technical resources, personal support and time available in the attempt, to develop a generic aircraft conceptual design stability and control methodology? Torenbeek [41] comments on the overall tail-sizing problem by referring to only the conventional tail-aft configurations (TAC): on "*During the preliminary design stage, the tail surfaces may present one of the most difficult problems in the dimensioning of the main parts of the aircraft, and this, in turn, may lead to many iterations.*" Acknowledging the challenge of the research undertaking, no compromise in scope and complexity has been allowed for. A saying comes to mind: "*... things should be as simple as possible, but no simpler...*".

Scope and complexity of the present research undertaking is seen as *catalyst opportunity*, which translates into a chance to evaluate the feasibility of a generic-type design methodology or parts of it. Clearly, the true complication of the task has been hidden in the inclusion of asymmetric aircraft configurations in addition to the range of symmetric aircraft layouts. Table 2.3 lists aircraft configurations which are identified to be of direct relevance for the development of a generic methodology.[9]

The asymmetric aircraft configuration type has been selected to be the *benchmark vehicle* for the development of the generic stability and control methodology. Technically, asymmetric aircraft configurations resemble the most demanding aircraft type with an unmatched performance potential. Asymmetric aircraft types, in particular the OFWC, are the single correct choice for minimum wave drag and minimum induced drag due to lift. In addition, the structural efficiency of the OFWC is superior due to its span-loader concept and volumetric efficiency. However, the real complications are their inherent stability and control characteristics. In contrast to symmetric aircraft types, asymmetric aircraft represent highly coupled systems due to inertia coupling and aerodynamic coupling effects. Figure 2.7 illustrates the performance and stability and control aspects of asymmetric aircraft. For introductory reading, Nelms [42] has produced an excellent summary of oblique-wing technology programmes. A more recent summary of oblique flying wing studies is presented by Li, Seebass, and Sobieczky [43].

[9]As implied by the term 'generic', the configuration selection is easily expandable.

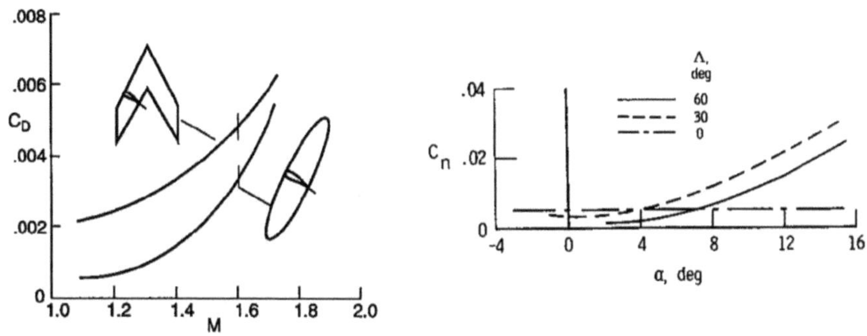

Fig. 2.7 Comparison of a sweptback and oblique wing (left) [44] and untrimmed yawing moment coefficient at unity load factor for different wing sweep angles of the AD-1 research aircraft (right) [45]

It has been the aim of the present research undertaking, to develop a *generic* stability and control methodology. The term 'generic' implies, that the asymmetric aircraft type is considered to be the most general aircraft arrangement, whereby symmetric types represent rather 'simplified' or special cases where certain simplifying assumptions are acceptable during the early design phase. Clearly, functionality of the methodology concept for asymmetric aircraft ascertains functionality for the range of symmetric aircraft. An important by-product of this approach is the capability, to enable handling of asymmetric flight conditions of symmetric and asymmetric aircraft configurations, a non-typical ability for a conceptual design method.

Figure 2.8 qualitatively illustrates the coupling between preparatory work (knowledge utilisation) required and development time necessary. As a fact, the quality of preparatory work needs to be particularly high for the present research undertaking, leading to an intrinsically broad scope of the investigation. The maximum of 'Preparatory Work Quality' obtainable is constrained by the time span available and by the availability of high-quality information resources. The 'Can Do Limit' qualitatively characterises the minimum quality and time requirement, below which development of a generic methodology concept is thought not to be possible at all. In the present context, the enabling constituents like information sources and time available have been judged healthy, to surpass the 'Can Do Limit' or critical-mass towards the research objectives defined.

It has become clear during the early stages of the research undertaking, that successful development of a generic stability and control methodology will demand systematic technology utilisation work by seeking an understanding of past, present, and future technology developments. In particular, it has been a primary aim to acknowledge the pool of case studies available from the past to the present, representing the anatomy of successes and failures. Figure 2.9 lists case studies selected for the construction of a comprehensive knowledge baseline.

Criteria for the selection of a particular aircraft case study has been its overall degree of significance towards construction of the representative generic aerospace

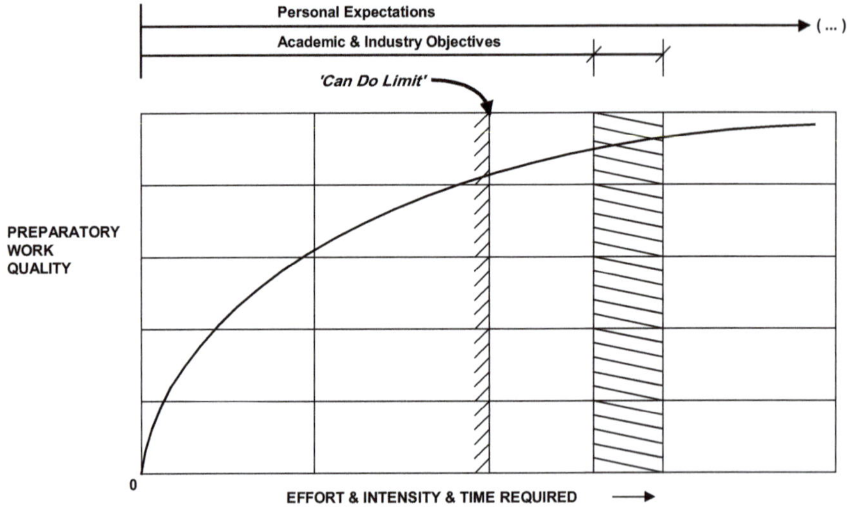

Fig. 2.8 Coupling between minimum preparatory work required and synthesis work to construct a generic stability and control methodology

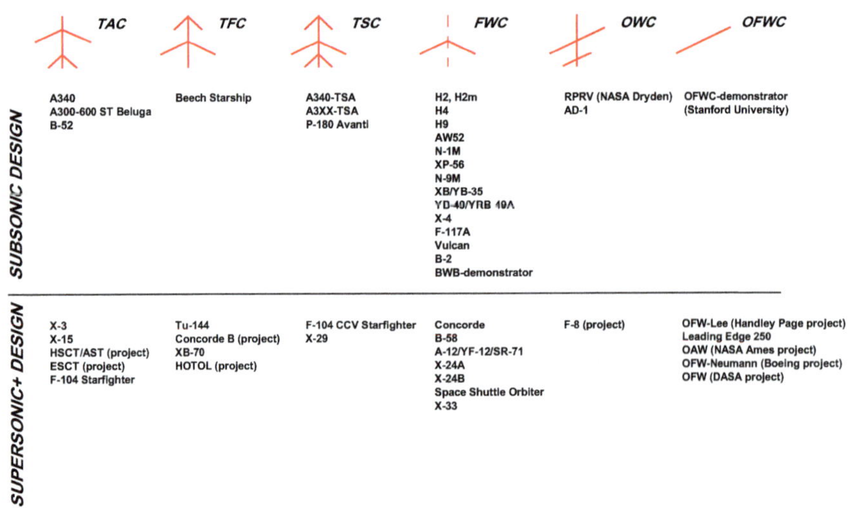

Fig. 2.9 Representative case studies selected for assemblage of a conceptual design knowledge baseline

vehicle design knowledge baseline. Obviously, the meaningfulness of a case study is dependent on the quality and richness of information available. This illustrates the importance of the knowledge utilisation activities, ranging from the search for documentation to expert advice.

Table 2.4 Organisation-scheme of knowledge utilisation activities towards conceptual design parameter reduction

1	Data-Base System (DBS) [computer-based]	Section 2.5.1
2	Report: *'Stability and Control Characteristics of Subsonic, Supersonic, and Hypersonic Aircraft Configurations'*	Section 2.5.1
3	Knowledge-Based System (KBS) [computer-based]	Section 2.5.2
4	Report: *'Aircraft Configuration Characterisation For Project Flight Mechanics'*	Section 2.5.2
5	Final design parameter reduction process	Chapter 4

Aerospace engineering design can be broadly idealised as consisting of three components which are, however, largely interrelated, being conceptual design, preliminary design, and detail design. Miles and Moore [36] rightly argue that "*The area which to date has been resistant to the introduction of computer systems is conceptual design.*" As a fact, computer systems dominate preliminary design and detail design to a far greater degree compared to the conceptual design level. Ideally, a combination of a *Data-Base System* (DBS) containing information on existing designs, and a *Knowledge-Based System* (KBS) with knowledge about the design process, coupled to analysis packages organized in a multidisciplinary synthesis system, should provide the designer with a great deal of assistance at all stages. However, the elements usually missing in conceptual design methodologies are, in particular, an up-to-date DBS and KBS for making data, information, and knowledge readily available for design-decision making.

The following two sections present the aircraft conceptual design DBS and KBS, as developed for the present research undertaking. The main research body is built on both systems. Table 2.4 summarises the knowledge organising activities performed in sequence for the current research undertaking, enabling a final design parameter reduction process for method construction.

2.5.1 *Aircraft Conceptual Design* Data-Base System (DBS)

The first step in utilising existing aerospace vehicle design knowledge has been an extensive literature survey, which in itself has been an ongoing effort throughout the research period. Source for accessing normal and radical design knowledge have been (a) public domain literature,[10] (b) institution and company internal sources,[11] and (c) expert advice. For efficient handling of design related data, information, and knowledge, a dedicated computer-based aircraft conceptual design *Data-Base*

[10]Public domain literature consists of books, periodical articles, grey literature, handbooks, standards, citation searches, and the world-wide web.

[11]The author utilised, to a large degree, company internal information and industry specialist advice.

System (DBS) has been set up. Appendix A.1 presents the literature DBS file-structure. This system handles disciplinary and inter-disciplinary literature relevant for conceptual design (methodologies, flight mechanics, aerodynamics, etc.), interview-protocols, aircraft case study information (descriptive-, historical-, numerical information on conventional and unconventional aircraft configurations), simulation and flight test information, etc. The overall requirement for the creation of the DBS has been simplicity in construction, maintenance, and operation, to comply with the underlying research project time constraints.

The comprehensive database prepared by Lorell and Levaux [46, 47] contains information on most winged and related military and commercial aircraft R&D programs undertaken by U.S. aerospace contractors after World War II. However, this database has been set up to evaluate primarily the role of prior fighter, bomber, and related research and development (R&D) experience among prime contractors in promoting successful R&D programmes. The descriptive, historical, and numerical information contained in the database itself is not of immediate relevance for aircraft conceptual design, whereby the information provided in the main body of the two references certainly is. Mason [48] offers an interesting compilation of aircraft conceptual design information sources, which have been particularly helpful in the initial phase of DBS construction. Overall, the author is not aware of any published aircraft conceptual design database similar to the DBS, offering the scope and contents as demanded by the present research undertaking.

A detailed description of the DBS is beyond the scope of the present discussion. The system has become a steadily growing, comprehensive, and effective working tool. Clearly, the quality of such system is only as good as the degree of completeness, actuality, and familiarity by the user. The DBS has matured to be the central instrument for managing aerospace vehicle design data, information and knowledge. However, the true potential of this system for utilising design knowledge[12] has been opened up by proceeding as follows:

1. availability of a *reference list* containing meaningful entries;
(*DBS*)
2. availability of these references as a *hardcopy* on the table;
(*DBS*)
3. utilisation of time to *absorb* the information and knowledge;
(*DBS*)
4. *review, select, classify, subtract, and document* the knowledge provided;
(*DBS*)
5. *extraction, combination and utilisation* of the knowledge in a pre-defined manner.
(*KBS*)

The first four steps are handled within the DBS. The DBS has been put to use to provide in an intermediate step (step four) suitably selected, structured, and

[12]In the present context, the term knowledge implies data and information.

condensed aircraft conceptual design knowledge in form of a report with the title '*Stability and Control Characteristics of Subsonic, Supersonic, and Hypersonic Aircraft Configurations*' [49]. The table of contents of this document is reproduced in Appendix A.2. The research goal, to develop a generic stability and control methodology for conceptual design application, requires to account for as many design-related interactions as necessary, since the rationale for the evolution of aircraft is diverse as a quick browse through aviation history reveals. The aerospace vehicle design disciplines identified relevant, see Fig. 2.6, and the representative case studies of design ingenuity selected, see Fig. 2.9, both elements need to be appreciated mutually, to efficiently serve the design understanding where innovation provided answers to otherwise troublesome problems. The DBS and the report, both together embody a knowledge- and technology baseline attained, which is considered state-of-the-art for the current research undertaking.

The document *'Stability and Control Characteristics of Subsonic, Supersonic, and Hypersonic Aircraft Configurations'* is subdivided into two main sections.

(a) *Aircraft Configuration Independent*
(b) *Aircraft Configuration Dependent*

Part (a) describes physical phenomena and constraints independent on the aircraft configuration, whereby section (b) concentrates on aircraft configuration dependent design considerations as illustrated along selected aerospace case studies. The configuration-independent metrics discussed are:

Stability and Control; Control Effectors; Aeroelasticity; Airframe-Propulsion Interactions; Certification Requirements;...

The configuration-dependent metrics discusses case studies along the following scheme:

Manufacturer; Initiation; Mission Objectives; Number Built; Historical Perspective; General Arrangement; Configuration Evolution; Aerodynamic Design; Stability and Control Design; First Flight Dates; Flight Testing; Specification; Summary.

This strategy has enabled the author to live through and comprehend selected aircraft programmes and projects in an expedient manner, thus acknowledging technical and non-technical design-reasoning.

Summarising, all five knowledge utilisation steps have been followed and completed to a representative and satisfactory degree within the time span allocated. Step five itself has been organised outside the DBS. Clearly, it is the process of knowledge extraction, -compilation, and -provision into an organised and concise format, which finally makes aircraft conceptual design knowledge available 'at the fingertips' for knowledge generation and finally problem solving activities. For this purpose, a simplified *Knowledge-Based System* (KBS) has been constructed as detailed in the following section.

2.5.2 *Aircraft Conceptual Design* Knowledge-Based System (KBS)

The aircraft conceptual design *Knowledge-Based System* (KBS), as developed for the present research undertaking, has to be considered an early development-version of a fully operational design KBS. Without reiterating the capability of exemplary KBSs [50], the system developed here is a 'manual' system in contrast to the ideally automated KBS.[13] However, independent on the degree of automation, both systems have in common that knowledge itself is the focus and that the knowledge acquisition activity is recognised as being one of the most problematic areas of KBS development [51, 52]. Clearly, it is the knowledge acquisition and knowledge utilisation activity, where the priorities for the present aircraft conceptual design KBS have been laid due to time constraints imposed. The implementation of the KBS-typical computer overhead, required to convert the manual-type KBS into an automatic KBS, has been considered secondary and therefore excluded from the present research undertaking.

The primary objective of developing the dedicated aircraft conceptual design KBS has been, to make relevant normal and radical design knowledge effortlessly available. The particular strength of the system manifests, in that it enables the user to advance his/her understanding with respect to the variety of aircraft configurations by identifying aircraft configuration commonalties and peculiarities. This feature has been empowered by placing particular emphasis on consistently grouped aircraft configuration-specific design knowledge. As a result, design detail, for example longitudinal stability, can be compared between the range of aircraft configurations as defined in Table 2.3. This approach finally enables a reliable and trust-worthy generic aircraft configuration parameter identification process.

The aircraft conceptual design KBS is subdivided into two main sections[14]:

(a) *Longitudinal Motion*
(b) *Lateral/Directional Motion*

Each motion is subdivided into:

– Flight Character	(Design Constraining Flight Conditions: trim, control, stability)
– Aerodynamic Character	(Stability and Control Derivatives: u, u/t, $w(\alpha)$, w/t (α/t), …)
– Flow Character	(Flow Phenomena: tuck, pitch-up, non-linearity, …)

[13]In general, the capabilities of the ideal KBS are not achievable yet with current technology. Thus, even industrial KBS have only reached a rudimentary level of sophistication.

[14]It has been found necessary for clarity reasons, if possible, to separate physical phenomena into longitudinal and lateral/directional motion. In case of physical coupling effects, clear reference is made.

– Additional Grounds (landing gear location, geometry limitations, c.g. range, …)

The 'living-character' of the KBS is ensured by permitting unconstrained knowledge entries into the organizing scales or categories as gained through the knowledge generation activities. Clearly, the DBS and the KBS are both living systems, which have matured towards fully functional and practical means throughout the research period. Appendix A.3 reproduces the contents of the aircraft conceptual design KBS.

2.6 Summary of Results

This chapter reviews the primary constituent essential for the aircraft conceptual design activity, being utilisation of design knowledge. In this context, it is of particular value to examine design office nature, which has evolved flying creatures far beyond today's human engineering ability in multiple areas. Emulation of nature's evolutionary process is of engineering interest, except for the evolutionary time span. Nature's largest flying animal, *Quetzalcoatlus Northropi*, is understood to inherit today's most advanced technology achievements, and thus can serve as a design case-study. Man-made design evolution and design revolution is primarily dependent on design knowledge available. Clearly, the composition of knowledge, its utilisation and generation are pivotal to enable development of a generic methodology concept. Howe accentuates such comprehension by saying that "*Life and money are both too short to have to repeat the learning process in each generation.*" [53]

The research strategy selected mirrors the above understanding by placing strong emphasis on systematic and thorough knowledge utilisation. The primary[15] literature familiarisation period provided a perspective on the original contribution the research might make to aerospace science, in particular to the conceptual design of aerospace vehicle control effectors. Completion of the search has convinced the researcher that not only is the study distinctive and different from previous research, but that it is worth doing. Clearly, no known party is working on a stability and control or other methodology of generic character.

The multi-disciplinary integrity of the subject made the development of a computer-based *Data-Base System* (DBS) and a manual *Knowledge-Based System* (KBS) obligatory. Both systems appear to be original in contents and scope for the conceptual design arena, in that they provide relevant design knowledge 'at the fingertips'. Bushnell [54] supports such approach by saying: "*... educate yourself about everything that has been done related to the problem, ... the best way of being creative is to violate the assumptions...*". Summarising, the objective of the knowledge utilising activity has been, to augment knowledge-generation and

[15]The literature survey consists of a primary and secondary search. The primary search is an initial familiarisation period whereby the secondary search endures throughout the research period.

problem-solving activities towards construction of the generic conceptual design stability and control methodology.

Chapter 3 presents an assessment of relevant elements of the aircraft conceptual design process, in particular design and certification requirements, aircraft synthesis procedures, configuration aerodynamics and stability and control in design.

References

1. Santayana, G., "Flux and Constancy in Human Nature," *Life of Reason*, Vol. 1, Ch.XII, 1905–1906.
2. Lilienthal, O., "Der Vogelflug als Grundlage der Fliegekunst," First Edition, *R. Gaertners Verlagsbuchhandlung*, Hermann Heyfelder, Berlin, 1889.
3. Rich, B.R. and Janos, L., "Skunk Works – A Personal Memoir of My Years at Lockheed," First Edition, *Little Brown and Company*, 1994.
4. Kroo, I., "Reinventing the Airplane: New Concepts for Flight in the 21st Century," Presentation, December 1995.
5. Davies, R.E.G., "Supersonic (Airliner) Non-Sense – A Case Study in Applied Market Research," *Paladwr Press*, 1998.
6. MacCready, P.B., "Natural and Artificial Flying Machines," Vol. IX, No. 3, *Technical Soaring*, Adapted from a Presentation at the Symposium on Perspectives in Fluid Mechanics, California Institute of Technology, 12 January 1985.
7. Rayner, J.M.V., "Vertebrate Flight: A Bibliography," Department of Zoology, University of Bristol, Last Update 5 October 1998.
8. Rayner, J.M.V., "A Bibliography of the Pterosauria," Privately Published, School of Biological Sciences, University of Bristol, 1 June 1998.
9. Herzog, K., "Anatomie und Flugbiologie der Vögel," First Edition, *Gustav Fischer Verlag Stuttgart*, 1968.
10. Saunders, M.W., "Flight and Nature," Second Edition, Printed in Great Britain, 7 Leazes Avenue, Chaldon, Surrey, 1972.
11. Chow, K., "Micro Air Vehicles – Technology Review and Design Study," M.Sc. Thesis, College of Aeronautics, Cranfield University, September 1999.
12. Ellington, C.P., van den Berg, C., Willmott, A.P., and Thomas, A.L.R., "Leading-Edge Vortices in Insect Flight," Vol. 384, No. 19, *Nature*, 26 December 1996, pp. 626–630.
13. Alexander, R.M., "Smokescreen Lifted on Insect Flight," Vol. 384, No. 19, *Nature*, 26 December 1996, pp. 609–610.
14. Winn, A., "Buzzing Along," Flight International, 23 December 1998–5 January 1999, p. 46.
15. Langston, W., Jr., "Pterosaurs," 224(2), *Scientific American*, February 1981, pp. 92–102.
16. MacCready, P.B., "QN – The Time Traveler™," NASM Research Report, 1985, pp. 185–194.
17. MacCready, P.B., "Of Birds, Bees, and Airplanes," 0278-6648/0500-0029, *IEEE Potentials*, May 1987.
18. Wellnhofer, P., "Die große Enzyklopädie der Flugsaurier," Third Edition, *Mosaik Verlag*, 1993.
19. Cowley, M., "Quetzalcoatlus Northropi," Model Aviation, August 1986, pp. 84–91.
20. Brooks, A.N., MacCready, P.B., Lissaman, P.B.S., and Morgan, W.R., "Development of a Wing-Flapping Flying Replica of the Largest Pterosaur," AIAA Paper 85-1446, *AIAA 21st Joint Propulsion Conference*, Monterey, CA, 8-11 July1985.
21. Hoey, R.G., "Research on the Stability and Control of Soaring Birds," AIAA Report 92-4122-CP, AIAA, 1992.
22. Ashkenas, I.L. and Klyde, D.H., "Tailless Aircraft Performance Improvements With Relaxed Static Stability," NASA Conractor Report CR-181806, NASA, March 1989.

23. Anon., "Negative Stability Tests Aid Airbus A3XX Design," Flight International, 15–21 September 1999, p. 12.
24. Nickel, K. and Wohlfahrt, M., "Tailles Aircraft in Theory and Practice," Translation of the First Edition 'Schwanzlose Flugzeuge' from 1990 by E.M. Brown, *Edward Arnold*, 1994.
25. Cunningham, J., "Quetzalcoatlus – Nature's Largest Flying Machine," Flapping Wings, *The Ornithopter Society Newsletter*, Winter 2000.
26. Hoey, R.G. and Gabor Miklos, G.L., "Genetic Aspects of Bird Evolution," Personal Communication, 4 March 1999.
27. Kroo, I., "Tailless Aerospace Vehicle Design - Recent Experiences," Chapter 7 of *Aerodynamics and Aeroacoustics*, K.Y. Fung, March 1993.
28. Darwin, C., "The Origin of Species," *Gramercy Books*, 1979.
29. Connor, S., "Man of the Millennium," *The Independent*, Tuesday Review, 24 August 1999.
30. Wood, R.M. and Bauer, X.S., "A Discussion of Knowledge Based Design," AIAA Paper 98-4944, *7th AIAA/USAF/NASA/ISSMO Symposium on Multidisciplinary Analysis and Optimization*, St. Louis, MO, 02-04 September 1998.
31. Gill, P.E., Murray, W., and Wright, M.H., "Practical Optimisation," Tenth Printing, *Academic Press*, 1995.
32. Crispin, Y., "Aircraft Multicriteria Optimisation Using Simulated Evolution," ICAS Paper 94-1.5.2, *32nd Aerospace Sciences Meeting and Exhibit*, Reno, NV, 10-13 January 1994, pp. 1520-1526.
33. Gorn, M.H., "The Universal Man – Theodore von Kármán's Life in Aeronautics," Smithsonian History of Aviation Series, *Smithsonian Institution Press*, 1992.
34. Brézillon, P., "Successes and Failures of KBSs in Real-World Applications," Knowledge-Based Systems 10, *Elsevier Science B.V.*, 1998, pp. 253–257.
35. Davenport, T.H. and Prusak, L., "Working Knowledge – How Organisations Manage What They Know," *Harvard Business School Press*, 1998.
36. Miles, J. and Moore, C., "Practical Knowledge-Based Systems in Conceptual Design," First Edition, *Springer-Verlag London Limited*, 1994.
37. Bavitz, P.C., "So You Want to be An Aerospace Vehicle Designer?!," Aerospace America, February 1964.
38. Vincenti, W.G., "What Engineers Know and How They Know It – Analytical Studies from Aeronautical History," John Hopkins Studies in the History of Technology, *The John Hopkins University Press*, 1993.
39. Hoey, R.G. (former USAF flight test engineer and lifting body project manager) and Chudoba, B., Personal Communication, 23 March 2000.
40. Scott, B., "Inside the Stealth Bomber – The B-2 Story," *TAB/AERO Books*, 1991.
41. Torenbeek, E., "Synthesis of Subsonic Airplane Design," 6th Printing, Delft University Press, *Kluwer Academic Publishers*, 1990.
42. Nelms, W.P., "Applications of Oblique-Wing Technology - An Overview," AIAA Paper 76-943, *AIAA Aircraft Systems and Technology Meeting*, Dallas, TX, 27-29 September 1976.
43. Li, P., Seebass, R., and Sobieczky, H., "The Oblique Flying Wing as the New Large Aircraft," ICAS 96-4.4.2, *20th Congress of the International Council of the Aeronautical Sciences*, 08-13 September 1996.
44. Jones, R.T., "Wing Theory," *Princeton University Press*, 1990.
45. Sim, A.G. and Curry, R.E., "Flight Characteristics of the AD-1 Oblique-Wing Research Aircraft," NASA TP-2223, NASA Ames Research Center, Moffett Field, CA, 01 March 1985.
46. Lorell, M.A., "Bomber R&D Since 1945 – The Role of Experience," Prepared for the United States Air Force, *RAND*, 1995.
47. Lorell, M.A. and Levaux, H.P., "The Cutting Edge – A Half Century of U.S. Fighter Aircraft R&D," Prepared for the United States Air Force, *Project Air Force*, RAND, 1998.
48. Mason, W.H., "Aerospace Vehicle Design Information Sources," *VPI Aerospace Vehicle Design Program and the NASA/USRA Advanced Design Program*, Department of Aerospace and Ocean Engineering, Virginia Polytechnic Institute and State University, 15 April 1999.

49. Chudoba, B., "Stability and Control Characteristics of Subsonic, Supersonic, and Hypersonic Aircraft Configurations," CoA Report NFP0103, Department of Aerospace Technology, College of Aeronautics, Cranfield University, April 2001.
50. Davis, R., "TEIRESIAS: Applications of Meta Level Knowledge," Part 2 of Davis, R. and Lenat, D.B., "Knowledge Based Systems in Artificial Intelligence," *McGraw-Hill*, 1982.
51. Welbank, M., "A Review of Knowledge Acquisition Techniques for Expert Systems," Martlesham Consultancy Services, British Telecom Research Laboratories, Ipswich, 1983.
52. Greenwell, M., "Knowledge Engineering For Expert Systems," *Ellis Horwood*, Chichester, 1987.
53. Howe, D., "Substitution of Sophistication for Common Sense," Airworthiness Aspects of New Technologies, *I.Mech.E. Seminar*, Bristol University, 20 November 1996.
54. Bushnell, D.M. (Chief Scientist, NASA Langley) and Chudoba, B., Personal Communication, College of Aeronautics, Cranfield University, 8 December 1998.

Chapter 3
Assessment of the Aircraft Conceptual Design Process

3.1 Introduction

Goldin rightly asks "*What tools do you need to stay aggressive* ..." while justifying NASA's *Faster, Better, Cheaper* approach [1]. With a similar intention, this chapter reviews essential constituents and peculiarities of the aircraft conceptual design process relevant in the context of developing a generic stability and control methodology. Examined are the problems of airworthiness, approaches to aircraft conceptual design synthesis, the methodology of aerodynamic project prediction, and finally stability and control in design. The peculiarities and challenges specific to the aircraft conceptual design process are highlighted for each of the above subject-matters, leading to the formulation of a requirement catalogue. After setting the scene with the formulation of development requirements, Chap. 4 outlines in detail how to satisfy those pre-defined requirements with respect to the development of a generic stability and control methodology for aircraft conceptual design.

3.2 Interrelationship Between Aerospace Vehicle Design and Airworthiness

Certification of advanced aircraft[1] is a challenge for those in the area of aeronautical engineering and science, as well as for those in the field of certification of aerospace vehicle design. Clearly, the aviation experimentalist must be prepared to explore the unknown in the field of certification. The present research investigation is particularly concerned with airworthiness issues. At first, the discipline of stability and control has a strong certification related focus on safety. Secondly, generic

[1]The expression advanced aircraft implies conventional aircraft and unconventional (novel) aircraft types.

B. Chudoba, *Stability and Control of Conventional and Unconventional Aerospace Vehicle Configurations*, Springer Aerospace Technology,
https://doi.org/10.1007/978-3-030-16856-8_3

aerospace vehicle design implies design of advanced flight vehicles. This fact represents a particular certification challenge due to the coupling of rational technical facts with irrational opinion-based fears about novelty.

Howe [2] narrates his concerns by saying "*Aviation history is littered with the debris of sophisticated projects which failed because inadequate initial thought was given to fundamental airworthiness considerations.*" Wood examines [3] many examples of projects from the depressing record of British aircraft developments since 1945, that suffered from requirements that were too restrictive, too ambitious or were changed during development. Conceding the organisational context and the political parameters shaping the organisation's options, it is a primary task of the aircraft conceptual design team to balance the narrow path between mundane technical conservatism and the implications of an overly radical approach.

3.2.1 Principles of the Certification Process

The following two definitions of airworthiness are a mandate for aircraft of conventional and unconventional configuration layout and of subsonic to hypersonic design.

> A declaration by an appropriate authority that an aerospace vehicle design is proven as fit for flight. [4]
>
> The acceptable safety standard of an aerospace vehicle designed and built according to applicable requirements, when operated within its intended environment and within its quantified and declared limitations, and maintained in accordance with procedures acceptable to the responsible authority. [5]

The following characterises the general elements of the certification process. It should be noted that aviation history shows several civil, military and experimental aircraft programmes, where the sequence and contents of the individual certification steps may deviate from the list below.

(i) An **organisation** to procure the design from those who perform the design.
(ii) **Airworthiness criteria** (design regulations, certification requirements, airworthiness code, airworthiness regulations, or airworthiness standards) of fitness for flight which are acceptable to the certification authority.
(iii) **Operational requirements** and finally a **specification** for the design.
(iv) A flight vehicle **design**.
(v) An **assessment** of that design by the certifying authority against the design requirements.
(vi) A **conclusion** by the authority that the design has (or has not) met the design requirements.
(vii) A **certification** (or not) of the design to be fit for flight.

Clearly, the legal responsibility for airworthiness, the Certificate of Airworthiness (C of A) rests with (a) the airworthiness authority, (b) the aircraft manufacturer, and (c) the operator. Overall, the primary objectives of airworthiness applicable to aircraft are, at first the safety of the people on the ground and on board, and secondly environmental aspects.

Without reiterating more detail in the present context, Bradshaw outlines [4] how the above key elements (i)–(vii) of airworthiness may be developed into general principles for the certification process by inclusion of decision paths and feed back loops. Overall, it has to be the intention of the certification process to assess any aircraft programme from an entirely objective perspective, allowing and supporting novelty when compliant with safety. Clearly, when discussing airworthiness requirements, it is essential to harmonise the interests and constraints given by the people involved from design to operation of an aircraft and the people who regulate the airworthiness activities. This does not imply that it is solely the imposition of the requirements which make an aircraft safe. As the following sections outline, some of the concepts behind airworthiness requirements may require modification.

3.2.2 Some Limitations of Airworthiness Codes

The aerospace vehicle designer has to make the correct choice of airworthiness regulations to which the aircraft will be designed due to their far-reaching influence on overall design and operation. Regulations have been introduced for all classes of aircraft, specifying certain minimum standards to which the design must conform. Table 3.1 summarises the important codes of regulation for design and fitness for flight. It should be noted that the US Mil Specs (MIL-F-8785C) and US Mil Stan (MIL-STD-1797A and B) are referring to *Flying Qualities of Piloted Vehicles*, being of particular interest for the present research investigation. This section describes some general limitations of airworthiness codes, whereby Sect. 3.2.3 is concerned with some of the airworthiness codes in more detail.

Table 3.1 Overview of selected aerospace vehicle design codes of airworthiness

Civil aircraft	BCAR (British Civil Airworthiness Requirements)
	FAR (Federal Aviation Requirements)
	JAR (European Joint Aviation Requirements)
	TSS (Concorde TSS Standards)
	Tentative Airworthiness Standards for Supersonic Transports (‘White Book’ for US HSCT)
Military aircraft	UK Def. Stan. (DEF-STAN 00-970)
	US Mil Specs (MIL-F-8785C)
	US Mil Stan. (MIL-STD-1797A and B)

The airworthiness codes listed in Table 3.1 have been developed, formulated and applied over many years. By their very nature, the following design-related problems arise with such collection of mature wisdom:

(i) For those intended for state-of-the-art subsonic civil production aircraft, the underlying historic assumptions of existing airworthiness regulations penalise the combination of safety and economics.
(ii) Changes of regulations during the development programme subsequently disturb the balance between initial design objectives and actual design decisions.
(iii) Consequent application of a single airworthiness code can result in over-specification and over-design, with attendant cost and schedule overruns.
(iv) The interrelationship of the aerospace vehicle design processes and operational requirements has not yet matured to the degree, to sufficiently utilise safety and economic advantages.
(v) As subsequently used for advanced technology and experimental aircraft, parts of the existing airworthiness regulations are not appropriate at all to authorise a renaissance of traditional design trends leading to design novelty.

Saha [6] identifies the concept behind all airworthiness regulations as related to the performance of an aircraft. The standard approach is to ensure a given level of design incident probability by defining a performance margin over the assumed datum performance. However, the performance margins taken into account for today's regulations are defined on the basis of aircraft of the forties and early fifties. Clearly, the underlying assumptions, when compared to today's design standard, differ in the areas of propulsion, speed and range, aircraft size and weight, wing sweep, flight-control system architecture, etc. Frustration materialises when realising, that technological innovations and improvements in aviation can only be translated into true economic gains through the regulatory framework. Clearly, the underlying performance margins of today's codes of airworthiness are obsolete.

The extent of application of airworthiness codes is discussed by Coburn [7]. Over the past twenty years, several military aircraft programmes have successfully adapted existing commercial aircraft to meet specific military operational mission requirements and vice versa. Prominent examples include military derivatives of the B707 and DC-10 developments, and conversely the civil derivatives of the initial military B747 development. Clearly, each procurement agency must make hard decisions about the issue of airworthiness regulations, in particular the degree to which such demanding and costly codes need to be imposed.

Fickeisen remarks [8], that until the time period of the 1970s, "*... commercial airplane design and operational requirements and processes were handled as largely separate entities. ... There was communication but only limited interactive coordination.*" Clearly, since the 1970's a higher level of design-operational interrelationship has been established, a development which needs constant recognition and integration effort. Examples of significant design-operational interrelationships for aerospace vehicle design are aircraft icing and steep approach. Bradshaw concludes [4], that "*Although the principles at the core of the certification process are simple enough, it is likely that for an advanced* [technology or] *experimental aircraft, the means for establishing a safe design might not be readily available within existing codes and procedures.*" Clearly, for the advancement of aerospace vehicle design, a significant factor is the choice and extent of application of codes of airworthiness regulations.

3.2.3 Airworthiness Codes and Design Philosophy

Chapter 2 has identified engineering as often a heuristic skill practiced before 'all the facts are in'. Engineering decision-making is characterised by estimating risks, balancing budgets and deadlines. However, the harmonisation process of the engineer's design approach and design philosophy with the regulating airworthiness body and certification code has overruling importance, as technology is designed, tested, built, and set into operation. Pinkus et al. [9] identify three basic principles that together characterise the interrelationship between the aerospace vehicle design activity and airworthiness: "... *competence, responsibility, and Cicero's Creed II (an updated version of engineering's oldest ethic, to 'insure the safety of the public').*" Black [10] emphasises the importance of this interrelationship by saying: "*As the science of air travel has expanded the safety of the vehicle has more or less kept pace. ... The importance of continuity in development and safety cannot be overstressed.*"

Advancements in aerospace vehicle technology and increasing complexity of multidisciplinary design integration have placed a challenging demand on the description of airworthiness codes, which themselves lag behind current needs. The following is primarily concerned with the ambitious task of certifying advanced conventional and unconventional aerospace vehicle configurations. Coburn recapitulates in [7], that from the aircraft manufacturer's perspective the civil and military engineering end product, the aerospace vehicle design and system engineering development process are similar, although there is a basic difference in approach to aerospace vehicle design. At first it is instructive to contrast the military versus the civil aerospace vehicle design philosophy, as mirrored in the existing airworthiness codes.

The highly demanding and diversified military air operations have triggered unique requirements which are subsequently imposed on the design specifications. Military aerospace vehicle design specifications have evolved from experimental flight testing and flight mishaps, to achieve a level of detailed specification of every aspect of the design. As a result of testing and modifications, the design specifications are constantly refined within the constraints of "... *literally hundreds of imposed MIL-SPECs* ..." [7], resulting in a highly iterative process between the developer and the military customer, see also Whitford [11]. The consequence is a significant increase in scope, complexity, and cost of the military aerospace vehicle design activity. Further effects include over-specification and over-design with attendant cost and schedule overruns, because many military programmes do not mandate complete use of MIL-SPECS design methods. On the other hand, one of the strengths of airworthiness codes like MIL-F-8785C [12] and MIL-STD-1797A [13] is the quantification of the available design space.[2] Note that if a civil design satisfies the MIL-Specs, then it is very likely that the design fulfils most of the civil requirements (may result in over-design).[3]

The civil design approach rather guides the aircraft manufacturer, but it does "... *not impose the myriad of detailed design specifications that the military design process does.*" [7] Clearly, such a design approach offers a greater flexibility in the design process due to a rather qualitative description of the design space. However, these requirements do not cover new developments in civil aviation. Airbus aircraft are certified with reference to JAR-25 (subpart B-Flight). As an example, the mechanism required to certify the specific control philosophy engaged in Airbus aircraft (FBW system incorporating C* law) requires the introduction of so-called Special Conditions (SC), which are an annexation to the standard JAR-25 document for certification of advanced technology features. Clearly, SCs are not generally valid design and certification schemes; they change depending on the type of aircraft and technology features included.

Turning now to the certification problem of advanced or unconventional aircraft types. Here, the significant factor is the choice of and extent of application of codes of design and airworthiness requirements. When recalling that the form of the aircraft may be seen as a function of the requirements that it is designed to satisfy, then it becomes clear that civil airworthiness codes place preference towards the conventional or classical tail-aft configuration (TAC) layout. Clearly, successful certification of advanced aircraft will depend on the design, the assessment methods, and the airworthiness code. As illustrated above, parts of the existing airworthiness codes are not applicable as state-of-the-art certification—and design guidelines for advanced military, civil, or experimental aircraft types. Bradshaw [4] identifies two options for proceeding towards the C of A for advanced aircraft types in today's environment:

---

[2]It should be mentioned that MIL-F-8785C is relevant for open-loop aircraft, whereby MIL-STD-1797A has been specifically developed to give rules for military fly-by-wire (FBW) aircraft.

[3]The UK Def. Std. and US Mil Specs assume a classical second order like aircraft behaviour. Both codes are appropriate to produce a safe aircraft with good flying qualities, but without the need to imply good handling qualities (the quest for good handling qualities is clearly a different matter). In contrast, the US Mil Std. tries to define requirements for stability augmented aircraft, to enable the design of safe aircraft with good handling qualities, an approach which does not seem to work well. Clearly, problems with these certification codes have been recognised and there is still no clear way to go. Overall, aircraft design is primarily concerned with guidance by the UK Def. Std. and US Mil Specs, because handling quality issues are a secondary issue in aircraft design.

1. "*Use from another code those requirements which are suitable.*"
2. "*Draw up new requirements specifically for the project*" [e.g. introduction of SCs].

He rightly suggests that the "... *establishment of new requirements should be as early as possible in the projecti life span, ideally the definition phase. ... it is essential that they* [requirements] *are identified early and are brought to the notice of both engineering and project management within the design organisation and the certificating Authority. In this way, the risk that tasks necessary to provide evidence of proof of design not being identified or not completed will be reduced.*"

The following case study deserves special mention in the present context. The Anglo-French Concorde resembled a unique opportunity not only to advance science and engineering, but also to question and advance design certification. A new certification document had been devised for Concorde which was thought to aid the visibility and traceability of evidence of proof of design. In the early years of Supersonic Commercial Transport (SCT) aircraft development, much thought was given to transformation of airworthiness requirements to cater to the new needs of the SCT. Rech and Leyman [14] recall that "... *the differences from existing practice were so great that they decided to write a completely new set of rules specifically aimed at supersonic transports.*" In particular flying quality related unknowns during transonic penetration and during extended supersonic cruise were of concern for an aircraft type embodying several advanced features.[4] An interesting outcome was, that comparatively few new requirements were needed to deal with supersonic flight as such. "*At least as many changes arose because of the new shapes of the aircraft and there have, of course, been the usual problems involved in distinguishing between a new requirement needed because of the nature of the vehicle and one stemming from the continuing March of safety improvement.*" [10]

The British and French collaborated in publishing the 'Concorde TSS Standards' [16], subsequently the only existing airworthiness code formulated exclusively for a SCT, being Concorde. The SCT airworthiness requirements, as proposed in the United States, were published in the form of a book entitled 'Tentative Airworthiness Standards for Supersonic Transports', covering both operations and airworthiness [10]. However, this airworthiness code has never been put to practice. One of the principle differences between the proposed US airworthiness code and the TSS Standards can be seen in that the US document is set out in the form of amendments to the current FAR Part 25, whereas the TSS Standards have been developed with the "*underlying concept of 'graceful degradation' based on a*

[4]Historically, E.P. Warner has the distinction of having first embodied flying quality requirements into a specification that could be applied to a new aircraft design [15].

probability approach rather than a 'single failure' philosophy [which] *was and is innovative and useful*" [14]. To put this into context, the world's first SCT aircraft, the Tu-144 received a NLGS C of A in 1977.[5]

3.2.4 AeroMech *Development Requirements—* Airworthiness

Aviation history reveals numerous cases, where known configurations confirmed physical effects during the flight test phase as 'known unknowns', but surprised with 'unknown unknowns'. In contrast, advanced aircraft types inherit this surprise factor from design to flight testing to a far greater degree, a fact often justifying a flying demonstrator. The differences in approach to demonstrate and test the C of A of civil aircraft and military aircraft are discussed by Coburn [7]. To complement the flight test issue, the aircraft conceptual design environment is obliged to minimise this surprise factor during the initial design phases. Clearly, a strong correlation needs to be established between aircraft conceptual design and the final steps of aircraft development, being the flight test and certification phase. Burdun and Parfentyev [18] and Burdun et al. [19] propose a technique for fast quantitative analysis of a 'virtual test article' with possible applications to aerospace vehicle design. Obviously, simulation techniques[6] will gain relevance to support the aircraft conceptual design risk reduction tasks, presuming that the analytical modelling complexity and computational turn around times can show consistency with conceptual design demands.

The current research investigation has devised a simulation technique for sizing the vehicle's control effectors without the need to map the entire flight envelope. As will be described later, the variety of design-critical flight conditions which show relevance for the design of control effectors are reviewed. Clearly, it is an objective of the current research investigation, to integrate design-relevant flight test and certification aspects into conceptual aerospace vehicle design. This intention is motivated by the fact, that stability and control is, to a large degree, a certification-related discipline.

[5]It should be recalled that Tupolev had tremendous experience in developing high-speed military aircraft. The Tu-144 was one of the first Soviet aircraft designed purely for civil applications. Consequently, the passenger comfort levels were only marginally beyond those of a bomber with the high maintenance profile and fuel consumption of a combat aircraft. The latest SCT activities refer back to 1993, where an agreement was signed to make airworthy a Tu-144 (termed Tu-144LL) to be used as a research vehicle for a future US supersonic transport [17].

[6]The engineering simulator at aircraft conceptual design level is one of the means to couple design with flight test and certification requirements.

Table 3.2 *AeroMech* development requirements—*Airworthiness*

Priority	Development requirement
1	Establish the connection between stability and control during conceptual design and flight test, both disciplines being safety dominated. [Conceptual design and flight test are the only disciplines throughout the vehicle design process, which have to consider the total flight vehicle].
2	Establish a platform to quantify the usually qualitative civil certification requirements and take those constraints and requirements into account during the conceptual design phase.
3	Strive for independance with respect to the existing airworthiness codes. [Such flexibility allows investigation of the effects of airworthiness codes on design and enables airworthiness code modifications, if necessary].
4	Establish clear design- and certification-related configuration optimisation criteria with the aim, to augment the usually ad hoc optimisation procedures dependent on the intuitive skills of the aircraft designer.

Table 3.2 lists the airworthiness development requirements for *AeroMech*, which ensure compatibility with the new generation of *Class V* synthesis systems, see Sect. 3.3.2.[7]

3.3 Aircraft Conceptual Design Synthesis

The aircraft and aerospace vehicle design process is conveniently broken down into the following three sequential phases: conceptual-, preliminary-, and detail design, followed by production, flight test and operation. For a detailed definition of those phases see e.g. Anderson [20] and Raymer [21]. Moore rightly recalls in [22] that most configuration synthesis occurs at the conceptual and preliminary design phases. Clearly, the boundary between the two phases is not distinct, "... *but conceptual design extends from the development of requirements to the determination of a vehicle concept and a size estimate, while preliminary design carries the design to the point where there is a geometric definition of the vehicle which will support the design of actual hardware*" [22]. Clearly, the design space is defined at conceptual design level, whereby a first design solution is proposed at preliminary design level. As a rule of thumb, it can be assumed that around 80% of the aircraft configuration is determined at the conceptual design phase alone. The present research investigation is therefore primarily concerned with the conceptual design phase. As identified in Chap. 1, this phase can be seen as the most important but least well understood phase of the aerospace vehicle design process.

[7]It should be recalled that *AeroMech* has stand-alone execution capability. Integration of *AeroMech* into *Class IV* and *Class V* synthesis systems represents, per development objective, no difficulty.

Construction of a generic stability and control method for conceptual design application requires thorough insights into the philosophy, limitations and potential of vehicle synthesis environments. It needs to be decided first, which generation and type of synthesis system the stability and control module will be developed for. It is obvious that the functional contribution of such a method can only unfold its potential, when harmonisation with the chosen synthesis system type is secured right at the beginning. A detailed discussion of vehicle design synthesis environments is not a subject of the present research investigation. However, results of a functional survey of flight vehicle synthesis environments are presented to augment the formulation of development guidelines for the generic stability and control methodology *AeroMech*.

3.3.1 Characteristics of the Conceptual Design Phase

During the conceptual design phase, principal design-decisions are made leading to aircraft configuration choice, -shape, and -size. This involves deciding which parameters need to be used to describe the design. Overall, such decisions will, to a large extent, determine the final cost of the aircraft. Hollowell and Bitten [23] illustrate, that "... *conceptual design tends to be very fast paced, with the total time from start to transition into preliminary design typically measured in months*". However, the overall design-criticality of the conceptual design phase is not apparent when reviewing the typical size of conceptual design teams.

The conceptual design process is an intense learning phase for the project team due to its iterative process structure, where feedback loops and successive refinements are involved. The following comments by Hollowell and Bitten [23] deserve special attention in this context.

> The final quality of the design is directly related to how much the design team learned during the conceptual design process. We have observed that how much the team will learn is a strong function of how much of the design space is explored. This, in turn, places a premium on being able to iterate through the design cycle rapidly. During the process the chief designer/engineer will have many questions that must be answered by the various specialists very quickly in order to keep the process on schedule. If a specialist does not provide the answer when it is needed, then the chief designer/engineer will quite often make the best decision he can based on past experience, without input from the specialist. When this happens, the specialist has lost his opportunity to positively influence the design. This is an extremely important observation relative to the application of optimization techniques to the conceptual design process. TIME IS OF THE ESSENCE!

The above mentioned emphasis on quick design-responses during the conceptual design phase results in the use of simplified analysis techniques. Clearly, it is a distinct characteristic of the conceptual design phase, that correct trends are more important than absolute accuracy. To further complicate the issue, the aerospace vehicle designer is often more interested in the interactions between the disciplines that the methods apply to, than in the individual methods themselves. However, the challenge remains to consistently engage simplistic models via first-order or highest-of-importance disciplines and parameters to correctly forecast multidisciplinary gross phenomena for the total flight vehicle. This then correctly identified starting point or baseline design needs to be refined in accuracy via successive iterations to further assess the impact (trends and sensitivities) of new technologies on the design, ultimately helping to provide the basis for the allocation of engineering development resources.

3.3.2 *Classification and Characterisation of Vehicle Synthesis Efforts*

Literature shows numerous classification schemes, to distinguish the multitude of vehicle analysis methods and vehicle synthesis environments. Classification schemes adopted have been according to:

- *Development Period* (pre-computer, CAD-era, etc.);
- *Application Type* (aircraft, aerospace vehicles, fighter, supersonic vehicle, hypersonic vehicle, etc.);
- *Development Phase* (conceptual design, preliminary design, detail design, manufacture, single- and multi-level analysis/synthesis, etc.);
- *Modelling Complexity* (empirical, analytical, numeric, parametric, response surface, optimiser types, etc.).

Using the *development period* alone as a classification scheme in not thought to be useful, although the individual design approaches may have originally evolved throughout a distinct era. Several of the 'dated' methodologies are in use today and are even further-developed for a variety of reasons. Classification according to the targeted vehicle *application type* clearly works for some of the methodologies considered. However, the development intention towards the generic methodology type prohibits such grouping. Classification according to *development phase* involves considerable uncertainty due to non-regulated definitions and understanding of the separate development phases. The classification scheme selected in the present context distinguishes the multitude of vehicle analysis and synthesis approaches according to their *modelling complexity*, expressing their limitations and potential.

The following categorises five different classes of vehicle design sophistications. It should be noted that the range of vehicle synthesis approaches, from *Class I* to *Class IV*, is in use today. *Class V* synthesis systems are under development, and are not yet considered operational.

3.3.2.1 *Class I* Synthesis Approach (Early Dawn)

From the earliest days until around 1905, aerospace vehicle design was characterised by the experimentalist, operating via intuition, limited or no experience, and sometimes via the trial & error approach [24]. With some exceptions,[8] there was obviously no common systematic approach to aerospace vehicle design, an era where intuition and common sense had to prevail.

It was this early period, where the margin between success and failure was virtually non-existent. It should be recalled that the *Class I* approach to aerospace vehicle design is still practiced today by the majority of aeromodel builders. The experience gained may result in a sound physical understanding of the main design variables. Clearly, such background should not be underestimated.[9]

3.3.2.2 *Class II* Synthesis Approach (Manual Design Sequence)

The period from 1905 until around 1955 is characterised by methods based on a mixture of fundamental principles and empirical data. Lovell recalls that during the early period, "... *due to the relative simplicity of design, development and manufacture, the specification and evaluation of costs was not a significant feature of the design process*", an issue which soon had to change [27]. The evolving manual but practical design sequences, as exemplified by the standard aerospace vehicle design handbooks, were based on limited analytical derivations.

Many of today's embedded analysis techniques were initially devised within this period, with the objective to bridge the gap between physical observation and practical implementation. Aerospace vehicle design was typically done by means of parametric studies, involving the interpretation of a series of plots to locate the appropriate configuration. The methods by Mises [28], the rapid sizing method by Loftin [29], and the graphic method by Cherry and Croshere [30] belong in this category. Countless hours of research were dedicated to developing parametric data bases that could be applied to new aerospace vehicle designs. Some sort of 'mechanical' automation of analysis sequences had been devised with special purpose slide rules, nomographs and overlay methods, allowing for an initial rapid evaluation and parametric analysis of new aircraft concepts, see Driggs [31].

Overall, design synthesis itself was entirely manual with the result that only a small number of design options could be calculated by a finite number of people. "*The usual procedure has been to rely on experience and quick 'back of the envelope' studies to get into the 'ball park' and then conduct a series of two- or three-dimensional parametric studies to establish a refined conceptual design*", Jensen et al. [32]. Clearly, the outstanding benefit of *Class II* design methods is the degree of physical design transparency they offer. The ability to gain explicit insights into the design-effects of each separate discipline involved enables the designer to develop a feel for the sensitivity of the individual design parameters on

---

[8]The most prominent early systematic researchers are Otto Lilienthal [25] and the Wright brothers [26].

[9]Probably every aircraft designer may have gained experience with the *Class I* aircraft design approach during his/her design career.

the overall configuration, including an appreciation of how their assumed level of technology actually influences the results. Clearly, the succeeding *Class III–Class VI* design methodologies are developed based on *Class II* methods. Today, *Class II* methods are in use to a certain degree within the general aviation industry and the aircraft home-builder scene.

3.3.2.3 *Class III* Synthesis Approach (Computer Automation)

The advent of multiprocessing computers around 1955 enabled the automation of isolated *Class II* analysis methods, enabling less reliance on judgement and more reliance on detailed design tradeoffs. "*Initial computer applications were confined to aspects of structural analysis and wing design. There was some resistance to the use of computers in initial project design because of the complex decision-making process involved. However they enabled more detailed analyses to be made and hence allowed a greater range of carpet plots with additional overlays to be prepared to show the effects of configuration variables on performance*" [27]. Such accelerated analysis reduced design cycles significantly, thus allowed a more detailed exploration of the design space in the time given. However, design synthesis was still performed manually with discipline-specific (stand-alone) software programs, separately developed by specialists in different departments in the aircraft industry.

Further automation and enhancement of the design-estimation quality was made possible with the advancement of computers and numerical mathematics, which enabled solving of more demanding problems. The application of numerical optimisation techniques was first attempted in structural design [33]. The capability to simulate the actual physics of the problem by solving the governing mathematical equations resulted in one-dimensional (single-point) optimisation capability.

3.3.2.4 *Class IV* Synthesis Approach (Multidisciplinary Integration)

> There followed a number of attempts at applying numerical optimisation techniques to the initial layout of aircraft. These had little success because of the gross simplifications made in the aerospace vehicle design synthesis in order to obtain a solution within a reasonable time using the limited computing power available in the early 1960s. … The main reasons for this position were the continuing difficulties of interpreting and visualising the resulting configurations, and the lack of confidence in the mathematical algorithms used for obtaining optima. … To try to circumvent this problem a Latin Square technique[10] was used. [27]

However, the introduction of computer graphics capabilities in the 1960s [35] initiated the trend, to assist the designer by integrating the stand-alone analytical

[10]The 'Latin Square technique' resembles a least-squares fitting procedure and is described by Healey et al. [34].

programs into a complete computerised design system, consisting of libraries of analysis programs. Using computer graphics lessened the difficulties of communication between man and computer. At first, this allowed participants to share common aircraft databases with the advantage of reduced data transcription errors. The demand to interchange design information between different departments was addressed with the introduction of a first generation of Multi-Variate Optimisation (MVO) techniques during the 1970s.

Substantial increase in computing power with reduced computing cost converted the computer from an exotic research facility to the every-day tool for the designer. Acceptance of computer-aided design and manufacture (CAD/CAM) systems "... *removed much of the prejudice against automated design methods*" [27]. Development of more robust optimisation algorithms resulted in more complex design synthesis systems for conceptual design application. The problems of integrating MVO and the resulting increase in design complexity were discussed in the mid 1970s. MVO enabled economical multidimensional design optimisation that resulted in substantial improvements in design cycle time and quality of the design.

The activity of centralising design authority from the functional, disciplinary groups to configuration management, enabled increased problem visibility of the design process across the disciplinary lines up to managerial levels. This approach eliminated the traditional decentralisation and its associated problems. Advanced generations of computer systems have enabled the first steps towards true multi-dimensional (multi-point) optimisation capability, still with little physical insight into the multidisciplinary coupling effects. Today's latest generations of such integrated systems offer a potentially valuable guide in selecting the overall vehicle configuration for more detailed consideration, see Van der Velden [36].

Clearly, such ability is in its infancy today with limited or no capability to reliably assess advanced vehicle layouts. "*Although these systems are expensive and complex, they are essential in a commercially competitive manufacturing environment*" [37]. It can be summarised, that today's computational systems for conceptual design evaluation are software limited rather than computer limited. Table 3.3 assembles an overview of former and contemporary *Class IV* design synthesis systems, given with their acronyms, originator, and field of application.

3.3.2.5 *Class V* Synthesis Approach (Generic Design Capability)

Configuration independent (generic), more rigorous disciplinary engineering analysis methods are linked to a sophisticated design synthesis framework, which include not only calculation but project management procedures as well. This enables consideration of a wider range of design alternatives and provides truly robust mathematical optimisation capability. "*Once robust design methods are developed within the technical discipline, multidisciplinary design problems can be attacked*" as expressed by Burgess [38].

Table 3.3 Aircraft and AEROSPACE vehicle *Class IV* synthesis systems

AAA [39]	Advanced Airplane Analysis	DARcorporation	Aircraft
ACDC [40]	Aircraft Configuration Design Code	Boeing Defense and Space Group	Helicopter
ACDS [41]	Parametric Preliminary Design System for Aircraft and Spacecraft Configuration	Northwestern Polytechnical University	Aircraft and AeroSpace Vehicle
ACES [42]	Aircraft Configuration Expert System	Aeritalia	Aircraft
ACSYNT [43]	AirCraft SYNThesis	NASA	Aircraft
ADAM [44]	(–)	McDonnell Douglas	Aircraft
ADAS [45]	Aircraft Design and Analysis System	Delft University of Technology	Aircraft
ADROIT [46]	Aircraft Design by Regulation Of Independent Tasks	Cranfield University	Aircraft
ADST [47]	Adaptable Design Synthesis Tool	General Dynamics/ Fort Worth Division	Aircraft
AIDA [48]	Artificial Intelligence Supported Design of Aircraft	Delft University of Technology	Aircraft
AircraftDesign [49]	(–)	University of Osaka Prefecture	Aircraft
APFEL [50]	(–)	IABG	Aircraft
AProg [51]	Auslegungs Programm	Dornier Luftfahrt	Aircraft
ASAP [52]	Aircraft Synthesis and Analysis Program	Vought Aeronautics Company	Fighter Aircraft
ASCENT [53]	(–)	Lockheed Martin Skunk Works	AeroSpace Vehicle
ASSET [54]	Advanced Systems Synthesis and Evaluation Technique	Lockheed California Company	Aircraft
AVID [55]	Aerospace Vehicle Interactive Design	N.C. State University, NASA LaRC	Aircraft and AeroSpace Vehicle
AVSYN [44]	?	Ryan Teledyne	?
BEAM [56]	(–)	Boeing	?
CAAD [57]	Computer-Aided Aircraft Design	SkyTech	High-Altitude Composite Aircraft
CAAD [35]	Computer-Aided Aircraft Design	Lockheed-Georgia Company	Aircraft
CACTUS 58	(–)	Israel Aircraft Industries	Aircraft

(continued)

Table 3.3 (continued)

CADE [59]	Computer Aided Design and Evaluation	McDonnel Douglas Corporation	Fighter Aircraft (F-15)
CAP [54]	Configuration Analysis Program	North American Rockwell (B-1 Division)	Aircraft
CAPDA [60]	Computer Aided Preliminary Design of Aircraft	Technical University Berlin	Transonic Transport Aircraft
CAPS [61]	Computer Aided Project Studies	BAC Military Aircraft Devision	Military Aircraft
CASP [62]	Combat Aircraft Synthesis Program	Northrop Corporation	Combat Aircraft
CASTOR [63]	Computer Aircraft Synthesis and Trajectory Optimization Routine	Loughborough University	Transonic Transport Aircraft
CDS [64]	Configuration Development System	Rockwell International	Aircraft and AeroSpace Vehicle
CISE [65]	(–)	Grumman Aerospace Corporation	AeroSpace Vehicle
COMBAT [66]	(–)	Cranfield University	Combat Aircraft
CONSIZ [67]	CONfiguration SIZing	NASA Langley Research Center	AeroSpace Vehicle
CPDS [68]	Computerized Preliminary Design System	The Boeing Company	Transonic Transport Aircraft
DesignSheet [69]	(–)	Rockwell international	Aircraft and AeroSpace Vehicle
DRAPO [70]	Definition et Realisation d'Avions Par Ordinateur	Avions Marcel Dassault/Breguet Aviation	Aircraft
DSP [71]	Decision Support Problem	University of Houston	Aircraft
EASIE [55]	Environment for Application Software Integration and Execution	NASA Langley Research Center	Aircraft and AeroSpace Vehicle
ESCAPE [62]	(–)	BAC (Commercial Aircraft Devision)	Aircraft
ESP [72]	Engineer's Scratch Pad	Lockheed Advanced Development Co.	Aircraft

(continued)

Table 3.3 (continued)

FASTPASS [73]	Flexible Analysis for Synthesis, Trajectory, and Performance for Advanced Space Systems	Lockheed Martin Astronautics	AeroSpace Vehicle
FLOPS [74]	FLight OPtimization System	NASA Langley Research Center	?
FPDB & AS [75]	Future Projects Data Banks & Application Systems	Airbus Industrie	Transonic Transport Aircraft
FPDS [76]	Future Projects Design System	Hawker Siddeley Aviation Ltd	Aircraft
FVE [77]	Flugzeug VorEntwurf	Stemme GmbH & Co. KG	GA Aircraft
GASP [78]	General Aviation Synthesis Program	NASA Ames Research Center	GA Aircraft
GPAD [79]	Graphics Program For Aircraft Design	Lockheed-Georgia Company	Aircraft
HASA [80]	Hypersonic Aerospace Sizing Analysis	NASA Lewis Research Center	AeroSpace Vehicle
HESCOMP [40]	HElicopter Sizing and Performance COMputer Program	Boeing Vertol Company	Helicopter
HiSAIR/ Pathfinder [81]	High Speed Airframe Integration Research	Lockheed Engineering and Sciences Co.	Supersonic Commercial Transport Aircraft
Holist [82]	?	?	?
ICAD [44]	Interactive Computerized Aircraft Design	USAF-ASD	?
ICADS [83]	Interactive Computerized Aircraft Design System	Delft University of Technology	Aircraft
IDAS [84]	Integrated Design and Analysis System	Rockwell International Corporation	Fighter Aircraft
IDEAS [85]	Integrated DEsign Analysis System	Grumman Aerospace Corporation	Aircraft
IKADE [86]	Intelligent Knowledge Assisted Design Environment	Cranfield University	Aircraft
IMAGE [87]	Intelligent Multi-Disciplinary Aircraft Generation Environment	Georgia Tech	Supersonic Commercial Transport Aircraft
IPAD [44]	Integrated Programs for Aerospace-Vehicle Design	NASA Langley Research Center	AeroSpace Vehicle
MacAirplane [88]	(–)	Notre Dame University	Aircraft
MIDAS [89]	Multi-Disciplinary Integrated Design Analysis & Sizing	DaimlerChrysler Military	Aircraft

(continued)

Table 3.3 (continued)

MIDAS [90]	Multi-Disciplinary Integration of Deutsche Airbus Specialists	DaimlerChrysler Aerospace Airbus	Supersonic Commercial Transport Aircraft
MVA [91]	Multi-Variate Analysis	RAE (BAC)	Aircraft
MVO [92]	MultiVariate Optimisation	RAE Farnborough	Aircraft
ODIN [93]	Optimal Design INtegration System	NASA Langley Research Center	AeroSpace Vehicle
OPDOT [94]	Optimal Preliminary Design Of Transports	NASA Langley Research Center	Transonic Transport Aircraft
Paper Airplane [95]	(–)	MIT	Aircraft
PASS [96]	Program for Aircraft Synthesis Studies	Stanford University	Aircraft
PIANO [97]	Project Interactive ANalysis and Optimisation	Lissys Limited	Transonic Transport Aircraft
POP [98]	Parametrisches Optimierungs-Programm	Daimler-Benz Aerospace Airbus	Transonic Transport Aircraft
PrADO [99]	Preliminary Aircraft Design and Optimisation	Technical University Braunschweig	Aircraft and AeroSpace Vehicle
PreSST [100]	Preliminary SuperSonic Transport Synthesis and Optimisation	DRA UK	Supersonic Commercial Transport Aircraft
PROFET [50]	(–)	IABG	Missile
RCD [53]	Rapid Conceptual Design	Lockheed Martin Skunk Works	AeroSpace Vehicle
RDS [101]	(–)	Conceptual Research Corporation	Aircraft
Rubber Airplane [96]	(–)	MIT	Aircraft
SENSxx [98]	(–)	DaimlerChrysler Aerospace Airbus	Transonic Transport Aircraft
SSP1 [40]	System Synthesis Program	University of Maryland	Helicopter
SSSP [102]	Space Shuttle Synthesis Program	General Dynamics Corporation	AeroSpace Vehicle
SYNAC [103]	SYNthesis of AirCraft	General Dynamics	Aircraft
TASOP [104]	Transport Aircraft Synthesis and Optimisation Program	BAe (Commercial Aircraft) LTD	Transonic Transport Aircraft

(continued)

Table 3.3 (continued)

TRANSYN [105]	TRANsport SYNthesis	NASA Ames Research Center	Transonic Transport Aircraft
TRANSYS [106]	TRANsportation SYStem	DLR (Aerospace Research)	AeroSpace Vehicle
VDEP [107]	Vehicle Design Evaluation Program	NASA Langley Research Center	Transonic Transport Aircraft
Vehicles [108]	(–)	Aerospace Corporation	Space Systems
VizCraft [109]	(–)	Virginia Tech	Supersonic Commercial Transport Aircraft
WIPAR [110]	Waverider Interactive Parameter Adjustment Routine	DLR Braunschweig	AeroSpace Vehicle (Waverider)
X-Pert [111]	(–)	Delft University of Technology	Aircraft
(–) [112]	Dialog System for Preliminary Design	TsAGI	Transonic Transport Aircraft
(–) [113]	Hypersonic Aircraft Conceptual Design Methodology	Turin Polytechnic	AeroSpace Vehicle
(–) [114]	Design Methodology for Low Speed High Altitude UAV's	Cranfield University (Altman)	Unmanned Aerial Vehicles
(–) [115]	Preliminary Design of Civil Transport Aircraft	ONERA	Transonic Transport Aircraft
(–) [116]	Numerical Synthesis Methodology for Combat Aircraft	Cranfield University (Siegers)	Combat Aircraft
(–) [117]	Synthesis Model for Supersonic Aircraft	Stanford University (Van der Velden)	Supersonic Commercial Transport Aircraft
(–) [118]	Spreadsheet Analysis Program	Loughborough University	Aircraft

Note Each synthesis system quoted is referenced with one representative source only

Industry and research institutions have taken only a first step in the direction of this higher level of sophistication when compared to *Class IV* design methods. Burgess continues that "*... these capabilities will come together to enable designers to fully explore the synergistic relationships that exist in aircraft ...*".

Class V systems enable detailed engineering analysis and synthesis of a user-defined aircraft, capacitating true inverse design capability.[11] The generic design capability facilitates the initial configuration selection and definition phase. Consequently, consistent vehicle configuration comparisons are made possible for a class of vehicles, where the ultimate performance may hinge on numerical subtleties. Such design resolution and capability is not seen with today's *Class IV* design methodologies. In contrast to *Class IV* design methods, the gross design variables are not frozen before the potential benefits of multidisciplinary interaction effects are explored. Clearly, reduced-risk advanced aerospace vehicle designs can only be proposed when concentrating on generic analysis methods rather than on experience based statistics, as seen with most *Class IV* approaches.

Optimisation history needs to be rigorously made transparent to the engineer, visualising the design-influence of the variety of disciplines and their assumed level of technology on the aircraft. This capability is of paramount importance, to ensure transparency of the underlying physics and finally to enable so-called 'sanity-' or 'reasonableness checks'. Inclusion of a Technical Competition Analysis System (TCAS) and an aerospace specific Knowledge-Based System (KBS) truly accelerates the learning process of the conceptual design team. The learning capability of the KBS provides a dynamic and intelligent design architecture with growth potential, to support decision making while having design information available 'at the fingertips'.

Finally, coupling of the conceptual design synthesis environment with a dedicated conceptual design engineering simulator enables inclusion of certification related constraints and -requirements at the earliest stage possible. The rapid conceptual design objective is further supported by performing wind tunnel tests using stereolithography (SLA) models. The cost and time associated with SLA wind tunnel testing fits within the conceptual design phase and supplements synthetic aerodynamic modelling activities during configuration trades.

It is not entirely coincidental, that nearly all *Class IV* synthesis systems are capable of dealing with the typical XB-47/B-707 type aircraft arrangement only. The success of the tail-aft configuration (TAC) stems in part from its *non-integrated* arrangement, because size and position of the wing, fuselage, empennage and engines are independently variable within wide limits before such arrangement becomes inappropriate. It can be concluded that *Class IV* type systems "... *operate by sizing the main components independently, and 'assembling' them into very simple geometric relationships. The next challenge for configuration design synthesis is to tackle designs with a high degree of structural* [and other] *integration, with realistic and complicated relationships between airframe and non-airframe components*" as expressed by Moore [22]. Clearly, most *Class IV* systems have not the competence to become *Class V* methods. Only a minority of systems like SENSxx [39] and PrADO [40] can be considered state-of-the-art *Class IV* design environments with development potential towards *Class V* synthesis methods.

[11]The inverse design problem quests: Which design characteristics produce the desired results?

3.3.3 AeroMech *Development Requirements*—Synthesis System

As the above classification illustrates, vehicle design has always involved some degree of optimisation, either accomplished 'manually' or numerically. Overall it is the conceptual design environment where computational prototyping has the potential for making the biggest contribution in engineering design. Kroo further recalls that such prototyping "*... is also farthest from being realized*" [41]. Currently utilised tools for vehicle analysis and synthesis range from prepared cookbook charts to complex analysis systems, see Table 3.3. However, the next-generation of generic *Class V* computational design methods, capable of dealing with conceptual-level design decisions, are in the process of being formulated. Referring back to Chap. 2 it can be concluded, that each conceptual design study (from unmanned aircraft to single-stage-to-orbit vehicles) is inclined to be unique,[12] although *they all tend to include the same core disciplines*. The survey of *Class I* to *Class IV* design synthesis formulations, ranging from aircraft to aerospace applications, confirms this fact. It is this common feature of the multitude of vehicle synthesis systems, which justifies the approach of the present research investigation, to strive towards the generic design capability as exemplified by the next-generation *Class V* synthesis systems, see Chudoba [42].

The development requirements summarised in Table 3.4 are specified for *AeroMech*, as they are requested by the new generation of *Class V* synthesis systems.[13]

3.4 Methodology of Aerodynamic Project Predictions

"*Aerodynamics is understanding what is happening!*" There exist intimate coupling effects between any aerospace vehicle's stability and control characteristics and its aerodynamic drivers. Those aerodynamic drivers are functioning in response to the overall aircraft layout. Clearly, any vehicle's stability and control analysis-results and successive design-decisions are dependent on the quality of the aerodynamic input data. Cook [43] accentuates this fact by saying that "*Probably the most difficult task confronting the flight dynamicist is the identification and quantification of the aerodynamic description of the aeroplane ...*". This demand poses a particular challenge on the aerodynamic modelling process, because the design of aerodynamic control effectors (CE) is performed in the non-linear areas of the flight envelope.

[12]Conceptual design studies tend to deviate significantly from each other due to: (a) distinct mission requirements, (b) special objectives, goals, and requirements, (c) dissimilar length of evaluation time available, and (d) various technologies being applied.

[13]It should be recalled that *AeroMech* has stand-alone execution capability. Integration of *AeroMech* into *Class IV* and *Class V* synthesis systems represents, per development objective, no difficulty.

Table 3.4 *AeroMech* development requirements—*Synthesis system*

Priority	Development requirement
1	Establish a generic stability and control methodology with the ability to function in *Class III* to *Class IV* vehicle synthesis systems. However, the primary development aim targets a supplementary analysis/design module for *Class V* synthesis systems with adequate growth potential.
2	Capability to assess stability and control characteristics in the multidisciplinary design context against design guidelines and/or quantified certification requirements.
3	Establish a status to consider stability and control as an evenly matched conceptual design discipline consistent with the classical design disciplines.
4	Transform the classical stability and control analysis function into an inter-disciplinary design function when embedded into a synthesis environment. In this mode, the synthesis system is functioning as the organisation concept and the analysis module is the estimation concept.
5	Construct a simple, physically correct, and robust modular analysis method. Compatibility with *Class V* synthesis systems requires derivation from first principles, configuration independent formulation, simplicity as far as useful, small computing time, iterable modelling structure, consistency and correctness rather than absolute accuracy, method transparency.
6	Comprehend only those influencing design variables with a strong interdisciplinary effect on aircraft design with respect to control effector sizing (provision of minimal information to make the idea work).
7	Ensure multi-fidelity estimation capability with succeedingly increasing modelling complexity. At first, determine the design space available using minimum modelling complexity, then proceed to a design proposal via enriched vehicle's main characteristics and geometric dimensions. The multi-fidelity estimation capability demands consequent use of consistent calculation routines.
8	Identify and visualise the primary physical interrelations between the disciplines of relevance (physical coupling effects, transparency, sensitivities, etc.).
9	Overall, the research investigation shall concentrate on the method development issue rather than on the programming overhead.

Clearly, the unfortunate coupling of non-linear aerodynamic modelling-demands with sparse conceptual design data-availability is the primary reason, why the stability and control discipline has been traditionally considered at a rudimentary level only during early conceptual design. Torenbeek [44] makes known the overall challenge by saying: "*During the preliminary design stage, the tail surfaces may present one of the most difficult problems in the dimensioning of the main parts of the aircraft, and this, in turn, may lead to many iterations.*" The following emphasises the importance of aerodynamic modelling in the conceptual design environment, as contrasted by today's prevailing practice to concentrate primarily on high-fidelity aerodynamic estimation, by itself non-practical for initial design. This is an area that seems to have been overlooked in technology development programmes particularly in Europe and the USA in recent years.

3.4.1 Configuration Aerodynamics

DEFINITION: *Configuration Aerodynamics considers overall flow phenomena present for the integral flight vehicle of a particular configuration and concept.*

Nicolai and Carty stress in [45] the importance of the aerodynamics group during the vehicle development process in saying that "... *they 'own' the OML (Outer Mold Line) through their control of the configuration aerodynamic characteristics and performance.*" Anderson classifies the contributors to aerodynamic understanding in [46] to be the 'three dimensions', (a) pure experiment, (b) pure theory, and (c) computational fluid dynamics (CFD[14]). He rightly concludes that CFD "... *nicely and synergistically complements the other two approaches of pure theory and pure experiment, but it will never replace either of these approaches.*"

From conceptual design to detail design, all 'three dimensions' contribute towards building a more complete understanding of a vehicle's aerodynamic behaviour. However, the processes involved and the understanding prerequisite to evaluate configuration aerodynamics at the conceptual design level are distinctly different compared to configuration aerodynamics evaluation during detail design. The present research investigation is primarily concerned with one of the 'three dimensions' mentioned above, being computational aerodynamic methods, which represent the 'work horse' for *Class IV* and *Class V* conceptual design synthesis methodologies.

The stability and control aircraft conceptual design evaluation phase has to guarantee a trimmed, stabilised (static and dynamic) and controllable vehicle. The technical means to realise the above requirements are manifold. On modern relaxed static stability aircraft, all three functions (trim, stability, and control) have to be accomplished with provision of adequate control power, just to emphasise its overall importance. In the context of configuration aerodynamics, special emphasis needs to be directed towards aerodynamic control effectors (CE) as an integral element of the total flight vehicle. Aerodynamic CEs are classical contributors to control power[15] during atmospheric flight. Ross and Thomas [47] and Thomas [48] have surveyed experimental data on the aerodynamics of controls in the light of future needs. They conclude that gaps exist in control-related knowledge for advanced aerospace vehicle design, an insight still valid today. "*The advent of Active Control Technology means that the aerospace vehicle designer needs as much, if not more, knowledge of control characteristics, with more emphasis on maximum control power and actuating force or moment than for the previous generation of aircraft*" [47].

[14]The broad area of CFD programs shows a range of approaches with varying complexity. In the present context we associate with CFD physically correct computational estimation schemes, where the latest generation of CFD methods is computer- rather than method limited.

[15]Control Effector (CE) types are generally of aerodynamic and propulsion (thrust vectoring) nature.

The dependence of vehicle design on aerodynamic CE characteristics has increased, particularly for advanced vehicles (TAC, TFC, FWC, …), for Control Configured Vehicles (CCV), and for vehicles where the flight control system (FCS) aims to be adaptive. Much of the systematic testing of conventional CEs rests with aerodynamic handbook methods, which are discussed in Chap. 4. More recent CE data have been acquired on a rather *ad hoc* basis for particular configurations only. Overall, the majority of non-generic aerodynamic estimation methods in use today for conceptual design work are not well-suited or suitable at all, to determine CE characteristics and configuration aerodynamics for today's and tomorrow's advanced vehicle concepts. Clearly, the present research investigation aims to advance CE-specific and configuration aerodynamics related understanding, with particular emphasis on utilising appropriate generic aerodynamic modelling methods suitable for conceptual design applications.

Clearly, building of configuration aerodynamics related understanding obligates all three dimensions, as mentioned above. The configuration aerodynamics discipline demands, according to Mason [49], the following ability from the practitioner:

- Develop an understanding to form a 'mental model' of each flow field or concept against which to gauge computational, theoretical, and experimental 'reality' [*physical insight*].
- Establish an understanding that computational, theoretical, and experimental tools must be used together [*engineering judgement*].
- Appreciate analytical theory because analytical formulas provide insight on the role of key parameters (aircraft were built before CFD) [*reduced order modelling*].

3.4.2 Status of Computational Aerodynamics for Conceptual Design

It should be recalled that the 'three dimensions' of fluid dynamics (experiment, theory, and computation) have resulted in three distinct classes of aerodynamic calculation methods: (a) *analytical*, (b) *semi-empirical/empirical*, and (c) *numerical.*[16] It is possible to categorise all existing aerodynamic calculation techniques into these three distinct classes (a more detailed survey is given in Chap. 4).

Today an imbalance can be observed throughout the *AeroSpace-Triangle*[17] between project-level aerodynamics and detail-level aerodynamics abilities, capabilities, and development effort with respect to configuration aerodynamics. Shevell

[16]CFD-type methods belong into the class of numerical methods.

[17]The *AeroSpace-Triangle* consists of Industry & Operator, Academia, and Research Institution.

comments on the apparent problem in his famous AIAA Paper [50] by saying: "*Early swept wing transport designers made remarkably successful use of inadequate theory and empirical data. This required a sound understanding of the physics of the flow, a talent that will always be important. CFD, by itself, may tend to discourage this since the computational art is largely devising acceptable grids and mathematical procedures. But to hypothesize designs with potentially better characteristics requires understanding the reasons for a less than adequate result .... Therefore, grasping the nature of a problem flow, and being able to judge what kind of changes in the configuration will improve the flow, will remain an important part of the applied aerodynamicist's skills. The design changes can then be evaluated relatively quickly with CFD procedures in many cases ... .*"

From Prandtl's lifting line model to Reynolds Averaged Navier-Stokes solver, "*... the fidelity of these methods has generally been constrained by the available computational capabilities ...*", as expressed by Kroo in [51]. "*Not unexpectedly, the time required for analyses seems to expand to fill what is available, and state-of-the-art aerodynamic analyses take days or weeks, despite the breath-taking progress in computer speed and capacity.*" As mentioned before, most aerodynamic method utilisation and method development activities in today's *AeroSpace-Triangle* concentrate on the well-justified effort towards high-fidelity flow solvers like full Navier-Stokes simulations. However, those modern high-order CFD approaches generally require the definition of geometry and other input at a level of detail inappropriate for conceptual design work. Even if we imagine the cycle-times of the most advanced CFD method to be compatible with conceptual design time-frames for multivariable and multidisciplinary optimisation, the input complexity naturally required to set-up and execute those methods clearly represents a show-stopper for conceptual design application.

Obert comments in [52], that "*For the sake of efficiency for each design problem a CFD method should be used with no higher degree of sophistication than is required to obtain the correct geometry that fulfils the design requirements.*" Today, the use of computational aerodynamics becomes routine once the configuration geometry is well defined. Clearly, advanced high-order CFD codes are normally used to assess only a few specific features of a new vehicle concept in order to stay within practical 'time-scales'. Consequently, the evaluation of configuration aerodynamics with high-fidelity methods carries a significant 'time-penalty'. The above argumentation justifies the author's disagreement with views like those expressed by Schmidt and Sacher in [53], to utilise high-fidelity modelling in conceptual design: "*... The message is that CFD and supercomputing has to be applied to a large extent during early conceptual and preliminary design stage where design freedom exists at low cost concerning changes with regard to the systems concept.*"

Summarising, two elements prevent high-fidelity CFD methods from being used at conceptual design level: (a) cycle time, (b) input data complexity. The present investigation is primarily concerned with project-level rapid turn-around configuration aerodynamics analysis. Computers have reached a point in their evolution that many, if not most, useful results for conceptual design application are obtainable with the current desktop computers available. It is therefore more appropriate to focus attention on the continued development of rapid turn-around lower-order computational methods (reduced-fidelity CFD methods and response-surface methods) with the idea of achieving useful results in very little time. Clearly, design synthesis and optimisation needs a balance of efficiency, correctness and accuracy, as opposed by high-fidelity detail design systems, where accuracy for *one* configuration may be the goal of years of research. Aerodynamic estimation methods relevant for conceptual design are discussed in Chap. 4.

3.4.3 Design Versus Analysis—Computational Aerodynamics in Vehicle Design

Mason [54] apprehends that "*Although computational fluid dynamics has become a major area of research, its use in the early stages of aircraft configuration development is not generally understood. An incredible variety of problems arise in advanced design, and this precludes the standard use of any simple, uniform procedure. Since the conceptual and preliminary design phases determine the basic configuration architecture, this is the area where improved design methods can make the biggest impact.*"

It is instructive to recall the aerodynamic design question, as expressed by Mason et al. [55]: "*The aerodynamic design question is typically posed at several levels, starting with some vague and general question about the 'best' shape of the airplane for a particular mission, and proceeds to more specific and detailed questions concerning* [e.g.] *the actual wing lines, subject to a large variety of constraints*." There are distinct differences between the *aerodynamic design mode* in contrast to the *aerodynamic analysis mode* to achieve the overall aerodynamic design goal.

In the classical *analysis mode* of design, the computational aerodynamics system is utilised to perform a synthetic wind tunnel evaluation campaign function for a pre-defined vehicle concept at an early time during the design cycle, before a real wind tunnel test is justifiable. Obviously, this strategy produces an improved final design at reduced cost in a shorter time period. In contrast, more refined designs emerge with shorter cycle times when the *design mode* of operation is explored. Mason et al. [55] illustrate the ultimate aerodynamic design scenario by stating "*Ideally, the aerospace vehicle designer would specify the aircraft mission (or missions) and a computer program would provide the detailed lines of the optimum aircraft. Such a smart computer program will not exist for some time.*"

Commonly, vehicle development proceeds along two successive steps, as reviewed by Mason et al. in [55] . As a **first step**, the gross features of the 'optimum' vehicle for a particular mission are predicted using either statistics or physically correct but simplistic calculation methods. Overall, the resulting start-configuration is defined by assuming an appropriate technology level. Typical aerodynamic gross features predicted are (i) aspect ratio, (ii) taper ratio, (iii) sweepback, (iv) thickness ratio, (v) Mach number, (vi) Reynolds number, target levels of (vii) maximum allowable drag coefficient, and (viii) cruise lift coefficient. In the **second step**, the design goals (a) minimum drag coefficient for cruise lift coefficient or alternatively (b) maximum lift coefficient for maximum allowable drag coefficient, and (c) the detailed geometry for the detailed aerodynamic vehicle definition have to be met, subject to geometric constraints like twist and root bending moment and aerodynamic requirements on performance and other flight conditions. Overall, the ideal but complex aerodynamic *design problem* has been reduced to a configuration defining step and a successive aerodynamic *analysis problem* (given is a start configuration, to find is the optimum configuration by consideration of selected constraints and requirements). This standard approach appears to be less vague and more manageable and is therefore predominantly in use today. Obert confirms in [52] such prevailing aerodynamic analysis practice: "*Although true design (i.e. inverse) methods are increasingly being developed, most designers use primarily analysis methods and vary the geometry based on experience in successive steps until a satisfactory geometry and pressure distribution are obtained. This applies in particular to three-dimensional geometries.*" This is precisely the mode in which today's high-fidelity CFD systems are used to design.

However, in aircraft conceptual design, the computational aerodynamics code can be used in a fundamentally different mode than to simply supplement wind tunnel testing. Operated in the *design mode*, such methods determine the optimum aerodynamic shape and performance directly to satisfy the design goal (objective function) specified. Two distinct design modes are feasible, being optimisation and the inverse method. The interested reader finds additional information about the design modes in reference [55]. It is a definite target of *Class V* synthesis systems, to utilise computational aerodynamic methods during conceptual design in the *design mode* rather the *analysis mode*.

The present section has illustrated the importance and challenge in obtaining high-quality aerodynamic data as input for stability and control design evaluations at conceptual design level. Clearly, the mode of operation of the aerodynamic estimation code (design- or analysis mode) at synthesis level is not of direct relevance in the present context to develop the generic stability and control methodology. The current focus, stability and control, remains locally an analyses procedure for the evaluation of combinations of input data like aerodynamics, weight, inertia, and geometry characteristics, against design- and certification requirements. However, in a more global sense, the compound aerodynamics and stability and control obviously can function in the direct design mode when integrated accordingly into a synthesis system. Consequently, the development or selection process of the generic aerodynamic modelling method needs to respect the requirements for its utilisation in the analysis and in particular the design mode.

Table 3.5 *AeroMech* development requirements—*Configuration aerodynamics*

Priority	Development requirement
1	Develop an intimate understanding of configuration aerodynamics flow phenomena in design and off-design conditions.
2	Familiarise and understand the origin, potential, and limitations of aerodynamic modelling methods for conceptual design application.
3	Aim for a single method with the following physical capabilities: Accurate and configuration independent (generic) prediction of aerodynamic coefficients and stability- and control derivatives, to enable prediction of performance, stability and control, and loads.
4	Deliver non-linear flow-solutions for the total aircraft configuration external shape during symmetric and asymmetric flight, low- and high angle-of-attack flight, implying symmetric and asymmetric flight vehicles.
5	Aim for a single method with the following computational capabilities: Rapid turn-around, input simplicity, robustness, PC as target computing platform, tabular output.
6	Quantify the quality of the estimation results (validation and calibration) to build confidence into the modelling process.

3.4.4 AeroMech *Development Requirements—* Configuration Aerodynamics

The purpose of Sect. 3.4 has been, to identify general requirements for non-linear aerodynamic prediction methods capable of handling complex configurations during the conceptual design phase (Table 3.5).

3.5 Methodology of Stability and Control Project Predictions

Wilbur Wright speaking before the Western Society of Engineers in September of 1901 expressed, that the "... *Inability to balance and steer still confronts students of the flying problem. ... When this one feature is worked out the age of flying machines will have arrived, for all other difficulties are of minor importance.*" [56]

In the 116 years since then, the 'age of flying machines' obviously has arrived. Still, we are confronted with an apparent weakness to reliably 'balance and steer'. Although not always recognised as such, stability and control is the single most critical requirement for flight safety. Therefore it has to be considered as a key discipline at the leading edge of aeronautics, in particular the aircraft conceptual design and future projects environment.

The challenges of stability and control at the aerospace vehicle design stage may be traced from before the first flight to the present day, and beyond. Renowned specialists McRuer and Graham conclude in [57]: "... *So, although we have progressed from tethered glider to hypersonic glider, we are still confronted by the same problem as the Wrights. But, now it ranges from hypersonic to subsonic speeds and its solution inherently requires systems which, in all their details, are beyond the ken of a single mind.*"

3.5.1 Classification of Flight Mechanics

DEFINITION: *Flight Mechanics—The applied science that deals with the problems of vehicle motion, flight conditions being taken into account.* [58]

One of the pre-conditions vital to enable systematic and effective development of flight mechanics, is the classification of the relevant phenomena into a specified and clear system. Kočka [59] describes the diversity and scope of the disciplines involved in flight mechanics as follows: "*This is an applied scientifical discipline which utilizes knowledge not only from mechanics of rigid and deformable bodies and from aerodynamics, but also from theories of propulsion, from theories of control, cybernetics, bio-mechanics, applied and numerical mathematics and system theories.*"

In the present context, the goal to formulate a classification scheme for flight mechanics lies in creating a basis, which supports the development of criteria for the quantitative evaluation of stability and control characteristics at aircraft conceptual design level. In the framework of the current research investigation, the classification of flight mechanics is limited to the mechanics of atmospheric flight. The classification scheme adopted combines and extends the groupings published in References [58, 59]. The resulting schematic diagram classifies the main parts of flight mechanics and is given in Fig. 3.1. Those areas are highlighted, which are considered to be of immediate relevance for the development of the generic stability and control methodology *AeroMech*.

With respect to the present research investigation, Table 3.6 classifies the elements of flight mechanics into the two categories 'relevant' and 'excluded'. It must be noted, that such selection does not principally discard those subject matters from being pertinent for future investigations.

The relevance of the subject matter 'handling qualities' during the conceptual design phase will be discussed in Chap. 4.

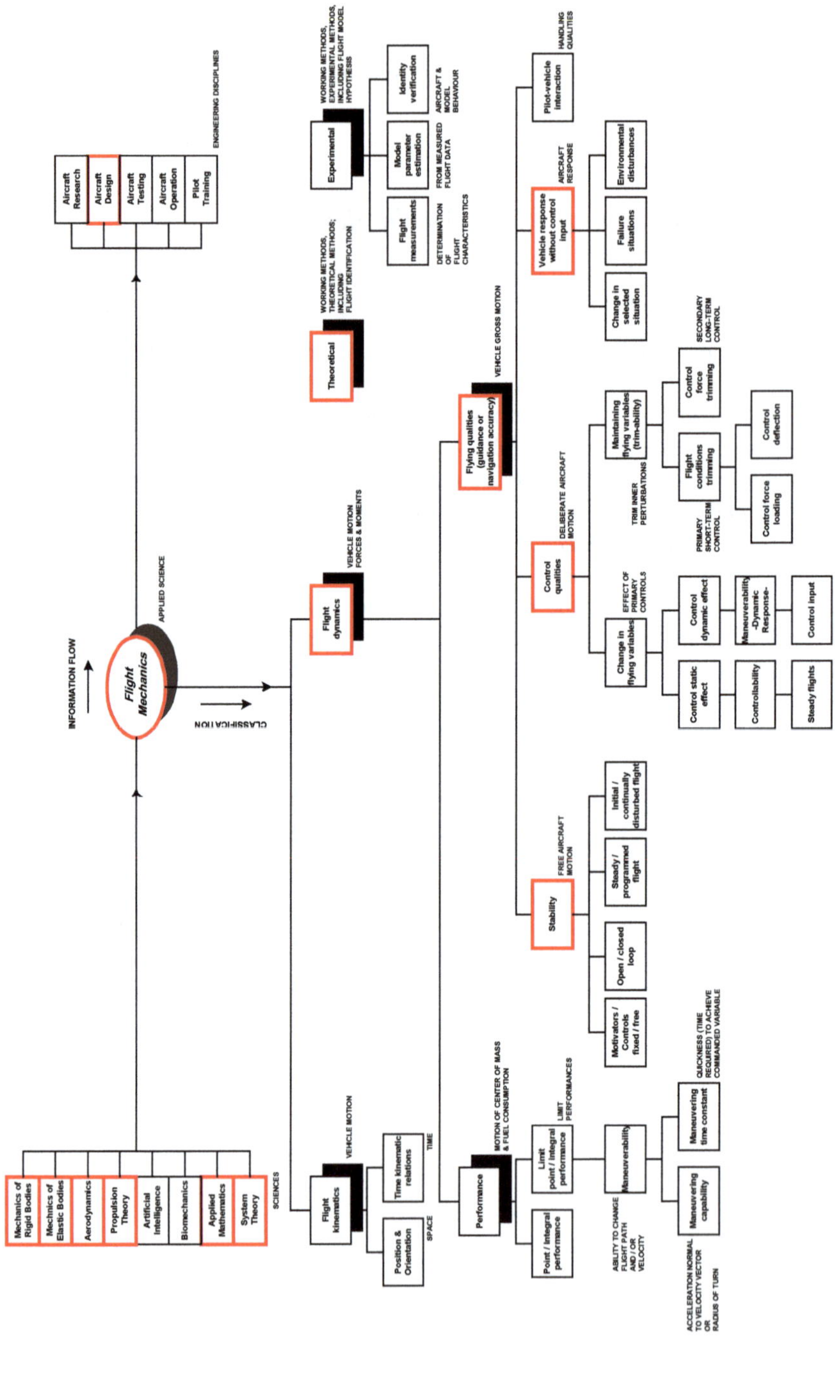

Fig. 3.1 Classification scheme for flight mechanics with subject matters relevant for stability and control at the design stage

Table 3.6 Relevant and excluded subject matters of flight mechanics at the conceptual design stage

Level	Relevant	Excluded
1	*Sciences*: Mechanics of Rigid Bodies; Aerodynamics; Propulsion Theory; Applied Mathematics	*Sciences*: Cybernetics; Biomechanics; Mechanics of Elastic Bodies; System Theory
	Engineering Disciplines: Aircraft Research; Aircraft Design; Aircraft Operation; Aircraft Testing	*Engineering Disciplines:* Pilot Training
2	(–)	*Flight Kinematics*: Position & Orientation; Time Kinematic Relations
	Flight Dynamics: Longitudinal and Lateral-Directional Vehicle Motion; Forces & Moments	(–)
	Theoretical: Working Methods; Theoretical Methods	*Theoretical*: Flight Identification
	(–)	*Experimental*: Flight Measurements; Model Parameter Estimation; Identity Verification
3	*Performance*: Maneuverability (-Capability, -Time Constant); Trim Drag Considerations	*Performance*: Point/integral performance
	Flying Qualities: Stability; Control; Response	*Flying Qualities*: Handling qualities

3.5.2 Confluence of Stability and Control Theory and Practice

For any type of aircraft, conventional or unconventional, a number of basic flying quality requirements have to be satisfied. Obviously, these requirements vary with the type of aircraft (military, civil, research) and from country to country. However, in all cases, the aim is to ensure that the vehicle is safe to fly and that it has desirable flying qualities. Abzug and Larrabee illustrate in their remarkable book [60], that history shows over and over again the neglect of stability and control fundamentals in otherwise excellent aircraft projects. "*If this work* [their book] *has any unifying theme it is the lag of stability and control practice behind currently available theory. Repeatedly, airplanes have been built with undesirable or even fatal stability and control characteristics out of simple ignorance of the possibility of using better designs.*"

The fundamentals of stability and control theory have been laid by the efforts of such men as Bryan, Routh, Lanchester, and Gates, just to mention some pioneer theoreticians and inventors from the early era. A remark from Root in a publication from 1935 has not lost any of its sharpness today; "*... those who have studied the theory realize its inherent complication so far as any rapid application to design is concerned*" [61]. Clearly, the subject is one in which it is easy to wander off into elaborate mathematics.

To apprehend such provocation in today's environment it has to be recalled, that the difficulties of this subject have mounted rapidly with the increase of speed and size and the severity of operational requirements for modern aircraft types. Lee identifies in his report from 1961 [62], that "*The aeroplane designer's approach to stability and control in the past has been conditioned largely by the historical development of the aeroplane.*"[18] Indeed, in our modern jargon, the originally stable airframe, capable of manual flight by a human pilot, has been transformed over the decades into an inherently indifferent or even unstable design of advanced configuration layout with powered controls, a flight control system, artificial feel, fly-by-wire signalling, and control allocation, just to mention some of the key technologies involved.

The above mentioned technical advancements experienced over the last decades justify the following question: Has the traditional approach to stability and control at the conceptual design stage reached its limits of applicability? The present research investigation answers the above question in the affirmative. Clearly, only minimum research effort has been invested into the development of working methods for conceptual design application. As a consequence, the theory and practice of stability and control at conceptual design lag behind modern performance design requirements. Clearly, the processes established to handle stability and control in today's conceptual design environments have not changed at all or not significantly since the paper was written by Root in 1935.

The underlying motivation for the design of an aircraft is to achieve an ever-increasing standard of performance. However, at the same time, an acceptable standard of flying qualities, thus safety, must be attained. An imbalance between overall aircraft performance and flying qualities repeatedly results in stability and control deficiencies for modern aircraft in the form of deficient control authority and control power, high trim drag levels, centre of gravity limitations, etc. There is still much that is not known about the make-up of good flying qualities for advanced technology aircraft, and new problems continually arise with the rapid pace of aircraft development.

Clearly, the theory, methods, and processes currently available to design for adequate flying qualities at the conceptual design level are not truly compatible with today's performance demands. This leads to the quest of a renewed confluence of theory and practice in the field of aircraft stability and control. McRuer and Graham describe in [57] the distinctive historical separation between the scientists/ theoreticians and tinkerers/inventors during the early eras of flight control development. The confluence of flight mechanics theory and practice is dated to around 1948 and was forced by the marked deficiencies in stability and control of the new jet aircraft, in particular the Northrop YB-49 control configured flying wing.

Flight mechanics at the detail design level provides us with "*... awesome capabilities to compute, simulate, and, sometimes, to confuse. ... As a consequence,*

[18]The above statement is valid when considering the processes involved for stability and control analysis during aircraft detail design.

the analysts' physical means now often exceeds his mental grasp, and what he can compute may far exceed his understanding or appreciation. This can lead to an excessively empirical approach to design which is similar to the one used by the tinkerers thirty or more years ago" [57]. While recognising that the above fear by McRuer and Graham has become reality in the conceptual design arena, the present research undertaking aims to construct the bridge from the theory of stability and control to the practical application for the conceptual design process with the consequent intent "... *things should be as simple as possible, but no simpler* ...."

3.5.3 Stability and Control at Conceptual Design Versus Detail Design

The earlier chapters have illustrated, that the design of controls is far from being straightforward. A great deal remains to be done particularly at conceptual design, before sizing of controls is a matter of calculation rather than of experience and experiment. Mason [63] depicts the problem as faced by the conceptual designer as follows: "*The Flight Control Guys (if they're even there ...): 'We need a complete 6 DOF, with an aero math model from –90° to +90° or else forget it.' The Conceptual Designers: 'Just use the usual tail volume coefficient.'*" The above extremist views represent a typical state throughout aircraft conceptual design environments. Obviously, a reasonable middle ground between those views is required.

Never before have we been presented with such tremendous wealth of data and information suitable for detail design of controls. In contrast, never before has it been necessary to approach any one of the primary design disciplines still so entirely *ad hoc* and *inconsistent*, as in the case of designing controls at the conceptual design level. The initial design of controls seems, with the present state of knowledge, to be very far removed from the 'exact science'as established and cultured in the mature world of modern flight dynamics at detail design. For the flight dynamicist, mathematical analysis is an absolutely essential servant, but it can be a highly dangerous master for the conceptual designer when loosing physical insights.

One of the main problems with the design of controls during the early design stage lies in their sensitivity to minute geometrical differences. As outlined in Sect. 3.4, the ability to quantify stability and control for design depends on the quality of the aerodynamic input available. The following chapters will identify, that utilisation of suitable generic aerodynamic estimation methods clearly justifies the development of a dedicated stability and control design algorithm for generic conceptual design application. While the aerodynamic estimation capability suitable for conceptual design is in process to evolve, it is straightforward to bring the traditional stability and control design-approach in line with today's technically demanding design applications, like control configured civil and military aircraft of conventional or unconventional configuration layout from subsonic to hypersonic speeds.

The inadequacy of the science of designing proper controls during conceptual design has far reaching effects, as those outlined in Chap. 1. Stability and control expert Blausey [64] comments on the typical industrial design practice for fighter aircraft, being representative, in part, for the development of civil aircraft. "... *The first steps in conceptual design are fuselage and wing sizing. ... Little or no thought is given to the empennage while this portion of the design process takes place. After the wing and fuselage are initially sized, the empennage is sized and added through a separate design effort. Stability and control requirements are considered one-at-a-time and the smallest empennage which meets all of the requirements is determined. Wing position on the fuselage and landing gear position are sometimes shifted during the empennage design process. At some point in the design process, and usually before the engineers are ready, management dictates a configuration freeze. After this time design changes are very difficult to make. However, small changes are possible. This is when wing strakes are reshaped, dorsal fins and ventral fins are added, wing and horizontal tail dihedral angles change, and wing fences, vortex generators, body strakes, fuselage plugs and wingtip extensions are added. These features usually appear when design deficiencies become evident after configuration freeze. Every last bit of control effectiveness is also squeezed out through leading and trailing edge flap deflection optimization. ... In the final stages of the design, stability and control takes on the dominant role in the aircraft development process.*"

McRuer and Graham talk about the dawning new fifth era of automatic flight control in their 1981 paper [57]. Cook accentuates the integration of stability and control into conceptual design by calling for the renewed confluence of theory and practice in his 1999 paper titled *'The New Age of Flight Control'* [65]. Cook's outlook into the future is motivation and justification enough for the present research investigation. Thus, special room is reserved to quote the author's mentor:

> So what of the future? It is no longer appropriate to regard the flight control system as a 'bolt-on' extra whose role is to rectify the legacy of stability and control deficiencies left by the airframe designer, and designers who continue to subscribe to this view simply have their 'heads in the sand'; and we all know about the flying qualities of the ostrich! The recognition by the Wright brothers of the critical importance of stability and control still applies today. The provision of good stability and control properties in an aeroplane must be at the top of the designer's agenda, for without that commitment the best laid plans will surely fail.

3.5.4 AeroMech *Development Requirements*—Project Stability and Control

The purpose of Sect. 3.5 has been, to identify general requirements for the development of a generic stability and control method for conceptual design application, see Table 3.7.

Table 3.7 *AeroMech* development requirements—*Project stability and control*

Priority	Development requirement
1	Lay down the detailed shortcomings of the traditional stability and control design sequence in project work with respect to today's and tomorrow's applications.
2	Develop an algorithm for longitudinal and lateral-directional control effector sizing of the aerodynamic and thrust vectoring type while taking static and dynamic conditions into account. Strive for minimum complexity while satisfying the primary research objective to develop a generic methodology.
3	Identify the minimum set of primary conceptual design variables and constraints like geometry parameters, flight conditions, failure conditions, and other subject matters of relevance.
4	Ensure the ability to execute the stability and control method in the stand-alone mode or as a robust module in a multi-disciplinary design methodology.
5	Discuss the relevance and implications of subject matters like flying qualities, handling qualities, control power and rest-stability, control allocation, flight control system, fuel transfer system, trim drag, etc., and integrate the relevant issues into the stability and control method.
6	Ensure physical transparency of design parameters involved to the designer with respect to stability, control, and trim. Consider the inclusion of reduced order models.
7	Establish the connection between conceptual design and flight test, to enhance safety and to accelerate the certification process.

3.6 Summary of Results

The inevitable provision of control effectors to any airframe has diminishing effects on aircraft performance. As a consequence, more refined stability and control methods applicable to the conceptual design phase will minimise the penalty on performance whilst ensuring a safe aircraft with adequate flying qualities.

The review of elements of the aircraft conceptual design process has to start with the airworthiness problem due to its intimate coupling with stability and control. The Certificate of Airworthiness (CoA) has, in general, to ensure the safety of the people on the ground and on board. Environmental aspects like noise around airports, sonic boom of supersonic aircraft, and finally atmospheric pollution, are fundamental design constraints. The need to strive for improved aircraft economics and improved safety standards demands advanced conceptual design processes, the assessment of advanced technology, and, as a consequence, flexible certification techniques. Clearly, the flying quality requirements for civil and military aircraft have become, to a large degree, inappropriate for modern technology aircraft. This applies in particular to those aircraft whose flying qualities depend primarily on a Flight Control System (FCS). Consequently, current certification requirements and processes need to be brought in line with today's and tomorrow's design trends, to enable an early and efficient assessment of design risks involved. This perspective is most critical for the emerging class of hypersonic flight vehicles.

Designing control effectors for good flying qualities and minimum trim drag, requires the designer to iterate between aerodynamic analysis, stability and control analysis, geometry modifications, etc. The stability and control method unfolds its true potential when integrated and automated in an aircraft conceptual design synthesis methodology. Then, design changes are economically evaluated to show their effect on flight performance and other global cost functions in the multidisciplinary context. For this reason, essential constituents and peculiarities of the aircraft synthesis design process have been analysed. The potential and limitations of past and present aircraft conceptual design processes are identified to assist the formulation of clear development targets for the development and integration of the generic stability and control methodology.

Particular emphasis is required for analysing the aerodynamic 'thumb-print' of the aerospace vehicle for stability and control design/analysis purposes at the conceptual design level. Control effectors are, in general, not designed at the cruise design point. Instead, they are sized in the 'grey-areas' of the flight envelope, where non-linear aerodynamics prevails. The long-term strategy is, to utilise the computer in the aerodynamic design mode compared to the usually practiced aerodynamic analysis mode. Utilisation of the computer to directly design and optimise the configuration is the most efficient use of computational aerodynamics in vehicle design. Clearly, it is essential to deliver trustworthy aerodynamic data, which is input for any follow on stability and control decision making.

Throughout most of the aerospace vehicle conceptual design environments, stability and control is considered a secondary rather than a primary design discipline. This fact is even more surprising when identifying its overall importance on flight safety, flight operation and certification. Clearly, there exists a wide discrepancy between the sophisticated approach of the modern flight dynamicist during detail design compared to the traditional approach in use during conceptual design. Realising the multitude of performance-driven advancements in the field of modern aerospace vehicle design such as relaxed static stability, fuel transfer, control allocation, and advanced configuration layouts, it becomes obvious that the traditional and well established stability and control methods in use have stagnated in their evolution over more than half of a century. It is the clear aim of the present research undertaking to bridge the gap between conceptual design and detail design work. Obviously, while accepting the research aim to develop a generic stability and control methodology, the effort required to simplify the problem and to strive for a generic solution has become a true challenge.

References

1. Goldin, D.S., "When The Best Must Do Even Better," Remarks by NASA Administrator to Employees of the Jet Propulsion Laboratory on the Future of NASA's Mars Exploration Program, Pasadena, CA, 29 March 2000.
2. Howe, D., "Substitution of Sophistication for Common Sense," Airworthiness Aspects of New Technologies, *I.Mech.E. Seminar*, Bristol University, 20 November 1996.
3. Wood, D., "Project Cancelled – A Searching Criticism of the Abandonment of Britain's Advanced Aircraft Projects," Second Edition, *Macdonald and Jane's London*, 1975.

4. Bradshaw, W., "Certification of Advanced Experimental Aircraft," ICAS Paper 86-3.4.3, *15th Congress of the International Council of the Aeronautical Sciences*, London, England, 07-12 September 1986, pp. 536–541.
5. McClymont, A.S., "Safety Regulations and Aircraft Performance," Airbus Industrie, Regulations and Strategies, AI/E, Seminar given at College of Aeronautics, Cranfield University, 8 March 1996.
6. Saha, R., "Current Airworthiness Requirements—A Need to Re-examine," ICAS Paper 84-4.3.1, *14th Congress of the International Council of the Aeronautical Sciences*, Toulouse, France, 10-14 September 1984, pp. 344–350.
7. Coburn, L.L., "Adapting Commercial Aerospace vehicle designs to Meet Existing and Future Military Aircraft Requirements," ICAS Paper 90-2.10.4, *17th Congress of the International Council of the Aeronautical Sciences*, 09-14 September 1990, pp. 1891–1898.
8. Fickeisen, F.C., "Interrelationships Between Commercial Airplane Design and Operational Requirements and Procedures," ICAS Paper 92-5.3.2, *18th Congress of the International Council of the Aeronautical Sciences*, Beijing, People's Republic of China, 20-25 September 1992, pp. 525–529.
9. Pinkus, R.L.B., Shuman, L.J., *et al*, "Engineering Ethics—Balancing Cost, Schedule, and Risk—Lessons Learned from the Space Shuttle," First Edition, *Cambridge University Press*, 1997.
10. Black, H.C., "The Airworthiness of Supersonic Aircraft," The Royal Aeronautical Society, Presentation given at the Lanchester College of Technology, Coventry, 21 February 1968.
11. Whitford, R., "Design for Air Combat," First Edition, *Jane's Publishing Company*, 1987.
12. Anon., "Flying Qualities of Piloted Airplanes—Military Specification," MIL-F-8785C, 1980.
13. Anon., "Flying Qualities of Piloted Vehicles—Military Standards," MIL-STD-1797A, 1990.
14. Rech, J. and Leyman, C.S., "A Case Study by Aerospatiale and British Aerospace on the Concorde," *AIAA Professional Study Series*, Library of Flight, 1994.
15. Abzug, M.J. and Larrabee, E.E., "Airplane Stability and Control—A History of the Technologies That Made Aviation Possible," First Edition, Cambridge Aerospace Series, *Cambridge University Press*, 1997.
16. Anon., "Concorde TSS Standards," Avion de Transport Supersonique, Supersonic Transport Aircraft, The Air Registration Board, 1969–1976.
17. Duffy, P. and Kandalov, A., "Tupolev—The Man and His Aircraft," First Edition, *Airlife Publishing*, 1996.
18. Burdun, I.Y. and Parfentyev, O.M., "Analysis of Aerobatic Flight Safety Using Autonomous Modeling and Simulation," SAE Paper 2000-01-2100, *2000 SAE Advances in Aviation Safety Conference and Exposition*, Daytona Beach, FL, 11-13 April 2000.
19. Burdun, I.Y., DeLaurentis, D.A., and Mavris, D.N., "Modeling and Simulation of Airworthiness Requirements for an HSCT Prototype in Early Design," AIAA Paper 98-4936, *7th AIAA/USAF/NASA/ISSMO Symposium on Multidisciplinary Analysis and Optimization*, St. Louis, MO, 02-04 September 1998.
20. Anderson, J.D., Jr., "Aircraft Performance and Design," First Edition, *WCB/McGraw-Hill*, 1999.
21. Raymer, D.P., "Aerospace Vehicle Design: A Conceptual Approach," Second Edition, *AIAA Education Series*, 1992.
22. Moore, W.F., "A Model for the Configuration Integration Process," AIAA Paper 95-3905, *Aircraft Engineering, Technology, and Operations Congress*, Los Angeles, CA, 19-21 September 1995.
23. Hollowell, S. and Bitten, R., "Application of Multidisciplinary Optimization to Conceptual Aerospace vehicle design at Rockwell International—A Status Report," AIAA Paper 92-1196, *Aerospace Design Conference*, Irvine, CA, 03-06 February 1992.
24. Bölkow, L., "Ein Jahrhundert Flugzeuge—Geschichte und Technik des Fliegens," First Edition, *VDI Verlag GmbH*, 1990.

25. Heinzerling, W. and Trischler, H. (Editors), "Otto Lilienthal—Flugpionier, Ingenieur, Unternehmer," *Bertelsmann Lexikon Verlag GmbH*, 1991.
26. Crouch, T., "The Bishop's Boys—A Life of Wilbur and Orville Wright," *W.W. Norton & Company*, 1989.
27. Lovell, D.A., "Some Experiences With Numerical Optimisation in Aircraft Specification and Preliminary Design Studies," ICAS Paper 80-2.4, *12th Congress of the International Council of the Aeronautical Sciences*, Munich, Federal Republic of Germany, 12–17 October 1980.
28. von Mises, R., "Fluglehre," First Edition, *Springer Verlag*, 1933.
29. Loftin, L.K., Jr., "Subsonic Aircraft: Evolution and the Matching of Size to Performance," NASA RP 1060, NASA Langley Research Center, 01 August 1980.
30. Cherry, H.H. and Croshere, A.B., Jr., "An Approach to the Analytical Design of Aircraft," SAE-TP-480169, Vol. 2, No. 1, *SAE Quarterly Transactions*, 01 January 1948.
31. Driggs, I.V., *"Aerospace Vehicle Design Analysis," Journal of RAeS*, February 1950, pp. 65–116.
32. Jensen, S.C., Rettie, I.H. *et al*, "Role of Figures of Merit in Design Optimization and Technology Assessment," Vol. 18, No. 2, *AIAA Journal of Aircraft*, February 1981, pp. 76–81.
33. Schmit, L.A., "Structural Design by Systematic Synthesis," *Proceedings of Second Conference on Electronic Computation ASCE*, 1960.
34. Healey, M.J., Kowalik, J.S. *et al*, "Airplane Engine Selection by Optimisation on Surface Fit Approximations," Vol. 12, No. 7, *AIAA Journal of Aircraft*, 1975.
35. Boyles, R.Q., "Aerospace Vehicle Design Augmented by a Man-Computer Graphic System," Vol. 5, No. 5, *AIAA Journal of Aircraft*, 1968, pp. 486–497.
36. Van der Velden, A., "Multi-Disciplinary Supersonic Transport Design," Chapter 17 in "New Design Concepts for High Speed Air Transport," Edited by Sobieczky, H., First Edition, CISM Courses and Lectures No. 366, *Springer*, 1997.
37. Jenkinson, L.R. and Rhodes, D.P., "Application of Spreadsheet Analysis Programs to University Projects in Aerospace Vehicle Design," AIAA Paper 95-3868, 1995.
38. Burgess, M.A., "The Future of Aerospace Vehicle Design," Aerospace America, September 1991, pp. 22–25.
39. Dirks, G.A. and Schneegans, A., "Scenario Based Aerospace Vehicle Design Using Knowledge Based Software Methods," ICAS Paper 5103, *22nd International Congress of Aeronautical Sciences*, Harrogate International Conference Centre, UK, 27 August-01 September 2000.
40. Heinze, W., "Ein Beitrag zur quantitativen Analyse der technischen und wirtschaftlichen Auslegungsgrenzen verschiedenster Flugzeugkonzepte für den Transport großer Nutzlasten," Ph.D. Thesis, Technische Universität Braunschweig, Institut für Flugzeugbau und Leichtbau, ZLR-Forschungsbericht 94-01, February 1994.
41. Kroo, I., "Computation-Based Design—A White Paper," Stanford University, 20 September 1996.
42. (a) Chudoba, B., "Aircraft Conceptual Design Tool Development—An Assessment of the Aircraft Conceptual Design Process," Report E2-086/2000, Future Projects and Technologies Department E2, Fairchild Dornier, 22 November 2000; (b) Chudoba, B., "Aircraft Conceptual Design Tool Development—Assessment of Fairchild Dornier's In-House Methods and Software Applicable to Aircraft Conceptual Design," Report E2-114/2000, Future Projects and Technologies Department E2, Fairchild Dornier, 22 November 2000; (c) Chudoba, B., "Aircraft Conceptual Design Tool Development—Specification for an Aircraft Conceptual Design Synthesis Methodology," Report E2-116/2000, Future Projects and Technologies Department E2, Fairchild Dornier, 22 November 2000; (d) Chudoba, B., "Aircraft Conceptual Design Tool Development—Evaluation of Aircraft Conceptual Design Synthesis Methodologies and E2 Development Proposal," Report E2-117/2000, Future Projects and Technologies Department E2, Fairchild Dornier, 22 November 2000.

43. Cook, M.V., "Flight Dynamics Principles," First Edition, *Arnold*, 1997.
44. Torenbeek, E., "Synthesis of Subsonic Airplane Design," Sixth Printing, Delft University Press, *Kluwer Academic Publishers*, 1990.
45. Nicolai, L.M. and Carty, A., "Role of the Aerodynamicist in a Concurrent Multi-Disciplinary Design Process," Paper RTO-4-1, *Symposium of the RTO (Research and Technology Organization) Applied Vehicle Technology Panel (AVT)*, Ottawa, Canada, 18–21 October 1999.
46. Anderson, J.D., "Computational Fluid Dynamics—The Basis With Applications," First Edition, *McGraw-Hill Inc.*, 1995.
47. Ross, A.J. and Thomas, H.H.B.M., "A Survey of Experimental Data on the Aerodynamics of Controls, in the Light of Future Needs," AGARD CP-262, September 1979.
48. Thomas, H.H.B.M., "The Aerodynamics of Aircraft Control—A General Survey in the Context of Active Control Technology," AGARD R-711, March 1983.
49. Mason, W.H., "AOE 4984 Configuration Aerodynamics Syllabus/Rules," Department of Aerospace and Ocean Engineering, Virginia Polytechnic Institute and State University, March 2000.
50. Shevell, R.S., "Aerodynamic Bugs: Can CFD Spray Them Away?," AIAA Paper 85-4067, *3rd Applied Aerodynamics Conference*, Colorado Springs, CO, 14-16 October 1985.
51. Kroo, I., "Aerodynamic Analyses for Design and Education," AIAA Paper 92-2664, *10th Applied Aerodynamics Conference*, Palo Alto, CA, 22-24 June 1992.
52. Obert, E., "The Role of Aerodynamics in Aerospace Vehicle Design and Operation," Delft University, November 1997.
53. Schmidt, W. and Sacher, P.W., "Applications of CFD Codes and Supercomputers to Aerospace Vehicle Design Activities," AGARD Report R-794, December 1993.
54. Mason, W.H., "Applied Computational Aerodynamics," Class Notes for AOE 4114, Department of Aerospace and Ocean Engineering, Virginia Polytechnic Institute and State University, 1995.
55. Mason, W.H., MacKenzie, D., Stern, M., Ballhaus, W.F., and Frick, J., "An Automated Procedure for Computing the Three Dimensional Transonic Flow Over Wing-Body Combinations, Including Viscous Effects," AFFDL TR-77-122, February 1978.
56. McFarland, M.W., "The Papers of Wilbur and Orville Wright—Volume One 1899–1905," *McGraw-Hill Book Co.*, 1953.
57. McRuer, D. and Graham, D., "Eighty Years of Flight Control: Triumphs and Pitfalls of the Systems Approach," Vol. 4, No. 4, Journal of Guidance and Control, July–August 1981, pp. 353–362.
58. Anon., "Recommended Practice—Atmospheric and Space Flight Vehicle Coordinate Systems," ANSI/AIAA R-004-1992, American National Standard, Approved February 1992.
59. Kočka, V., "Classification in Flight Mechanics," ICAS Paper 82-3.6.2, *13th Congress of the International Council of the Aeronautical Sciences, AIAA Aircraft Systems and Technology Conference*, Seattle, Washington, 22-27 August 1982.
60. Abzug, M.J. and Larrabee, E.E., "Airplane Stability and Control: A History of the Technologies That Made Aviation Possible," First Edition, Cambridge Aerospace Series Nr. 6, *Cambridge University Press*, 1997.
61. Root, L.E., "Dynamic Longitudinal Stability Charts for Design Use," Vol. 2, *Journal of Aeronautical Sciences*, May 1935, pp. 101–108.
62. Lee, G.H., "The Aeroplane Designer's Approach to Stability and Control," AGARD Report 334, April 1961.
63. Mason, W.H., "Control and Stability in Aircraft Conceptual Design," Class Presentation, Aerospace and Ocean Engineering, Virginia Tech, September 1999.
64. Blausey, G.C. (Stability & Control Expert, Lockheed Martin) and Chudoba, B., Personal Communication, Atlanta, USA, 28 November 1998.
65. Cook, M.V., "The New Age of Flight Control," Aerogram, Vol. 9, No. 4, College of Aeronautics, Cranfield University, December 1999, pp. 9–13.

66. Roskam, J., Malaek, S.M. *et al*, "AAA (Advanced Aircraft Analysis); A User-Friendly Approach to Preliminary Aerospace Vehicle Design," ICAS-Paper 90-2.10.2, *17th Congress of the International Council of the Aeronautical Sciences*, Stockholm, Sweden, 09-14 September 1990, pp. 1865–1879.
67. Newman, D. and Stanzione, K., "Aircraft Configuration Design Code Proof-Of-Concept: Design of the Crewstation Subsystem," AIAA Paper 91-3097, *AIAA Aircraft Design and Operations Meeting*, Baltimore, MD, 23-25 September 1991.
68. Yuan, L., "A Parametric Approach to Preliminary Design for Aircraft and Spacecraft Configuration," ICAS Paper 92-7.2.1, *18th Congress of the International Council of the Aeronautical Sciences*, Beijing, People's Republic of China, 20-25 September 1992, pp. 367–371.
69. Bargetto, R., Mazzetti, B., and Garbolino, G., "Aircraft Configuration Analysis/Synthesis Expert System: A New Approach to Preliminary Sizing of Combat Aircraft," ICAS Paper 88-1.11.2, *16th Congress of the International Council of the Aeronautical Sciences*, 28 August-02 September 1988, pp. 1645–1649.
70. Gelhausen, P., "ACSYNT—A Standards-Based System for Parametric Computer Aided Conceptual Design of Aircraft," AIAA Paper 92-1268, *AIAA Aerospace Design Conference*, Irvine, CA, 03-06 February 1992.
71. Heldenfels, R.R., "Integrated, Computer-Aided Design of Aircraft," AGARD CP-147-Vol. 1, June 1974.
72. Bil, C., "Development and Application of a Computer-Based System for Conceptual Aerospace Vehicle Design," Ph.D. Thesis, Delft University Press, 1988.
73. Alsina, J., Fielding, J.P., and Morris, A.J., "Progress Towards an Aerospace Vehicle Design Expert System," in "Computer Applications in Aerospace Vehicle Design and Operation," Editors Murthy, T.K.S. and Fielding, J.P., Computational Mechanics Publications, *Springer*, 1987.
74. Consoli, R.D. and Sobieszczanski-Sobieski, J., "Application of Advanced Multidisciplinary Analysis and Optimization Methods to Vehicle Design Synthesis," ICAS Paper 90-2.3.4, 17th Congress of the International Council of the Aeronautical Sciences, Stockholm, Sweden, 09-14 September 1990, pp. 458–467.
75. Anon., "AIDA: Artificial Intelligence Supported Design of Aircraft," Delft University, December 1999.
76. Park, C., "An Expert System for Instruction of Preliminary Aerospace Vehicle Design," AIAA Paper 90-3261, *Aircraft Design, Systems and Operations Conference*, Dayton, OH, 17-19 September 1990.
77. Ebeling, P. and Pfisterer, E., "The Use of Computer Aided Design Methods in Airborne Systems Evaluation," AGARD CP-280, September 1979.
78. Heilmann, A., "Kostenbezogenes Entwurfsverfahren—Dokumentation der Phase I," Dornier Bericht: FE 110-069/89, Dornier, February 1989.
79. Ladner, F.K. and Roch, A.J., "A Summary of the Design Synthesis Process," SAWE Paper No. 907, May 1972.
80. Straub, W.L., "Managerial Implications of Computerized Aerospace Vehicle Design Synthesis," Vol. 11, No. 3, *AIAA Journal of Aircraft*, March 1974, pp. 129–135.
81. Olds, J.R., "The Suitability of Selected Multidisciplinary Design and Optimization Techniques to Conceptual Aerospace Vehicle Design," AIAA 92-4791-CP, *4th AIAA/USAF/NASA/OAI Symposium on Multidisciplinary Analysis and Optimization*, Cleveland, OH, 21-23 September 1992.
82. Healy, M.J., Kowalik, J.S., and Ramsay, J.W., "Airplane Engine Selection by Optimization on Surface Fit Approximations," Vol. 12, No. 7, *AIAA Journal of Aircraft*, July 1975.
83. Frank, P., "Die Auslegung von Flugzeugen mit geringstem Antriebsbedarf," Ph.D. Thesis, Technical University Stuttgart, Germany, 1992.
84. Dror, B., "Development and Application of Computer Aided Design Technology at Israel Aircraft Industries," Auerbach Annual 1979 Best Computer Papers, New York/North Holland, 1979.

85. Herbst, W.B. and Ross, H., "Application of Computer Aided Design Programs for the Technical Management of Complex Fighter Development Projects," AIAA Paper 70-364, *Fighter Aircraft Conference*, 05-07 March 1970.
86. Haberland, C., Thorbeck, J. and Fenske, W., "A Computer Augmented Procedure for Commercial Aircraft Preliminary Design and Optimization," ICAS 84-4.8.1, *14th Congress of the International Council of the Aeronautical Sciences*, 10-14 September 1984, pp. 943–953.
87. Tambach, T., "Some Remarks on Computerized Aerospace Vehicle Design," Vorläufiger Entwurf, AL 10, Dornier GmbH, Immenstaad, July 1977.
88. Gerhardt, H.A., Letter to Dr. W. Schmidt at Dornier GmbH, Ref. 3811-80-65, Northrop Aerodynamics Research, 13 March 1980.
89. Simos, D. and Jenkinson, L.R., "Optimization of the Conceptual Design and Mission Profiles of Short-Haul Aircraft," Vol. 25, No. 7, *AIAA Journal of Aircraft*, July 1988, pp. 618–624.
90. Johnson, V.S., "A Computer-Assisted Process for Supersonic Aircraft Conceptual Design," AIAA Paper 85-4027, *AIAA Aircraft Design Systems and Operations Meeting*, Colorado Springs, CO, 14-16 October 1985.
91. Braun, R.D., Powell, R.W., *et al*, "Multidisciplinary Optimization Strategies for Launch Vehicle Design," AIAA Paper 94-4341, September 1994.
92. Woodford, S., "The Minimisation of Combat Aircraft Life Cycle Cost Through Conceptual Design Optimisation," Ph.D. Thesis, Cranfield University, College of Aeronautics, May 1999.
93. Anon., "CISE Methodology," Grumman Aerospace Engineering Report WT 997R-128, August 1976.
94. Wallace, R.E., "A Computerized System for the Preliminary Design of Commercial Airplanes," AIAA Paper 72-793, *4th Aircraft Design, Flight Test, and Operations Meeting*, Los Angeles, CA, 07-09 August 1972.
95. Buckley, M.J., Fertig, K.W., and Smith, D.E., "Design Sheet: An Environment for Facilitating Flexible Trade Studies During Conceptual Design," AIAA Paper 92-1191, *AIAA Aerospace Design Conference*, Irvine, CA, 03-06 February 1992.
96. Bernard, F., "DRAPO—Integral Computer-Assisted Design and Manufacturing System," ICAS Paper 80-3-1, *12th Congress of the International Council of the Aeronautical Sciences*, Munich, Federal Republic of Germany, 12-17 October 1980, pp. 112–122.
97. Marinopolous, S., Jackson, D., *et al*, "Compromise: An Effective Approach for Conceptual Aerospace Vehicle Design," AIAA Paper 87-2965, September 1987.
98. Bouchard, E.E., "Concepts for a Future Aerospace Vehicle Design Environment," AIAA Paper 92-1188, February 1992.
99. Szedula, J.A., "FASTPASS: A Tool for Launch Vehicle Synthesis," AIAA CP 96-4051, *6th Symposium on Multidisciplinary Analysis and Optimization*, 04-06 September 1996, pp. 655-663.
100. Dovi, A.R. and Wrenn, G.A., "Aerospace Vehicle Design for Mission Performance Using Nonlinear Multiobjective Optimization Methods," Vol. 27, No. 12, *AIAA Journal of Aircraft*, December 1990, pp. 1043–1049.
101. Wolf, G., "Future Projects Data Banks & Application Systems," Airbus Industrie, AI/ED-X, DGLR Workshop, 13 November 1995.
102. Bishop, A.W., Page, A.N. and By-Pass, B., "An Approach to Design Integration," AGARD CP-147, Vol. 1, June 1974.
103. Lang, M., "Programmdokumentation Flugmechanik," Document No. A55-20-110-1, Stemme GmbH & Co. KG, 1 February 1996.
104. Galloway, T.L., Hague, D.S., *et al*, "GASP: General Aviation Synthesis Program," NASA CR-152303, January 1978.
105. Batdorf, W.J., Holliday, J.F., and Peed, J.L., "A Graphics Program for Aerospace Vehicle Design—GPAD System," AIAA Paper No. 75-136, *13th Aerospace Sciences Meeting*, January 1975.

106. Harloff, G.J. and Berkowitz, B.M., "Hypersonic Aerospace Sizing Analysis for the Preliminary Design of Aerospace Vehicles," Vol. 27, No. 2, *AIAA Journal of Aircraft*, February 1990.
107. Dovi, A.R., Wrenn, G.A., Barthelemy, J.-F.M., "Multidisciplinary Design Integration System for a Supersonic Transport Aircraft," AIAA Paper 92-4841, *4th Symposium on Multidisciplinary Analysis and Optimization*, Cleveland, OH, 21-23 September 1992.
108. Lockwood, M.K., Petley, D.H., *et al*, "Airbreathing Hypersonic Vehicle Design and Analysis Methods," AIAA Paper 96-0381, *34th Aerospace Sciences Meeting and Exhibit*, Reno, NV, 15-18 January 1996.
109. Seffinga, B.F., "Development of an Interactive Computerized Aerospace Vehicle Design System," AIAA Paper 96-5529, October 1996.
110. Hollowell, S.J., Beeman II, E.R. and Hiyama, R.M., "Conceptual Design Optimization Study," NASA CR 4298, May 1990.
111. Wennagel, G.J., Mason, P.W., and Rosenbaum, J.D., "IDEAS, Integrated Design and Analysis System," TP-SAE-680728, *Society of Automotive Engineers*, October 1968.
112. Saggu, J.S., "Integration of Aerospace Vehicle Design and Manufacture Using Artificial Intelligence Paradigms," ICAS 92-7.2.2, *18th Congress of the International Council of the Aeronautical Sciences*, Beijing, People's Republic of China, 20-25 September 1992, pp. 372–380.
113. Tejtel, D., "Conceptual Aircraft Design Environment: Case Study Evaluation of Computing Architecture Technologies," AIAA Paper 98-4844, October 1998.
114. Anon., "MacAirplane—Version 3.0," AeroWorks, December 1999.
115. Kraus, W., "Zukünftige Gestaltung des Vorentwurfs bei DASA LM," DGLR Workshop, 13 November 1995.
116. Van der Velden, A., "Multi-Disciplinary SCT Design Optimization," AIAA Paper 93-3931, *AIAA Aircraft Design, Systems, and Operations Meeting*, Monterey, CA, 09-11 August 1993.
117. Hitch, H.P.Y., "Computer Aided Aircraft Project Design," Paper No. 469, *Aeronautical Journal*, 1977.
118. Kirkpatrick, D.L.I. and Larcombe, M.J., "Initial Design Optimisation on Civil and Military Aircraft," AGARD CP-147, October 1973.
119. Rau, T.R. and Decker, J.P., "ODIN: Optimal Design Integration System for Synthesis of Aerospace Vehicles," AIAA Paper 74-72, *12th Aerospace Sciences Meeting*, Washington, DC, 30 January-01 February 1974.
120. Sliwa, S.M., "Use of Constrained Optimization in the Conceptual Design of a Medium-Range Subsonic Transport," NASA TP 1762, NASA, December 1980.
121. Kolb, M.A., "A Flexible Computer Aid for Conceptual Design Based on Constraint Propagation and Component-Modelling," AIAA Paper 88-4427, 1988.
122. Kroo, I., "An Interactive System for Aerospace Vehicle Design and Optimization," AIAA Paper 92-1190, *AIAA Aerospace Design Conference*, Irvine, CA, 03-06 February 1992.
123. Simos, D., "Piano User's Guide—For Piano Version 3.6," Lissys Ltd., 2000.
124. Cachelet, J.P., "Pressto—A Programme for Preliminary Supersonic Transport Synthesis and Optimisation," College of Aeronautics, Cranfield University, 1998.
125. Raymer, D.P., "RDS: A PC-Based Aerospace Vehicle Design, Sizing, and Performance System," AIAA Paper 92-4226, 1992.
126. Anon., "Space Shuttle Synthesis Program (SSSP), Vol. 1, Part 1, Engineering and Programming Discussion—Final Report," NASA CR-114986, NASA, December 1970.
127. Lee, V.A., Ball, H.GG., *et al*, "Computerized Aircraft Synthesis," Vol. 4, No. 5, *AIAA Journal of Aircraft*, September–October 1967, pp. 402–408.
128. Cousin, J.M. and Metcalfe, M.P., "The BAe (Commercial Aircraft) Ltd Transport Aircraft Synthesis and Optimization Program (TASOP)," AIAA Paper 90-3295, *Aircraft Design, Systems and Operations Conference*, Dayton, OH, 17-19 September 1990.

129. Ardema, M.D. and Williams, L.J., "Automated Synthesis of Transonic Transports," AIAA Paper 72-794, *4th Aircraft Design, Flight Test, and Operations Meeting*, Los Angeles, CA, 07-09 August 1972.
130. Wolf, D.M., "TRANSYS: A Software System for Preliminary Design, Analysis and Evaluation of Space Transportation Systems," AIAA Paper 92-1193, *AIAA Aerospace Design Conference*, Irvine, CA, 03-06 February 1992.
131. Oman, B.H., "Vehicle Design Evaluation Program," NASA CR 145070, NASA, January 1977.
132. Gillam, A., "A Knowledge-Based, Extensible Architecture for Space System Design," AIAA Paper 92-1115, *AIAA Aerospace Design Conference*, Irvine, CA, 03-06 February 1992.
133. Goel, A., Baker, C.A., *et al*, "VIZCRAFT: A Problem Solving Environment for Configuration Design of a High Speed Civil Transport," MAD Center Report 99-06-01, Virginia Polytechnic Institute and State University, June 1999.
134. Eggers, T., "Aerodynamic Design of Hypersonic Waveriders for Space Transportation Systems," DLR, 1999.
135. Sharma, N. and Sim, M., "The Impact of New Software Engineering Technology on a New Aerospace Vehicle Design Methodology," SAE Paper 942189, *Aerospace Technology Conference and Exposition*, 1994.
136. Denisov, V.E., "Application of Methods and Tools for Computer-Aided Design in Investigation of Prospects for Civil Aircraft Progress," ICAS Paper 90-2.1.4, *17th Congress of the International Council of the Aeronautical Sciences*, Stockholm, Sweden, 09-14 September 1990, pp. 45–50.
137. Chiesa, S.G. and Maggiore, P., "Hypersonic Aircraft Conceptual Design Methodology," ICAS 94-1.4.2, *19th Congress of the International Council of the Aeronautical Sciences*, Anaheim, CA, 18-23 September 1994, pp. 1064–1074.
138. Altman, A., "A Conceptual Design Methodology for Low-Speed High Altitude Long-Endurance Unmanned Aerial Vehicles," Ph.D. Thesis, College of Aeronautics, Cranfield University, 2001.
139. Michaut, C., Cavalli, D., Huynh, H.T. and Le Thuy, H., "Preliminary Design of Civil Transport Aircraft," AIAA Paper 89-2152, *AIAA Aircraft Design and Operations Meeting*, Seattle, WA, 31 July-02 August 1989.
140. Siegers, F., "Conceptual design Synthesis and Optimization for New Generations of Combat Aircraft," Ph.D. Thesis, Cranfield University, 1996.
141. Van der Velden, A., "Aerodynamic Design and Synthesis of the Oblique Flying Wing Supersonic Transport," Ph.D. Thesis, Stanford University, Department of Aeronautics and Astronautics, Suddar 621, June 1992.
142. Jenkinson, L.R. and Rhodes, D.P., "Application of Spreadsheet Analysis Programs to University Projects in Aerospace Vehicle Design," AIAA Paper 95-3868, 1995.

Chapter 4
Generic Characterisation of Aircraft—Parameter Reduction Process

4.1 Introduction

The conceptual design parameters and design processes which are used to access the development of the generic stability and control method are identified and discussed in Sect. 4.4. Primarily, design related commonalties and peculiarities for the range of conventional and unconventional aircraft types are considered.

It must be noted, that no claim for completeness of the subject matters discussed can be made. The excessive scope of the research undertaking and the time limitation given, both warrant the above statement. Whilst this situation has been anticipated from the outset of the study, the quest for a generic method represents, indeed, the main research challenge. As a consequence, the research depth chosen for each individual subject matter has been adjusted to provide sufficient understanding to empower development of the generic stability and control method *AeroMech*.

Chapter 2 has identified four primary contributors to the design of controls: (1) ***geometry and mass properties***, (2) ***aerodynamics***, (3) ***stability and control***, and (4) ***flight evaluation expertise***. To recall, those four design contributions are conditioned in practice by the three categories (i) *INFRASTRUCTURE*, (ii) *OPERATION*, and (iii) *TECHNOLOGY*. Inevitably, such conditioning leads to a highly interwoven, thus difficult-to-quantify design influence, resulting in the before mentioned uniqueness of aerospace vehicle designs. Having assembled a generic set of design parameters and processes for the range of aircraft types under investigation in this chapter, Chap. 5 finally presents the generic stability and control methodology concept *AeroMech* with the key calculation algorithms.

B. Chudoba, *Stability and Control of Conventional and Unconventional Aerospace Vehicle Configurations*, Springer Aerospace Technology,
https://doi.org/10.1007/978-3-030-16856-8_4

4.2 Geometry and Mass Characterisation

4.2.1 Classification of Aircraft Configuration and Concept

The spectrum of aircraft can be conveniently broken down into four distinct classes being: (a) **subsonic designs**, (b) **transonic designs**, (c) **supersonic designs**, and (d) **hypersonic designs**. With this classification scheme adopted in the present context, fixed wing vehicles are grouped on the basis of their aerodynamic shape alone. It needs to be acknowledged, however, that those types of aircraft differ from one another not only via their design flow regime, but also in several other aspects including propulsion concepts and fuels, controls, materials, methods of construction, and finally modes of operation. Figure 4.1 presents the changing aerodynamic shape of the spectrum of aircraft[1] as illustrated originally by Küchemann and Bagley in [1].

Figure 4.1 vividly illustrates the statement '*form follows function*'. Independent on its overall mission objectives, any flying vehicle has to inherit the means to provide ***lift***, ***volume***, ***control***, and ***propulsion***. As visualised in Fig. 4.1, the higher the design speed of the flight vehicle, the more integrated or blended appear those four functions into the airframe. Conversely, slower flight speeds permit those functions to be rather separate and largely independent of one another. Overall, high functional integration of lift, volume, control, and propulsion into the airframe results in a particularly demanding and non-typical vehicle design process.

Before identifying those design parameters relevant in the present context, it is first necessary to characterise the flight vehicle's possible geometric design options. Clearly, the outer mold line of any flying vehicle, resembling either a subsonic-, transonic-, supersonic-, or hypersonic design, may be realised with an infinite number of design options. The grouping proposed categorises the flight vehicle's outer mold lines according to the understanding of aircraft configuration and aircraft concept.

DEFINITION *The **Aircraft Configuration** specifies the arrangement of the lift generating surfaces relative to the positioning, and/or number, and/or integration of the longitudinal control effector(s) (e.g., Tail-Aft Configuration [TAC], Three-Surface Configuration [TSC], Flying-Wing Configuration [FWC]).*

*The **Aircraft Concept** specifies, for a given aircraft configuration, possible permutations of either lift-, volume-, control-, and propulsion-generating contributors (i.e., possible wing concepts for a specific TAC are, high aspect-ratio wing, delta wing, variable sweep wing, etc.)*

[1]The axis of abscissa represents the flight Mach number, the axis of the ordinate the ratio of range versus Earth circumference.

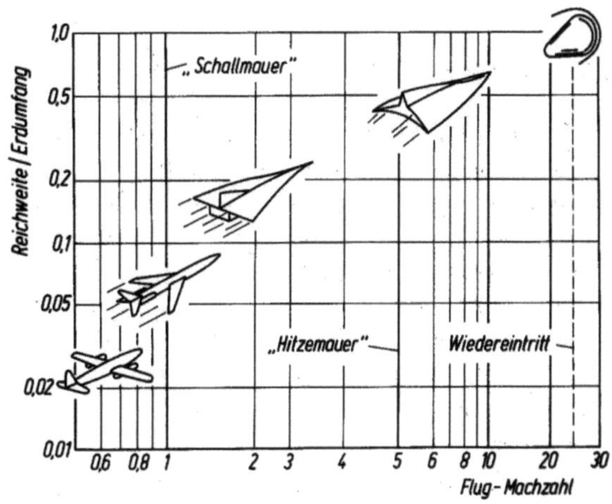

Fig. 4.1 The spectrum of aircraft and their changing aerodynamic shape with speed [2]

Figure 4.2 sketches this multi-dimensional aerospace vehicle design parameter space.[2] Although the evolution of flying vehicles has resulted in currently accepted design trends, the statement of the uniqueness of aerospace vehicle designs is legible when considering the far-reaching design terrain as implied by Fig. 4.2.

It should be noted that the problem of categorising aircraft types has always generated certain misapprehension. The most prominent example is certainly the confusion of the terms describing the Flying Wing Configuration (FWC). Nickel and Wohlfahrt devote a separate sub-chapter to this problem in [3]. Synonyms for FWC are tailless aircraft, flying wing aircraft, all-wing aircraft, wing alone aircraft, flying plank, etc. To complement such confusion, aircraft like Concorde and Vulcan are commonly considered delta wing aircraft, although resembling a FWC due to the lack of an auxiliary longitudinal control surface, but having a fuselage and a vertical fin.

Clearly, responsible for the apparent difficulty in defining a consistent terminology is the open-ended matrix of possible permutations of aircraft configurations and aircraft concepts. In the present context, the definition of the FWC and the other configurations, as introduced in Fig. 2.9, refer to the relation between the primary lifting surface and the controlling surface in the longitudinal plane. Design permutations like the absence/existence of the fuselage, vertical fins and other elements obviously may alter the aircraft concept choice but not the aircraft configuration itself.

A general classification strategy for aircraft configurations and aircraft concepts is proposed in Appendix A.4. This grouping scheme allows a consistent

[2]Figure 4.2 does not imply that the design parameter space is a continuum; in fact, several aircraft configurations and concepts belong into discrete groups (e.g., a wing may inherit either the variable sweep or fixed sweep concept).

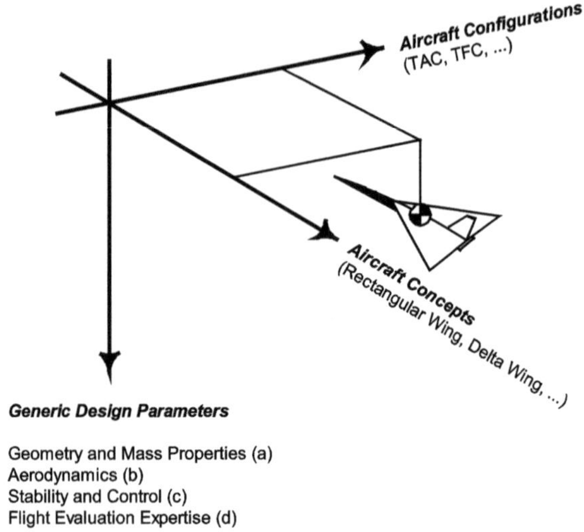

Fig. 4.2 Multi-dimensional aircraft configuration and aircraft concept design parameter space

characterisation of arbitrary aircraft types and their subassemblies.[3] The main benefit of this categorisation materialises when utilised as a 'virtual design-toolbox'. Each subassembly can be assigned to a set of physical characteristics. Then, different configuration- and concept-scenarios can be explored and a preliminary proposal can be mutated into a whole.[4] Still, any categorisation scheme, including the one proposed and used in the present context, will not be entirely fail-proof.

4.2.2 *Stability and Control Design Guide Parametrics*

The key requirement to enable the generic parameter reduction process is the availability of appropriate aerospace vehicle design information and design knowledge. Chapters 1–3 have documented the activities leading towards that aim. The Knowledge-Based System (KBS), outlined in Chap. 2, serves as the primary source. The process of identifying a generic set of design parameters (design-guide parametrics) primarily aims, to build-up consistent physical understanding of conventional and unconventional aircraft types. Clearly, such identification can never be complete nor is it intended to incorporate every design parameter into the method. As a consequence, the design guide parametrics has been kept as simple as possible by focusing primarily on the highest-of-importance parameters, in order to safeguard its relevance and transformability into an algorithm suitable for

[3]The classification scheme relates to the foundations laid for the KBS, see Sect. 2.5.2.

[4]An extended version of this *Virtual Toolbox* can be used for brainstorming sessions, to stimulate creaticity during the configuration definition phase.

conceptual design. A detailed discussion of the multitude of design interactions is beyond the scope of the current investigation. Occasionally selected case studies serve to illustrate only. For more detail, one has to refer to the KBS and the individual references quoted. The following discusses those *design interactions* and isolates those *design parameters*, which have a primary effect on aircraft Control Effector (CE) sizing. Sections 4.3–4.5 discuss several of the more important parameters and processes in more detail.

4.2.2.1 Mass Implications—Moment of Inertia and Centre of Gravity

Mass is a chief concern during the design and operation of any flying vehicle. Performance calculations typically are concerned with the moving mass point, which idealises the total aircraft. In contrast, the implications of mass on stability and control are more complex. The positioning of the centre of gravity relative to the neutral point and manoeuvre point of the vehicle significantly affects stability and control. The mass distribution throughout the vehicle determines the moments of inertia and consequently vehicle dynamics. To complicate the issues, there may be considerable differences in the total mass, the mass centre location, and the mass distribution throughout the mission profile, as fuel is consumed, as the optimum centre of gravity may be tracked with a fuel transfer system, as the sweep of the wing may be variable, as stores are released, etc.

Figure 4.3 subdivides the mass of rigid aircraft into the concepts of the mass point, moment of inertia, and centre of gravity.

The conception of the *mass point* will not be discussed in the present context, belonging to aircraft performance. The stability and control design implications and design parameters related to *moment of inertia* and *centre of gravity* are identified in the following. Roskam has evaluated in [4] the total mass change of aircraft and rockets over a 60 s period during the takeoff phase. He concludes, that because the mass change for aircraft is within about 5% of the initial mass over a 30–60 s period, the constant mass assumption is reasonable for stability and control conceptual design investigations.

The present investigation also assumes, that the mass distribution is constant during this 60 s time interval. Clearly, phenomena such as fuel transfer, fuel slosh, shifting payloads, wandering passengers, mass variance between takeoff and

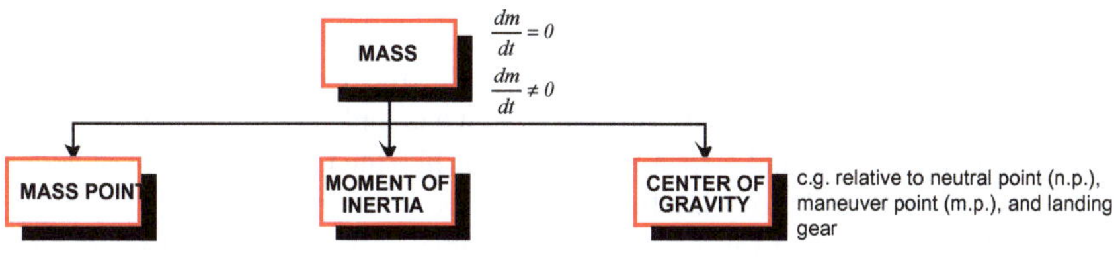

Fig. 4.3 Categorising of aircraft mass into the concepts of the mass point, centre of gravity, and moment of inertia

landing, are taken into account to alter the pre-defined centre of gravity position and the moments of inertia while ignoring the transition period.

Moment of Inertia
The total mass of the aircraft is given by the volume integral:

$$m = \int_V \rho_{a/c} dV \tag{4.1}$$

where $\rho_{a/c}$ is the aircraft mass density. The physical relation between vehicle mass and moment of inertia is given with the *inertia matrix* I_B, written in its most general form

$$I_B = \begin{bmatrix} I_x & -I_{xy} & -I_{xz} \\ -I_{yx} & I_y & -I_{yz} \\ -I_{zx} & -I_{zy} & I_z \end{bmatrix} \tag{4.2}$$

The elements of I_B are the moments and products of inertia,

$$\begin{aligned} I_x &= \int (y^2 + z^2)dm; \; I_y = \int (x^2 + z^2)dm; \; I_z = \int (x^2 + y^2)dm \\ I_{xy} &= I_{yx} = \int xydm; \; I_{xz} = I_{zx} = \int xzdm; \; I_{yz} = I_{zy} = \int yzdm \end{aligned} \tag{4.3}$$

Equation (4.2) is non-specific and complete in its formulation and applies to any flying vehicle, including the OWC and OFWC, where no plane of symmetry exists.

For the traditionally adopted symmetric aircraft layout, however, it is a usual assumption during conceptual design, that the *xz*-plane is a plane of symmetry (reference plane) and that the mass is uniformly distributed. As a result the products of inertia $I_{xy} = I_{yz} = 0$. If in addition the x axis is selected to be the *principal* axis, then the remaining product of inertia I_{xz} is also zero.

Figures 4.4 and 4.5 illustrate the body-fixed co-ordinate systems for those two engineering extremes as discussed in the present context.

Figure 4.5 presents the most general case of a geometric asymmetric aircraft layout in an asymmetric flight condition. The lack of geometric symmetry usually results in a non-uniform mass distribution. As a consequence, the principal x-axis of the aircraft may be inclined vertically and horizontally relative to the selected reference plane.

Goldsmith compares in [5] the moments of inertia for three types of aircraft about the principal axes of pitch, roll, and yaw. To complement such comparison, a lifting body aircraft [6] and a span loader aircraft like the OFWC have been added, see Table 4.1, where the roll inertia is taken to be unity in each case.

It can be observed that for the straight wing type of aircraft the inertias are all of the same general order, whereby the configuration extremes, the slender and spanloader type aircraft, have a large disparity between the inertias. It is well

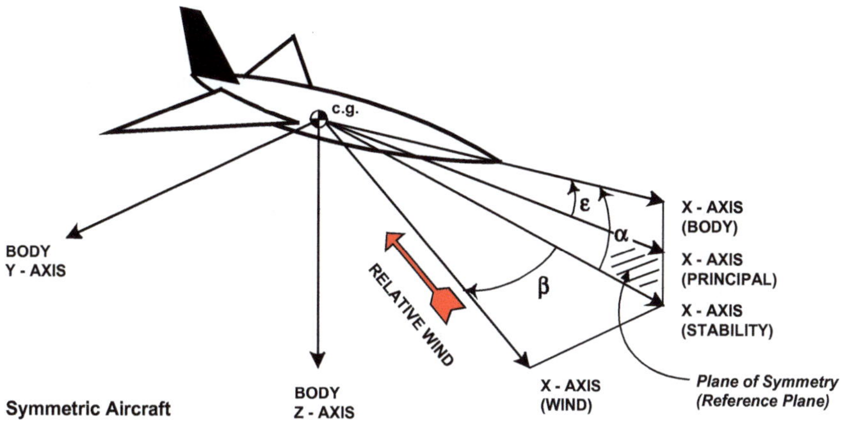

Fig. 4.4 Definition of aircraft axes and angles for the symmetric aircraft type, illustrated with operational asymmetry

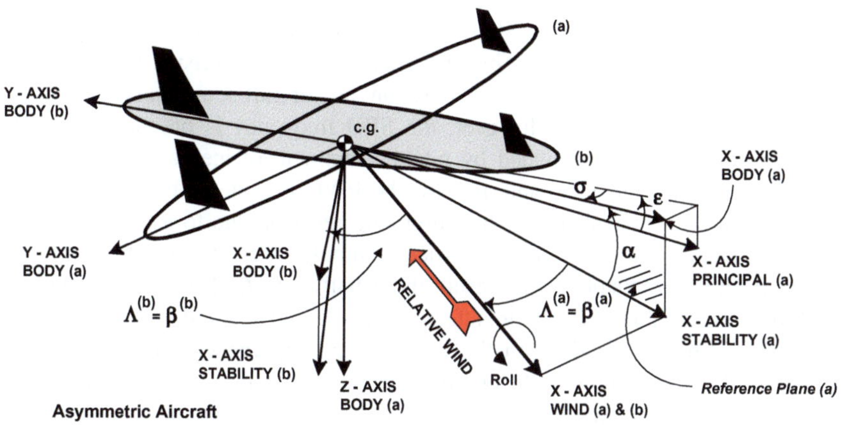

Fig. 4.5 Definition of aircraft axes and angles for the asymmetric aircraft type (OFWC), illustrated with operational asymmetry

Table 4.1 Moments of inertia about the principal axes of pitch, roll, and yaw

	Straight wing aircraft	Swept wing aircraft	Slender body aircraft	Lifting body aircraft	Span loader aircraft
Roll	1	1	1	1	1
Pitch	1	3.5	7	7.7	0.15
Yaw	2	4.5	8	8	1.2

Note Data adapted in part from Goldsmith [5]

known, that slender aircraft like Concorde tend to roll about their principal x axis of inertia rather than the wind axis. An excellent assessment of those phenomena is presented by Pinsker in [7]. In contrast, the FWC and in particular the OFWC, both tend to pitch about their principal y axes.

Figure 4.6 presents the *Moment of Inertia* design interaction and design guide parametrics. Four distinct physical design-related phenomena are of concern, being (a) dynamic modes, (b) inertia coupling, (c) principal axes inclination instability, and (d) spin resistance.

(a) ***Dynamic Modes***

The characteristics of the dynamic modes are in general related to the (i) aerodynamic outer mold line, (ii) the total vehicle mass, (iii) the centre of gravity position, and the (iv) moments of inertia. The following illustrates the effects of the moments of inertia on the dynamic modes along some selected case studies.

The phugoid oscillation is a lightly damped motion even for the TAC, and seems slightly less damped for the FWC due to their relatively low drag level, since drag is the chief means of energy absorption in this mode. Being a slow motion, the moments of inertia have little effect. For more detail see McRuer et al. [8].

The short period oscillation (SPO) of the C-5A Galaxy approaches in frequency that of the phugoid oscillation. This characteristic is not surprising since the phugoid oscillation frequency depends only on an aircraft's true airspeed and is invariant with aircraft size. However, the SPO frequency is proportional to the square root of a quantity divided by pitching moment of inertia, I_y. The SPO is highly damped for the TAC and also for the FWC in spite of the relatively low pitch-damping coefficient, C_{mq}. This surprising result is due to a combination of low static stability in pitch and the vertical damping force, Z_w, which absorbs the energy from the oscillation. As a result, the usually low I_y of the FWC makes the small numerical value of C_{mq} more effective when compared to the TAC. However, such behaviour is dependent on the centre of gravity range, thus is difficult to generalise. Lee et al. show in [9] the SPO characteristics for the X-33 lifting body, being the dynamic mode of importance in the longitudinal plane.

The tumbling oscillation is a dynamic instability of the FWC. Tumbling consists of a continuous pitching rotation about the lateral axis. The position of the centre of gravity and low values of the pitching moment of inertia, I_y, have a pronounced effect on the motion. Some descriptive information is given by Nickel and Wohlfahrt in [10].

The Dutch roll oscillation takes for inertially slender aircraft like Concorde the form of a low damped, constant amplitude roll oscillation. This phenomenon is called Gray's oscillation and has been extensively studied using the HP115, see Pinsker [7].

The roll mode time constant primarily depends on the moment of inertia in roll, I_x, and on the derivative describing the aerodynamic damping in roll. Inertially slender aircraft like Concorde have a small roll mode time constant. Those aircraft

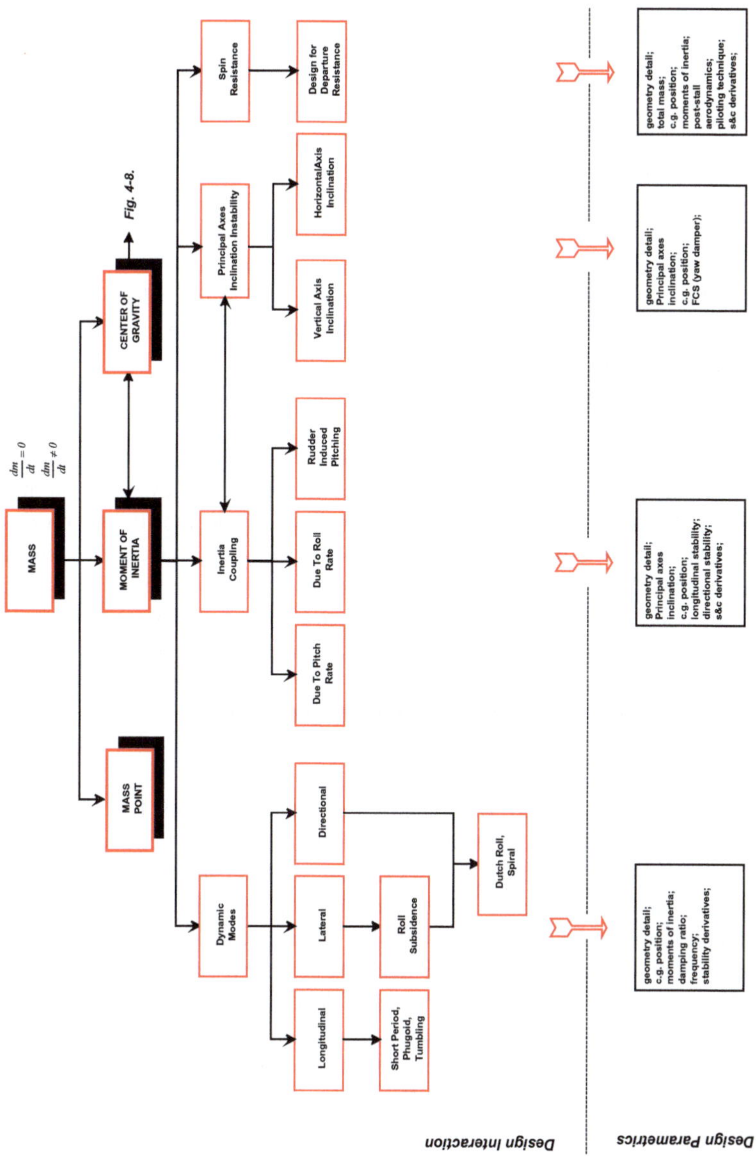

Fig. 4.6 Moment of inertia design interaction and design guide parametrics

develop a commanded steady state roll rate quicker than aircraft with a large roll mode time constant, see Pinsker [7].

The spiral mode is very slow to develop following a disturbance. Consequently, the coupled moments of inertia have little effect, see McRuer et al. in [8].

(b) ***Inertia Coupling***

The term 'inertia coupling' is somewhat misleading, because the complete problem is one of aerodynamic coupling as well as inertia coupling. For the AD-1 oblique scissors wing aircraft (OWC), the magnitude of the cross product of inertia, I_{xy}, is nearly as large as the roll inertia, I_x, at higher wing sweep angles. With a high value for I_{xy}, a pitch-roll coupling occurs. During the AD-1 research program, the aerodynamic cross-coupling derivatives were also determined, see Sim and Curry in [11]. However, these were relatively insignificant compared with the inertial effects. Abzug and Larrabee review in [12] the inertia coupling design evolution of aircraft, starting from the XS-1, P-80, X-3, F-100, YF-102, and F4D. An excellent physical explanation is given by Miller in [13].

(c) ***Principal Axes Inclination Instability***

Aircraft with masses high on the vertical tail, such as a T-tail, cause the principal x axis to be sloped nose-downward relative to the normal body axes. Sternfield reports in [14] that a principal x axis inclination relative to the flight path destabilises the Dutch roll oscillation. It has to be acknowledged, that accurate estimation of the moments of inertia and the position of the inertia axes remains a true challenge during conceptual design. Hallion reports in [15] of a rolling instability of the Douglas D-558-2 Skyrocket research aircraft induced by the inclination of the principal x axis. This phenomenon has been eliminated on modern high-speed aircraft with the use of yaw damping stability augmentation.

(d) ***Spin Resistance***

The spin is an uncontrolled rotation of a fully stalled aircraft. It is, as a consequence, not accessible to an analytical but only empirical assessment during design. The mass distribution of modern jet-engined aircraft has caused spins to be oscillatory rather than being a stable yawing motion. Modern aircraft have to inherit design detail, to make them spin resistant. The tragic accident of the YB-49 FWC, which resulted in renaming Muroc AFB into Edwards AFB [16], has been related to either tumbling or spinning of the FWC, see Pape and Campbell [17]. Northrop describes in his famous paper given at the 35th Wilbur Wright Memorial Lecture in London the spinning and tumbling characteristics of the FWC [18]. A more recent FWC project in the US, the Boeing BWB (Blended Wing Body), is undergoing tests to evaluate spin and tumble susceptibility [19, 20].

CENTRE OF GRAVITY

The choice of the operational c.g. range is the most important decision when designing for adequate stability and control characteristics. Figure 4.7 illustrates

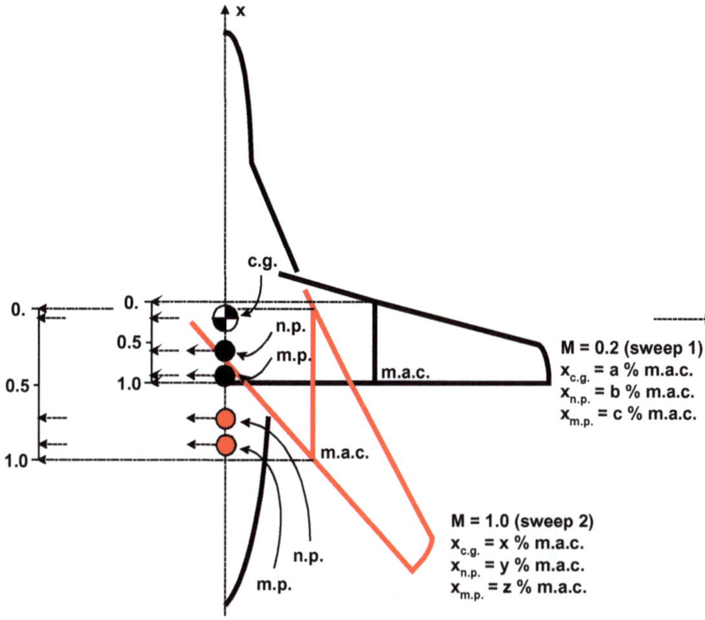

Fig. 4.7 Relative positioning of the c.g., n.p., m.p., and the m.a.c. positions for an aircraft with variable wing geometry

qualitatively the relationship between the c.g. position, n.p. position, m.p. position, and m.a.c. for two different flight conditions and wing sweep angles.

Figure 4.8 presents the *Centre of Gravity* design interaction and design guide parametrics. The following nine subject matters are relevant in the present context: (a) control element, (b) performance, (c) c.g. range and boundaries, (d) c.g. management, (e) CCV (Control Configured Vehicle), (f) moment of inertia, (g) lift element, (h) landing gear, and (i) propulsion element.

(a) ***Control Element***

The primary controls are the LoCE (longitudinal control effector), DiCE (directional control effector), and LaCE (lateral control effector). To satisfy pre-specified mission objectives and certification requirements, sufficient control power must be available to manoeuvre, stabilise, and trim the aircraft. The quantity of control power available depends on the physical control forces generated by the individual CEs (control effectors) and on the individual control-force displacement relative to the c.g. Clearly, the control power available of each individual CE depends on the c.g. position and on the flight condition. Harmonising the required and available control power with the c.g. range is a main objective of the current research undertaking.

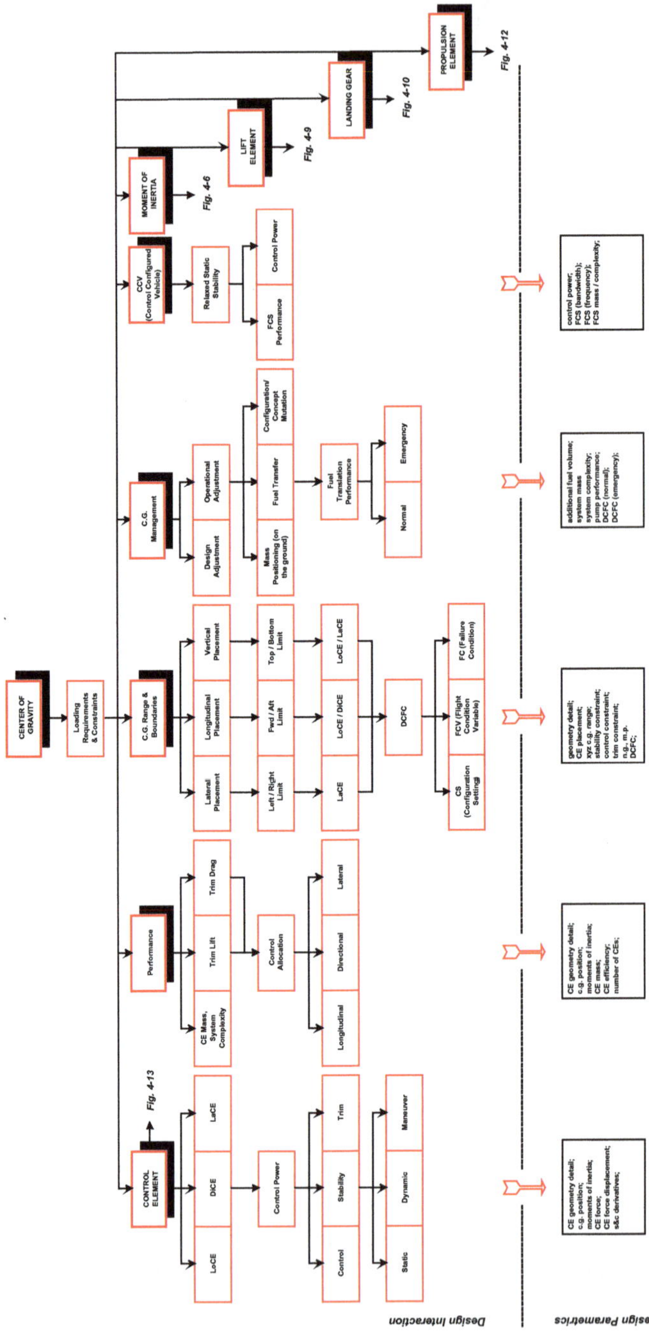

Fig. 4.8 Centre of gravity design interaction and design guide parametrics

(b) ***Performance***

Primary CEs have, in general, an adverse effect on aircraft performance, which manifests in additional mass, system complexity, trim lift, trim drag, and aircraft total drag. Those effects are in severity dependent on the choice of aircraft configuration and concept. The effect of trim lift on Concorde is discussed by Rech and Leyman in [21]. A deflection of FWC wing trailing edge elevons upwards with the intent to rotate the aircraft nose-up results, at first, in an appreciable wing lift loss (significant change of wing camber). This effect is called transient lift response and results in a small reduction in altitude of the c.g. It becomes more pronounced, if the c.g. moves closer the LoCE.

Trim drag arises if a CE deflection is used to generate moment equilibrium. This effect is most pronounced on supersonic aircraft, which experience a significant shift of the neutral point throughout the speed range. On Concorde, the c.g. is shifted with a fuel transfer system to reduce trim deflection of the elevons to zero during cruise. Sachs relates in [22] trim drag to the c.g. position.

The task to optimally schedule the deflection of redundant CEs is termed control allocation. Cameron and Princen describe in [23] the control allocation challenges for the current Blended Wing Body (BWB) FWC project.

(c) ***C.G. Range and Boundaries***

The available c.g. range is determined by operational loading demands, performance, and stability and control (safety) aspects. The c.g. range physically extends into three dimensions. The fuel transfer system of Concorde has to track the variable n.p. throughout the speed range in primarily the longitudinal sense. The narrow c.g. 'corridor' from 0.5 to 1.5 M is a characteristic of Concorde [21, 24]. The A300-600 ST Beluga has to balance a pronounced vertical c.g. displacement depending on loading [25], similar to the combination of the Space Shuttle Orbiter and the B747 Carrier Aircraft [26]. The vertical c.g. issue has been discussed by Weightman in [27]. Lateral c.g. displacements have to be trimmed on the BWB FWC with the LaCE.

The CEs have to be sized to comply with a range of design-critical flight conditions, called the Design-Constraining Flight Conditions (DCFCs). The DCFCs are specified by a c.g. position, configuration setting, flight condition variables, and failure conditions, see Sect. 4.5.

(d) ***C.G. Management***

Any aircraft is designed for a c.g. range which needs to comply with satisfactory flying qualities. During the design process, the aircraft is configured such, as to make provision for the c.g. range demanded. Freight positioning, variable seating layouts, and passenger seating are means to influence the c.g. range on the ground. Fuel transfer during flight is a standard feature on transonic transports [28] and Concorde [29], which enables to shift the c.g. into the optimum position relative to the n.p. and m.p. It should be noted that external changes of the aircraft configuration during flight indirectly alter the positioning of the n.p. and m.p. relative to the c.g. location. Variable sweep aircraft like the B-1 Lancer [30], the Tu-144 with the retractable canard [31], and the XB-70 with folding wing tips [32], belong in this category.

(e) ***CCV (Control Configured Vehicle)***

The CCV design approach capitalises on the potential of considering advanced flight control concepts during the conceptual design phase. The control functions resulting in significant performance improvements are described by Holloway in [33]. The control function of particular relevance in the present context is *augmented stability*. Relaxed static stability is considered the key technology, which triggered a renewed interest into the FWC, in particular the Boeing BWB project. Catalyst to this activity has been the report by Ashkenas and Klyde [34]. Relaxed static stability requires a balance between performance of the Flight Control System (FCS) and control power available.

The subject matters (f) moment of inertia, (g) lift element, (h) landing gear, and (i) propulsion element, are discussed in the separate sub-chapters.

4.2.2.2 Lift Element Integration

Figure 4.9 presents the *Lift Element* design interaction and design guide parametrics. The following three subject matters are relevant in the present context: (a) volume element, (b) control element, (c) propulsion element.

The relative positioning of the lift element (wing) to the control element, propulsion element, volume element (if a distinct fuselage is available), significantly affects the size of the CEs. Shifting the wing back or forward on the TAC has a large effect on the c.g. location and range, because the m.a.c. and n.p. move directly. The wing must be aligned with the fuselage frames on modern commercial transports. A shift-back or shift-forward of the wing by *n* frames relative to the fuselage affects in particular the LoCE and DiCE lever arm, the n.p. position, the wheel base, the ground loads on the nose and main gear, the c.g. limits, and possibly the fuel capacity in the belly tanks. Constraints for positioning the wing may be stability at high-speed and controllability at low-speed. Sanders [35] and Torenbeek [36] describe procedures for the TAC, as how to position the wing relative to the fuselage.

4.2.2.3 Landing Gear Integration

Figure 4.10 presents the *Landing Gear* design interaction and design guide parametrics. The following four subject matters are of relevance in the present context: (a) aerodynamics, (b) geometry limitation, (c) centre of gravity, (d) control element.

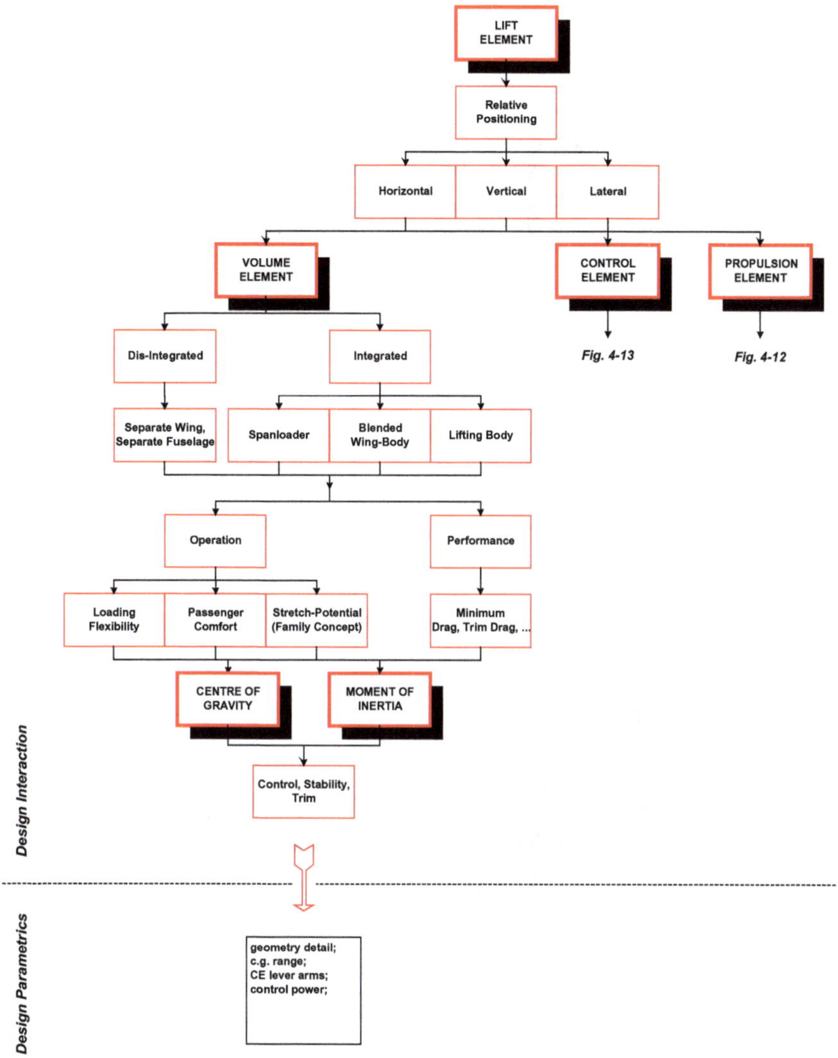

Fig. 4.9 Lift element design interaction and design guide parametrics

(a) *Aerodynamics*

The extension of the landing gear of any aircraft results in a significant increase in drag. Dependent on the aircraft configuration and concepts choice, the landing gear is expected to cause a nose-down trim change upon extension. Hoey recalls in [37], that those problems were particularly severe on the early lifting body aircraft like the M2-F2. Typical for the M2-F2, all three gears had to extend in less than one second. "*By delaying the landing gear extension until after flare completion, the pilots expected to be able to successfully land the airplane in the same manner as*

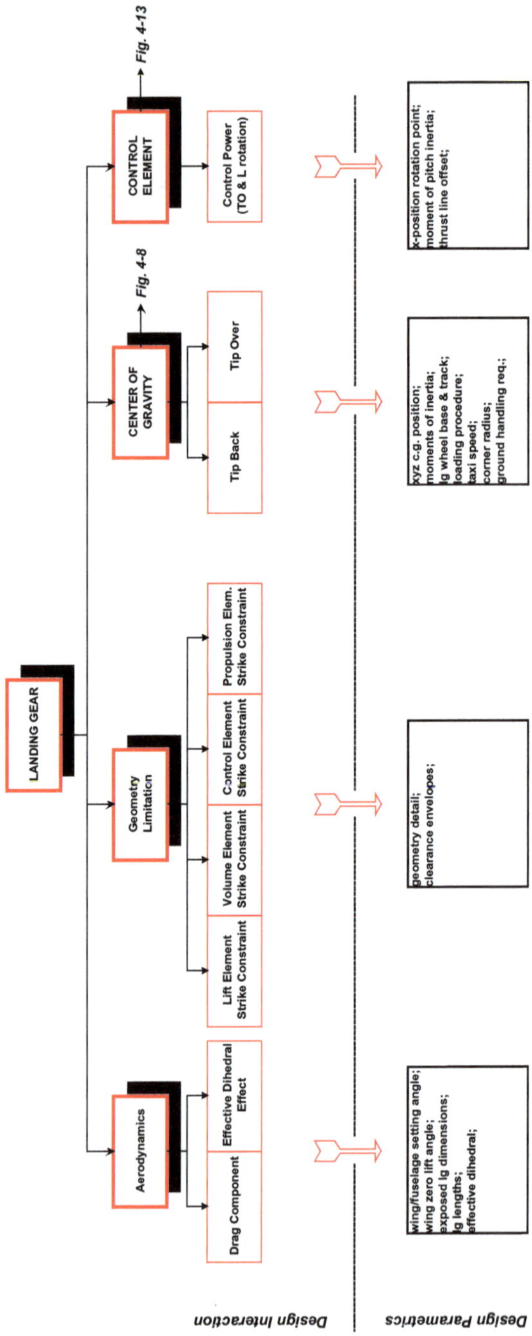

Fig. 4.10 Landing gear design interaction and design guide parametrics

the X-15." In comparison, the landing gear on the Space Shuttle Orbiter reaches the fully extended position within a maximum of ten seconds, see Jenkins [26]. The drag and trim change generated by the landing gear are of particular concern for the above vehicles due to their glider-like landing style. In contrast, commercial TAC transports have to fly with a high thrust setting when in landing configuration, resulting in particularly high noise levels. The landing gear length may be adjusted such, as to position the aircraft longitudinally on the ground to reduce the wing lift to approximately zero before take-off rotation. Apart from being a low drag case, this configuration setting was implemented during the design of the Convair B-58 Hustler. "*This is a favourable characteristic in strong cross winds because the undesirable dihedral effects of the delta wing are almost entirely avoided*" [38].

(b) ***Geometry Limitation***

The landing gear position and dimensions are restricted by the lift element, volume element-, control element-, and propulsion element ground clearance constraints. The landing gear height on Airbus- and Boeing-type passenger transports is determined, in part, by engine ground clearance. The wing dihedral is required to increase, to enable big fan engine capability. The swept, low positioned LoCE on such TAC types require dihedral, to prevent their tips from scratching the ground, see Greff [39]. On pusher aircraft like the Piaggio P.180 Avanti, the ventral fins are utilised to prevent prop damage in the event of over rotation on take-off or landing, see Sacco [40]. The US BWB FWC is a strongly geometry limited aircraft configuration. Due to the small CE moment arms longitudinally and the large wing chord, high-lift is predominantly produced via angle-of-attack, since flaps are difficult to trim. Without elaborating, the BWB requires a long main gear and is still angle-of-attack limited. For more detail see Wakayama and Kroo [41]. Figure 4.11 presents ground clearance envelopes of selected aircraft configuration types schematically.

(c) ***Centre of Gravity***

The centre of gravity position influences both the tip over and tip back susceptibility of the aircraft. The tip over case is dynamic during taxiing. The tip back situation is static on the ground during loading, or dynamic during take-off

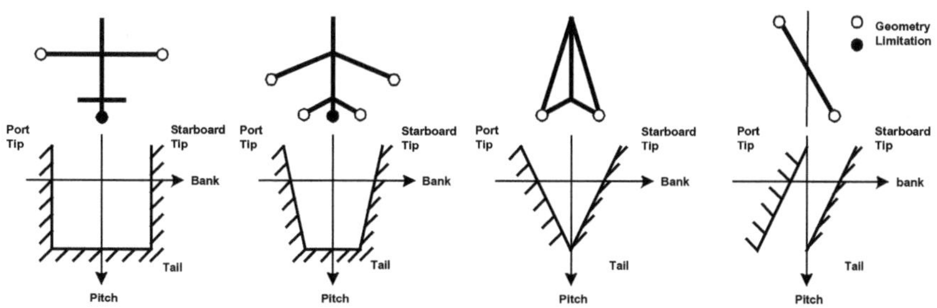

Fig. 4.11 Ground clearance envelopes qualitatively for the TAC, FWC, and OFWC

acceleration with full thrust setting. Twin-engined aircraft like the A330 and B-777, having thrust lines particularly low relative to the c.g., are sensitive to aft c.g. positions during the take-off run with high thrust settings. The OFWC demonstrator by Morris [42] has a four wheel landing gear system that prevents tip-over.

(d) ***Control Element***

The LoCE delivers the force to rotate the aircraft during the take-off rotation and landing de-rotation manoeuvre. Those flight cases clearly belong to the set of critical DCFC for the design of the LoCE. The main gear axel is the point of rotation of the aircraft. Control power available, location of the point of rotation, c. g. position, and the pitch inertia are interrelated during this DCFC (for more detail see Sect. 4.5). The Boeing B-52 Stratofortress is worth mentioning, because sizing of its LoCE and the positioning of the landing gear are de-coupled. The B-52 does not rotate during take-off, see Davies and Thornborough [43], but requires over-sized high-lift devices.

4.2.2.4 Propulsion Element Integration

Figure 4.12 presents the *Propulsion Element* design interaction and design guide parametrics. The following four subject matters are of relevance in the present context: (a) operation, (b) aerodynamics, (c) propulsion control system, (d) geometry limitation. Any propulsion system requires a *xyz*-placement of the thrust line(s) relative to the volume and/or lift element. Synergistic airframe-propulsion interactions and integrations are discussed by Yaros et al. in [44].

(a) ***Operation***

The modes of operation of the propulsion system are either symmetric thrust or asymmetric thrust conditions. The thrust condition directly influences controllability required, stability remaining, and consequently the ability to trim the aircraft. Again, the stabilising or destabilising effect of the propulsion system depends primarily on its aerodynamic influence on the n.p., m.p., and the c.g. location. The fatal accident of the jet-powered Horten Ho 9 FWC illustrates the effects of insufficient control power available. "*During slow flight with the landing gear lowered as well as the landing flaps extended, maintaining directional control for the aircraft with drag rudders only would not be sufficient if one of the turbojets were operating at 100% thrust. It was not enough to keep the aircraft in a straight flight path*" [45].

(b) ***Aerodynamics***

The integration of the propulsion system into the lift and/or volume element results in installation drag, installation lift, and installation moments. Those effects have, apart from their influence on flight performance, a pronounced effect on control, stability, and trim. The Northrop B-2 Spirit FWC presents a highly

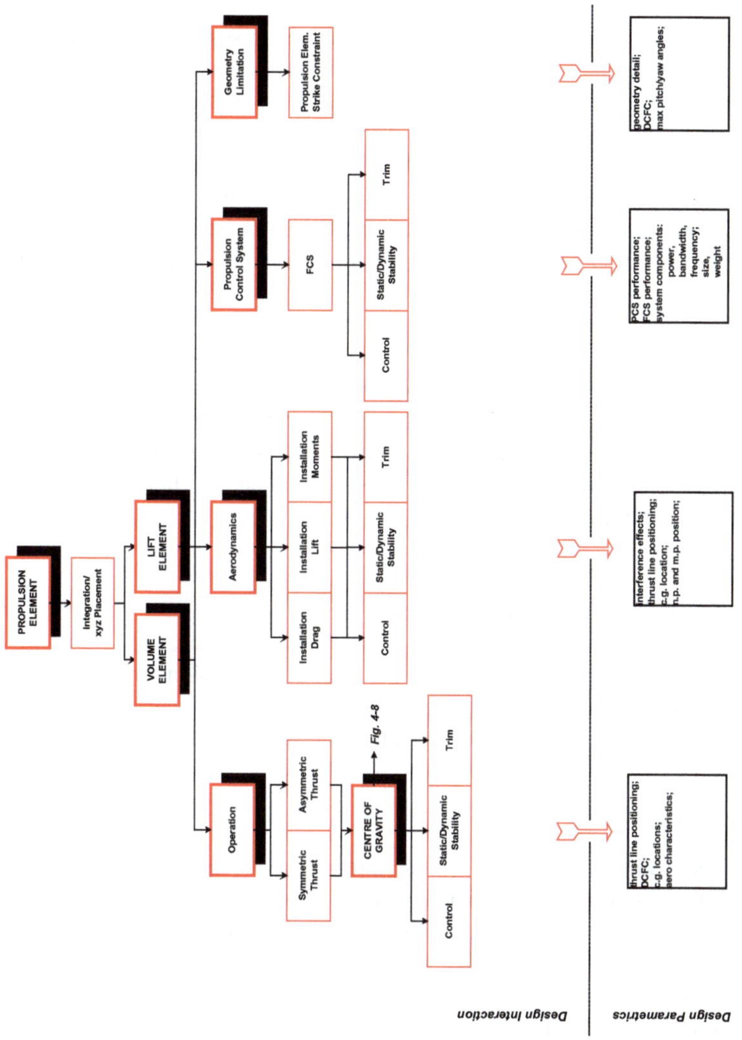

Fig. 4.12 Propulsion element design interaction and design guide parametrics

integrated propulsion system, which was part of the external aerodynamic design from the very beginning. "*Another challenge was the clear definition of an aero/propulsion bookkeeping system. ... The system is labeled 'aero/propulsion' and not 'thrust/drag', because on this aircraft, propulsion effects influence not only drag but also lift, pitching moments and rolling moments*" [46].

(c) ***Propulsion Control System***

The forces and moments generated by the propulsion system have a large impact on the FCS. The variable geometry propulsion system, which is the most demanding propulsion system, has been realised on aircraft like the Lockheed YF-12/SR-71 [47] and Concorde [21]. Aircraft like the second-generation SCT (Supersonic Commercial Transport) will have even stronger couplings between the PCS (Propulsion Control System) and the FCS. Unlike the YF-12/SR-71, XB-70, Tu-144, and Concorde, the next generation SCT[5] will be designed with relaxed static stability. "*The key issue here is the interaction and coupling of both systems*" [48].

(d) ***Geometry Limitation***

The integration of the propulsion element into the volume and/or lift element may be influenced by the strike constraint. Large span aircraft like the BWB FWC and the OFWC lack ground clearance with regard to lateral rotation. Lateral engine placement poses a particular challenge, possibly influencing dihedral, lateral positioning of the engines, landing gear height, etc.

4.2.2.5 Control Element Integration

Figure 4.13 presents the *Control Element* design interaction and design guide parametrics. The following five subject matters are of relevance in the present context: (a) control type, (b) control positioning, (c) control power, (d) FCS, and (e) operation.

(a) ***Control Type***

The CEs (Control Effectors) for the three axes are the LoCE (Longitudinal Control Effector), DiCE (Directional Control Effector), and the LaCE (Lateral Control Effector). The CE types are either of the aerodynamic type, of the reaction type (pitch-, yaw-, roll rockets) like on the North American X-15 [49], and/or of the thrust vectoring type like on the X-31 [50] and the VTOL aircraft types. The traditional aerodynamic CE concepts are either of the incidence control type like the variable incidence tailplane, of the camber control type like the fixed stabiliser-elevator combination, or a mixture of both principles like the trimmable horizontal

[5]Second generation SCT projects are the US HSCT (High-Speed Commercial Transport) and the European ESCT (European Supersonic Commercial Transport).

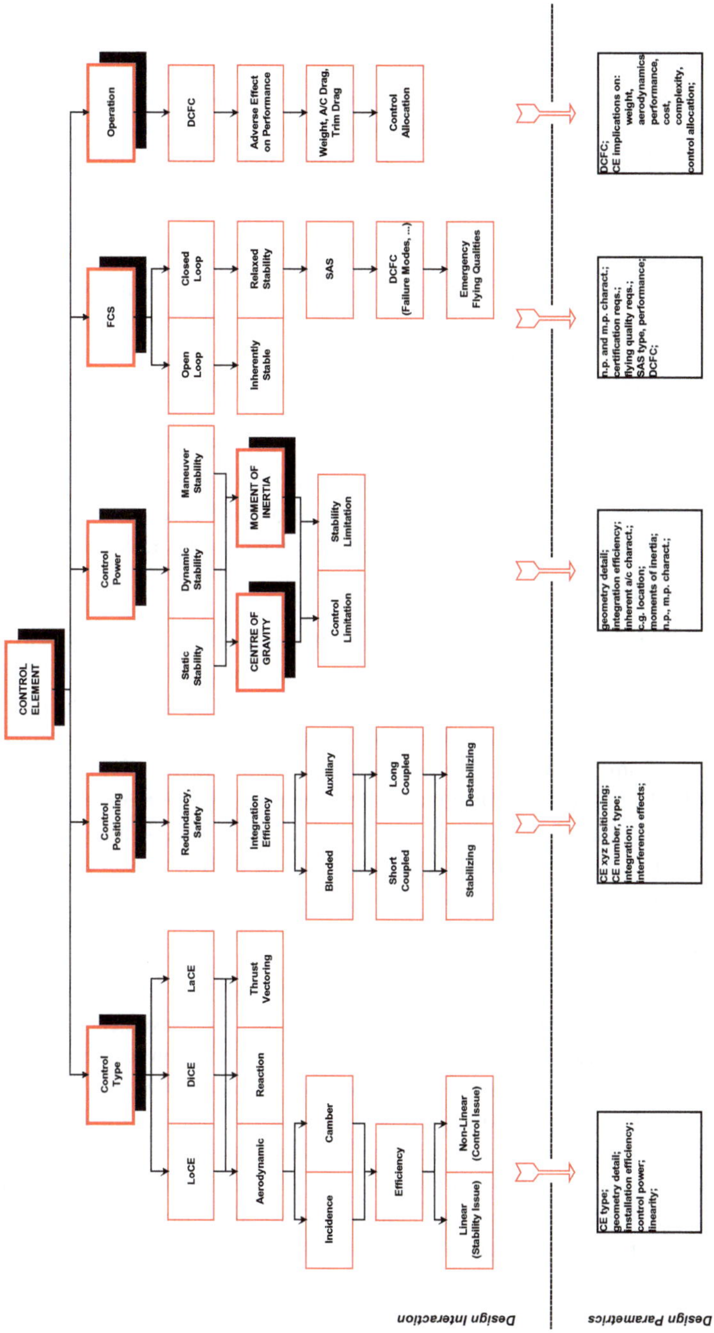

Fig. 4.13 Control element design interaction and design guide parametrics

stabiliser with elevator. The efficiency of those CEs varies significantly throughout the flight envelope.

(b) ***Control Positioning***

Control positioning means the arrangement of each individual CE, its distribution over the airframe, the positioning relative to the c.g., and the integration into the airframe. These factors altogether determine the stabilising or destabilising effects of the controls on the bare airframe.

(c) ***Control Power***

A certain amount of control power is generated by each individual CE. The sum of control power available for each axis needs to be harmonised with static-, dynamic-, and manoeuvre stability. The position of the c.g. and the ratios of the moments of inertia finally determine, if the aircraft is control or stability limited. The Lockheed Martin Skunk Works X-33 technology demonstrator design for the single-stage-to-orbit Venture Star is an impressive example of providing sufficient control power from hypersonic speeds to low subsonic speeds. The vehicle uses a range of aerosurfaces, engine thrust vector control, and reaction control system thrusters, dependent on the flight phase, see Lee, Chang, and Kaiser in [9].

(d) ***Flight Control System (FCS)***

The coupling of a bare airframe with a FCS not only provides the means to affect flying- and handling qualities, but to influence the choice of the primary configuration and concept design parameters (CCV). A closed-loop SAS (Stability Augmentation System) will be required in the case of a relaxed stable, indifferent, or unstable airframe. However, the demonstration of emergency flying qualities in the event of a FCS failure still need to be demonstrated. Clearly, the provision of a FCS is not a substitute for control power. The Grumman X-29 is a highly unstable aerospace vehicle design, demanding a fast and strong FCS in unison with adequate physical control forces available, see Moore and Frei [51].

(e) ***Operation***

As has been mentioned before, the necessity to control, stabilise and trim an aircraft has an adverse effect on aircraft performance. The installed CEs have to comply with the DCFCs (Design-Constraining Flight Conditions). Minimisation of the resulting performance penalties like additional mass, FCS power demands, aircraft overall drag, and trim drag, requires optimal allocation of the controls. The Piaggio P.180 Avanti is an overdetermined system where the CEs are staggered along the x axis, see Sacco [40]. In contrast, the redundant CEs on the US BWB are staggered along the y axis. Again, an efficient control allocation scheme is required to minimise actuator rate, hinge moments, and horsepower requirements, etc., see [23].

4.3 Configuration Aerodynamics Characterisation

When Professor J. Stollery presented The Sydney Goldstein Memorial Lecture with the title '*Aerodynamics, Past, Present and Future*' [52], he offered his personal view of the period 1930–2030. He summarised his perspective on future aerodynamics work, in particular the research and development needed for the range of flying vehicles, likely future projects, and the intellectual challenges involved. Overall, this presentation vividly indicates the scope of technical, scientific, and intellectual challenges lying ahead. To complement such vision, Prem forecasts in [53] an aerospace vehicle development scenario as reproduced in Fig. 4.14.[6]

Section 3.4 has outlined the shortcomings and the potential of configuration aerodynamics work during the conceptual definition phase. Clearly, current and future challenges, as identified by Stollery and Prem, *have to be addressed first at the conceptual design level.* The present chapter identifies modelling capability requirements and finally selects an aerodynamic estimation process, capable of delivering some of the aerodynamic information required by the generic stability and control methodology.

4.3.1 Configuration Aerodynamics Work During Vehicle Synthesis

4.3.1.1 Aerodynamic Estimation Techniques

The definition of an aerodynamic data base for use during the conceptual design of an aerospace vehicle is a task which must yield credible results in a timely manner. It has been mentioned in Sect. 3.4, that the 'three dimensions' of fluid dynamics (experiment, theory, and computation) overall result in three classes of aerodynamic calculation techniques: (a) *analytical*, (b) *semi*-empirical and *empirical*, and (c) *numerical.* As a consequence, it is possible to categorise all existing aerodynamic calculation methods into the following distinct classes, see Table 4.2.

The aerodynamic method development history is characterised by the use of analytically based linear and non-linear methods until the 1960s, requiring massive hand calculations at detail design level. This approach transitioned to Datcom-type handbooks in the late 60s, developed with the availability of extensive wind tunnel and flight test data. The early era of computational aerodynamics until the 1980s culminated in the development of, e.g., Digital Datcom and transitioned to linear- and potential method types. The Computational Fluid Dynamics (CFD) era from the 80s until today virtually swallows all resources, as expressed by Snyder in his remarkable paper from 1990 [74]. "*Linear methods are even further away from the*

[6]The time scale allocated obviously needs modification, but the development trends seem to be valid when reviewing the number of current US X-plane research programmes.

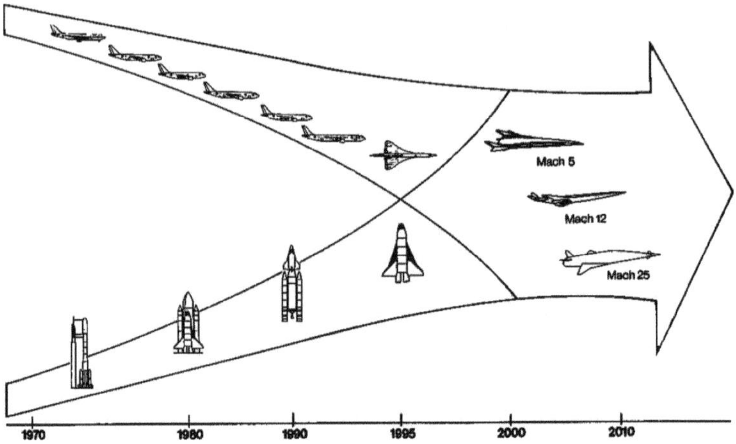

Fig. 4.14 Unification of aircraft and rocket developments [53]

cutting edge of technology than the full potential codes and thus have almost no technical glamour at all. As a result, interest in linear code development is at a low ebb. This is particularly unfortunate for those of us in the conceptual design business." All three approaches to computational aerodynamics are in use today.

(a) ***Analytical***

Numerous *analytical theories* have been developed since the beginning of manned flight, see Table 4.2. Those methods range from the most simple to the most complex with the aim to theoretically analyse two- and three-dimensional lifting systems. However, the solutions to the configuration aerodynamics problem are predominantly limited to incompressible, inviscid fluids. Shevell characterises in [75] the potential of analytical theory to aerospace vehicle design, in that "*... analytical methods plus wind tunnel studies allowed many airplanes to be developed and to meet the predicted performance with acceptable accuracy. Since the 1950's it has been correct to say that the experienced aerodynamicist could predict the drag and lift of a high subsonic speed transport airplane with analytical tools over almost all of the possible speed and angle of attack conditions. ... When flow separation was involved, as at the stall, or shock waves were present on the wing surface, the theories broke down. ... Unfortunately, for most aerospace vehicle designs these limited regions were the most important.*"

Classical analytical theories are still used during the conceptual design phase, to make a first order estimate of the forces and moments involved. Those theories are primarily employed, to identify the influence of gross geometric parameters used in the analytical formulations. Overall, the strength of analytical theory is the ability to gain an insight on the physical role of key parameters. Although analytical methods are based on physical reasoning, the modelling is restricted in practice to specific geometries and operational applications. As a consequence, the modelling

Table 4.2 Engineering techniques for configuration aerodynamics analyses

Analytical			Semi-empirical/empirical			Numerical		
Method	Investigator	Year	Method	Investigator	Year	Method	Investigator	Year
Lifting line theory [54]	Prandtl	1921	RAE standard method [60]	RAE staff	1940	Vortex lattice method [68]	Falkner	1943
Swept-wing theory [55]	Busemann	1935	Hoerner [61–63]	Hoerner	1951	Panel method [69]	Hess	1962
Swept lifting line theory [56]	Weissinger	1942	DATCOM [64]	Hoak	1960	Finite difference method (CFD) [70]	Adam	1975
Low aspect ratio wing theory [57]	Jones	1946	ESDU [65]	RAE	1963	Finite element method (CFD) [71]	Chung	1978
Loading function method [58]	Multhopp	1950	Schemenski [66]	Schemenski	1973	Finite volume method (CFD) [72]	Rizzi	1973
Modified lifting line method [59]	Küchemann	1952	Missile DATCOM [67]	Vukelich	1981	Spectral Method (CFD) [73]	Gottlieb	1977
(…)	(…)	(…)	(…)	(…)	(…)	(…)	(…)	(…)

flexibility of classical analytical methods must be considered non-generic and therefore not suitable in the present context.

(b) ***Empirical and Semi-Empirical***

The difficult balance between model complexity, calculation speed, and calculation accuracy has been classically satisfied with the use of empirical (database) methods and semi-empirical (engineering) aerodynamic methods. A definition of empirical is given with the following: "*Empirical knowledge or study is based on practical experience and observation rather than theories*" [76]. Snyder recalls in [74], that " ... *the 'semi' in semi-empirical methods means that the parameters used in the correlations were reasonable parameters based on the physics of the situation.*" He continues: "*In the development of semi-empirical methods, basic aerodynamic theory is used to make a first order estimate of the lift and drag and to define reasonable aerodynamic parameters to be used in the correlations. Then empirical corrections are made to the theory to produce good agreement with wind tunnel and flight test data.*"

The semi-empirical methods listed in Table 4.2 have in common, that " ... *when the geometric parameters used in an airplane design are significantly different than those in the data base that was used to develop the semi-empirical aerodynamic methodology, the methodology results are subject to question*" [74]. The data bases existing for the classical TAC are large, diverse, detailed, and accessible. As a consequence, a semi-empirical method like Digital Datcom (Data Compendium) may reach remarkable accuracy levels and short turn-around times for TAC, ideal for conceptual design work, see Blake [77] and Blake and Simon [78]. Razgonyaev and Mason compare in [79] several aerodynamic prediction methods with flight and wind tunnel test data available for the XB-70 tail-first configuration (TFC). "*Since DATCOM was designed to estimate aerodynamic derivatives of conventional configurations, some elements of the XB-70 geometry cannot be represented exactly, or sometimes cannot even be modeled at all. ... DATCOM provides no method for estimating aerodynamic derivatives of such configurations.*"

Clearly, empirical and semi-empirical aerodynamic estimation methods like Datcom and ESDU (Engineering Sciences Data) data sheets are restricted to conventional configurations only. Those methods do not function outside their underlying database. The development of new semi-empirical methods by incorporating corrections based on existing empirical data,[7] able to deal with conventional and as well unconventional aircraft configurations, will be extremely difficult due to the extensive engineering resources required. To recall, empirical corrections to theoretical equations will be required to reliably predict transonic effects, vortex stability, vortex breakdown, vortex interactions, control interference effects, attached flow predictions, etc., for the range of conventional and unconventional aircraft configurations and concepts. To conclude, the non-generic character of

[7]Much empirical data exists which is not included in low-order codes.

semi-empirical and in particular empirical estimation techniques disqualifies those methods from becoming the 'work horse' for the generic stability and control methodology *AeroMech*.

(c) ***Numerical***

Numerical estimation methods comprise linear methods like the vortex lattice- and panel method families, and non-linear CFD (Computational Fluid Dynamics) techniques, see Table 4.2. The broad term 'CFD' groups methods which use various approaches to the solution of the underlying non-linear fluid flow equations. Figure 4.15 illustrates the governing equations of numerical aerodynamic estimation techniques. Overall, the principals of mass, momentum, and energy are the generic basis of numerical techniques.

The governing equations of *non-linear CFD methods* are the following. The *Navier-Stokes* equations are coupled, non-linear partial differential equations. Neglecting viscosity (friction) yields the *Euler* equations, being of the coupled and non-linear partial differential equation type. Neglect of rotation yields the *Full Potential* equations, represented with non-linear partial differential equations capable of modelling compressible potential flow.

The governing equations of *linear numerical methods* are the following. Starting from the non-linear *Full Potential* equations, neglecting compressibility yields the *Prandtl-Glauert* equations, representing linearised potential equations. The flow is restricted to be inviscid, irrotational, linear, and is often assumed to be steady. These restrictions physically mean that important flow behaviour such as separation, skin-friction drag, and transonic shocks are not predicted. Items that are predicted include drag-due-to-lift (often called induced drag for subsonic flow, and vortex drag for supersonic flow), and wave drag. The equations are applicable for subsonic and supersonic flow excluding transonic flow. Compressibility effects are approximated with inclusion of the Prandtl-Glauert factor in the free-stream ($x-$)

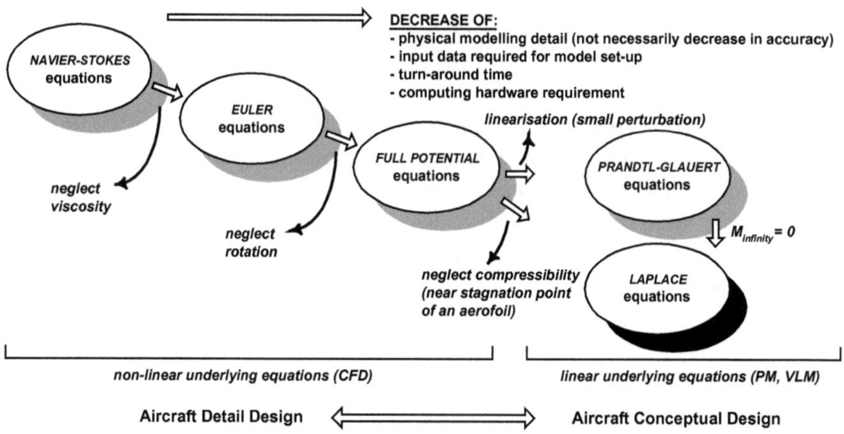

Fig. 4.15 The governing equations of numerical fluid-simulation methods

direction. This permits only application to slender aircraft where the perturbation quantities u, v, and w are small. Usually, compressibility inaccuracies are largest in the stagnation region due to the highest deviation from the free-stream velocity. For slender aircraft, the local Mach number becomes the free-stream Mach number due to the negligible influence of the stagnation point on the flow field. Clearly, those equations represent the simplest form of the fluid-flow equations that approximate compressibility effects. Being a linearised set of equations means, that quantities like air density and speed of sound are constant, whereby local Mach numbers become the free-stream Mach number. Neglect of free-stream- and local Mach number (compressibility) effects transforms the *Full Potential* equations into the *Laplace* and the *Wellen* equations. Transformation of the factor $\left(1 - M_\infty^2\right)$ in the subsonic *Prandtl-Glauert* equation to 1 yields the *Laplace* equation (linear potential equation), valid for the subsonic speed range. Transformation of the factor $(M_\infty^2 - 1)$ in the supersonic *Prandtl-Glauert* equation to 1 results in the *Wellen* equation, valid in the supersonic speed range.

CFD methods require the construction of a grid to fill the flow field volume of interest, resulting in a large number of mesh points. This consequently leads to a very large system of equations, posing particular demands on computational resources. As a consequence, CFD modelling requires tasks like grid generation, flow field discretisation, efficient solving of systems of equations, data storage and transmission, and computational flow visualisation. The resulting extensive demand for computational resources and the detailed geometry representation required[8] clearly disqualify high-order CFD estimation methods for conceptual design work.

Linear numerical estimation methods are most suitable for conceptual design work. Vortex lattice methods (VLM) study the mean geometric surface and panel methods study the geometric surface. Both methods circumvent the modelling task of the geometric volume, thereby avoiding much of the CFD-typical pre-processing complexity. However, linear estimation techniques do not estimate effects like boundary layer, wake roll up, transonic flow, or strong shocks. Instead, " *... they do provide reasonable estimates of the inviscid aerodynamics, including drag, for a large class of airplane geometries in both subsonic and supersonic flow*" [74]. Snyder proposes to augment those linear methods with semi-empirical models, resulting in non-linear potential methods.

The aerodynamic prediction techniques of choice in the present context are methods based on the linear potential equations. Linear analysis tools are well-proven for linear aerodynamics. However, aerodynamic control effectors are predominantly sized to comply with critical flight conditions at the boundary of the flight envelope. The resulting non-linear flow domination requires 'enhanced' linear tools (e.g., vortex lattice plus vortex model).

---

[8]During the conceptual design stage, only gross geometric parameters are available and are of relevance.

4.3.1.2 Computational Aerodynamics Integration Requirements

Rubbert and Tinoco describe in [80] the process of applying computational analysis to practical aerodynamic estimation. The process starts with the flow field around the flight vehicle. This physical situation is represented with a simplified mathematical model, which is solved numerically. Clearly, the estimation process is not finished when the results are examined. More importantly, the entire sequence (flow situation, physical model, mathematical model, numerical solution, result) requires an interpretation to provide the final aerodynamic solution. Mason remarks in [81]: "*Notice here that the numerical solution of a computational problem is a small part of the total engineering process. Successful aerodynamicists must master the entire sequence of steps.*"

Nicolai and Carty describe in [82] the integration of computational aerodynamics for conceptual design application into the multidisciplinary design environment RCD (Rapid Conceptual Design) at Lockheed Martin *Skunk Works*. Especially successful utilisation examples of computational aerodynamics for conceptual design application are given by Morris, which have culminated in the commercially available SWIFT foot-launch sailplane [83] and the NASA funded oblique flying wing (OFWC) small-scale demonstrator programme [84]. The activities by Morris, coupled with the Multi-Disciplinary Optimisation (MDO) expertise available at Stanford University, enabled his colleagues

Table 4.3 Priority list of functional non-linear aerodynamic prediction requirements

1	Must be responsive to quick geometry changes (setup time: for complete configuration 1–3 days maximum).
2	Possibility of automatic aircraft geometry modelling (simple input for mean surface geometry with controls).
3	Must not require detailed geometry (modelling with gross geometry parameters, no volume grids).
4	Interactive geometry editing and output viewing.
5	Able to handle a wide variety of aircraft geometries.
6	Single consistent calculation method instead of the multi-method approach.
7	Must run on workstations or equivalent PC's; modular characteristics.
8	Favorable run times (10 h maximum run time).
9	Output in tabular, plotted, and electronic format.
10	Trend information more important than absolute accuracy.
11	Effort should concentrate on subsonic and supersonic speeds.
12	Specific emphasis on non-linear flow phenomena for control surface sizing, in particular high angle of attack.
13	Easy adaption to a multidisciplinary optimisation environment.
14	Ensure calculation robustness for a legal set of input.
15	Keep basic method as simple, fast, and general (generic) as possible (avoid black boxes).
16	Understanding 'why' more important than 'what'.

Wakayama and Kroo [41] to actively advance the multidisciplinary configuration optimisation process of the US BWB flying wing program. Overall, those activities represent true state-of-the-art in aircraft conceptual design.

Table 4.3 lists non-linear aerodynamic prediction requirements relevant to stability and control work at conceptual design level.

4.3.2 *Identification of Gross Configuration Aerodynamics Parameters*

Having identified the challenge of configuration aerodynamic modelling during conceptual design in Sect. 3.4, the available modelling techniques have been reviewed and functional requirements have been formulated for a computational aerodynamic estimation method, see Sect. 4.3.1. However, before selecting a method suitable for integration into the generic stability and control methodology, it is required to secure a balance between the minimum aerodynamic data required and data finally available for the design of controls.

Any permutation of aircraft configuration and concept is tied to a specific set of aerodynamic flow behaviour and associated sensitivities, see Fig. 4.16.

The following examines configuration flow phenomena, the aerodynamics of control effectors, and finally stability and control derivative coefficients as the primary means to quantify aerodynamics. The primary source of information is the conceptual design KBS, which has been described in Sect. 2.5.

4.3.2.1 Configuration Flow Phenomena

During the conceptual design process of new aircraft types, only estimates of gross aerodynamic characteristics are attainable. Chronic data shortage coupled with permanent time pressure, high cost in manpower and computation time, those aspects clearly prevent the study of secondary flow effects. The relative design-influence of individual gross aerodynamic flow phenomena on the vehicle depends on the operational conditions encountered, see Table 4.4. Clearly, the challenge in

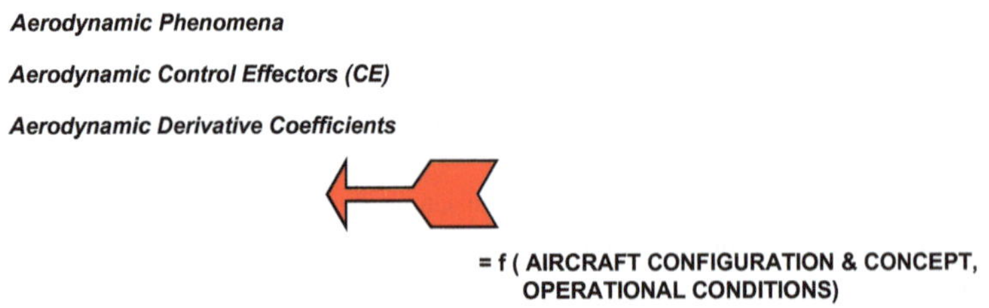

Fig. 4.16 Configuration aerodynamics dependency

Table 4.4 Flow phenomena dependency on operating conditions

Mission segment	Take-off, climb, cruise, descent, approach, loiter, landing, etc.
Aircraft setting	Centre of gravity position, high-lift setting, etc.
Aircraft state	Steady, perturbed
Ambient properties	Temperature, density, speed of sound, dynamic pressure, etc.

developing a generic method lies in identifying a set of gross aerodynamic flow phenomena, common to the range of aircraft configurations and concepts.

A comprehensive assessment of gross aerodynamic flow phenomena for the range of aircraft configurations and concepts, dependent on all possible permutations of operating conditions, requires examination of the entire flight envelope. Figure 4.17 indicates the magnitude of this multidimensional undertaking. Obviously, such an approach is not easyly accomplished during the conceptual design phase and an examination of such depth is well beyond the scope of the current research undertaking.

To recall, Fig. 4.17 illustrates the primary challenge of advanced design, being confronted with the risk of overlooking potential problem areas for reasons like (a) insufficient design experience, (b) lack of time to conduct in-depth studies, (c) lack of wind tunnel data, and (d) insufficient accuracy of available aerodynamic methods.

Table 4.5 identifies gross aerodynamic flow phenomena for the range of aircraft configurations and concepts from subsonic to hypersonic speed. The problem of designing aerodynamic controls manifests itself in the requirement to consider longitudinal and lateral-directional flow phenomena. Apart from mission performance, the flight conditions and associated flow phenomena relevant to the design of controls are primarily of non-linear flow character.

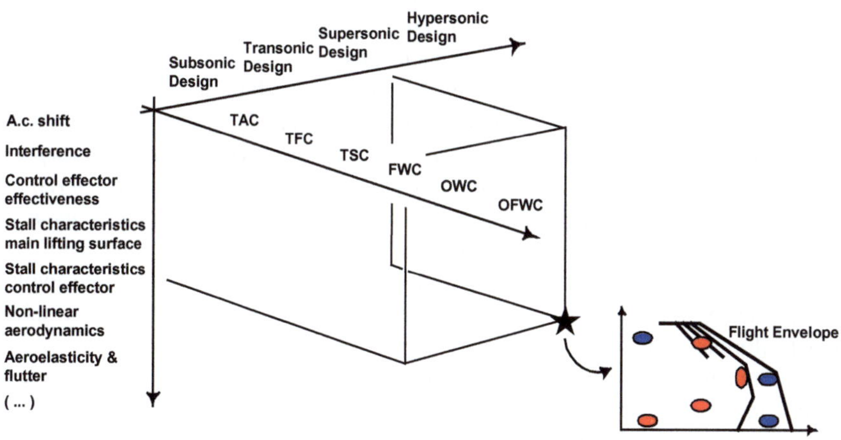

Fig. 4.17 Multi-dimensional dependence of aerodynamic flow phenomena

In addition to the static and dynamic derivative coefficients implied in Table 4.5, the aircraft conceptual design KBS (Sect. 2.5) discusses the following gross flow phenomena for selected case studies: non-linear aerodynamic behaviour, main lifting surface stall characteristics, CE stall characteristics, aeroelasticity, flutter, tuck, pitch-up, high angle-of-attack departure, interference effects, and apparent mass effects.

In classical conceptual design, the following parameters need to be assessed sequentially due to lack of data availability: (a) lift & drag, (b) moment & trim, (c) longitudinal control, (d) lateral-directional control, and (e) high-lift. It is the aim of the present research undertaking to avoid prioritising the above calculations, instead being able to estimate those quantities concurrently. To conclude, the aerodynamic estimation tool ideally aims for the challenging case of low-subsonics, transonics, manoeuvrability, and high angle of attack capability, to assure non-linear estimation capability.

4.3.2.2 Aerodynamic Control Effectors

Hancock introduces in [86] the subject matter 'aerodynamics of controls' by stating: "*Aerodynamics of controls are concerned with the understanding, both in qualitative and quantitative terms, of the aerodynamic loading induced on the surface of an aircraft configuration following the deployment of a control surface. It is necessary to know the overall forces and moments on the aircraft configuration in order to calculate aircraft response and to ensure structural integrity; it is necessary to know the loads on the control in order to design the actuation system.*"

Lifting surface trailing-edge flaps used as secondary aerodynamic high-lift devices and primary aerodynamic CEs (LoCE, DiCE, and LaCE) are basically one and the same device, controlling the circulation of the lifting surfaces. Hoerner vividly outlines the physical rationale of trailing-edge flaps in [62].

A review of the history of aerospace vehicle design and the design for stability and control (KBS see Sect. 2.5, Abzug and Larrabee [12] and Young [87]) reveals, that the technology and processes implemented to design aerodynamic control effectors (CEs) at conceptual design level have changed little in the last 100 years. Ross and Thomas challenge rebuttal with their remarkable paper from 1979 by stating, that "*The advent of Active Control Technology means that the aerospace vehicle designer needs as much, if not more, knowledge of control characteristics, with more emphasis on maximum control power and actuating force or moment than for the previous generation of aircraft*" [88].

Ross and Thomas structure the problem of aircraft control into three evolutionary stages. Figure 4.18 is directly reproduced from Ref. [88], since no additional information is required 40 years after publication.

With unaugmented aircraft, the aerodynamic CEs establish a direct link between control forces and the pilot. In the special case of controls freed, stability is even adversely affected. With augmented aircraft, the CEs are used to augment stability and controllability. "*Successful design now shows increased dependence on the*

Table 4.5 Identification of generic gross aerodynamic flow phenomena during aircraft conceptual design

Design analysis	Mission performance	Takeoff and landing performance	Manoeuvre performance	Envelope performance
Data requirement	The quantities to be predicted are range and payload.	The quantities to be predicted are control power available, stall margin, pitch and roll performance, engine out performance, power induced effects, and ground effect.	The quantities to be predicted are control power available to enable sufficient turn rate, roll performance, pitch agility, and finally excess power.	The quantities to be predicted are V_{min} and V_{max} and the maximum angle of attack α.
Aerodynamic parameter identification	The cruise condition is characterised by low angles of attack α and low control deflections δ. Primarily involvement of the longitudinal coefficients C_L, C_D, and C_M since sideslip angle β is equal to zero. Estimation of trimmed polars.	Those flight conditions are characterised by low speed, moderate to high angles of attack α, partial flow separations on primary lifting surfaces and CEs, large CE-deflections, the existence of sideslip angles β, and strong power induced interference effects.	The flight conditions of concern range from low subsonic to hypersonic speeds, from high to low angle of attack α, from attached to fully separated flow, large CE deflections, from low to high sideslip angles $_\beta$, dynamic and unsteady flow conditions, and power induced interference effects.	The flight conditions of concern range from low subsonic to hypersonic speeds, from high to low angle of attack α, from attached to fully separated flow, large CE deflections, from low to high sideslip angles $_\beta$, dynamic and unsteady flow conditions, and power induced interference effects
	Internal trim calculation not required, better done independent of predictive methods. Effort should concentrate on subsonic speeds, transonic too complex, little need for nonlinear effects supersonically (except aeroelasticity).	Involvement of the longitudinal coefficients C_L, C_D, C_M, and the lateral-directional coefficients C_Y, C_l, C_n.	Involvement of the longitudinal coefficients C_L, C_D, C_M, and the lateral-directional coefficients C_Y, C_l, C_n.	Involvement of the longitudinal coefficients C_L, C_D, C_M, and the lateral-directional coefficients C_Y, C_l, C_n.

Note Data adapted, in part, from Dorsett and Peters [85]

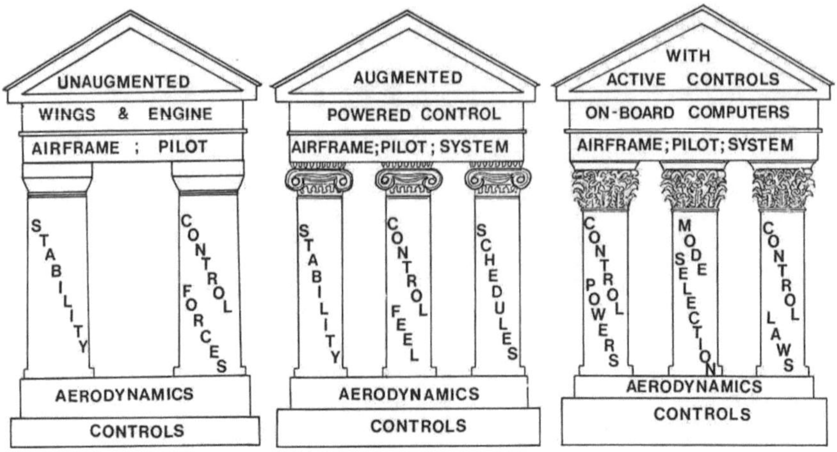

Fig. 4.18 Dependence of aerospace vehicle design on aerodynamic data and control data [88]

characteristics of the motivators and hence on the knowledge needed to achieve the desired characteristics." The third class of aircraft and primary focus in the present context are CCVs (Control Configured Vehicles), which depend on classical as well as advanced CE types. "*The dependence of the design on motivator characteristics has increased even further and may even dominate particularly in the design of an aircraft control system which aims to be adaptive*" [88]. It is interesting to note that for aircraft with active controls (CCV), stability is no longer a primary design issue. Adequate stability can be produced by having adequate control power available. Illustrative examples are FWCs like the Horten IX [45] and the B-2 [89], where directional stability is no longer generated with the installation of a vertical fin. Instead, design emphasis lies on generating adequate control power in pitch, roll, and yaw.

In summary, the dependence of aerospace vehicle design on CE characteristics has significantly increased for (a) unconventional or advanced aircraft, (b) CCVs, especially relaxed static stability vehicles, and (c) aircraft where the FCS aims to be adaptive. The results of extensive and systematic testing of conventional aerodynamic CEs rests with methods given within ESDU data items and Datcom handbooks, see Table 4.2. However, today's controls related data are acquired on a rather ad hoc basis for specialised aircraft applications only, without having yet culminated in design procedures for the next generation of aircraft types. An illustrative example is given by Moul et al. [90], where the stability and control characteristics of stealthy flying wing type aircraft are examined. Clearly, the apparent control-related knowledge gaps need to be filled using appropriate design tools.

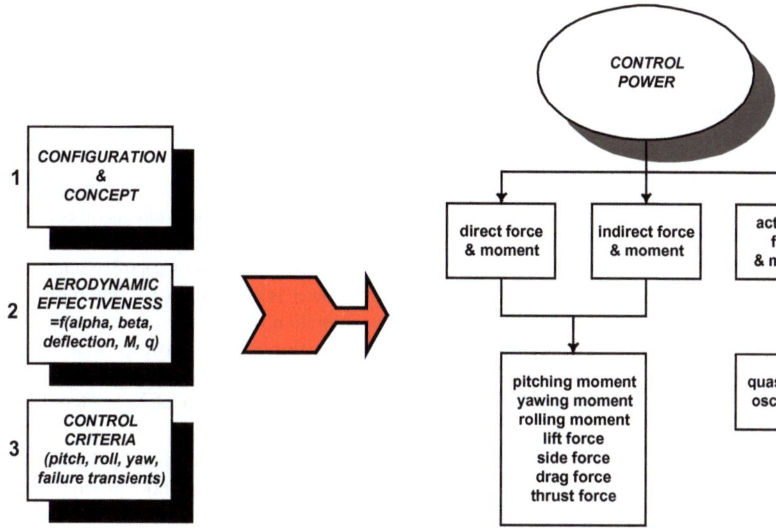

Fig. 4.19 Dependency of control power on configuration & concept, aerodynamic effectiveness, and stability and control criteria

The overall dependence of control power on the choice of the aircraft outer mold line, the effectiveness of the CE of choice,[9] and stability and control design and certification criteria, is shown in Fig. 4.19.

With reference to Figure 4.19, the aircraft response to deflection of aerodynamic CEs is conveniently subdivided into direct-, indirect-, and actuating forces and moments. Direct control effects are those intended to be generated by the CE, whereby indirect effects are largely undesired secondary coupling forces and moments due to unfavourable pressure distributions. The actuating forces are directly related to CE hinge moments, actuation power required, actuator size, -band width, and -frequency. Ross and Thomas present in [88] and Thomas in [91] an impressive survey of the characteristics of aerodynamic controls, a CE-related knowledge-base judged fully representative for the range of classical aerodynamic CEs in use today.

Referring back to Fig. 4.19, task 1 '*Configuration & Concept*' has been discussed in Sect. 4.2.1, whereby the subject matter of task 3 '*Control Criteria*' will be dealt with in Sect. 4.5. Task 2 '*Aerodynamic Effectiveness*' is assessed with the following. Aerodynamic control effectiveness is, by definition, a measure for the lift differential produced per degree deflection of an aerodynamic CE (camber-, or incidence change). Table 4.6 presents an overview only of primary design

[9]The term CE (Control Effector) is used throughout this report to describe all types of controls, including aerodynamic controls, thrust vectoring, thrusters or jets, etc.

Table 4.6 Design conditions and design parameters influencing the aerodynamic efficiency of control effectors

Effect	Physical explanation
Wing planform	The primary lift generating element influences aerodynamic control efficiencies to a large degree, in particular when the aircraft is wing dominated. The range of wing planform-permutation possible (primarily variation of aspect ratio versus leading-edge (LE) sweep) has governing effects on control efficiency. The wing aspect ratio, % of wing area LEX, LE sweep, and LE droop or crank, have strong effects on $C_{m\beta}$ (high angle of attack effect). Roll control power is affected by LE droop or crank, LE sweep, the incorporation of kinks or of a sawtooth. $C_{n\beta DYN}$ is affected by LE droop or crank, LE sweep, % of wing area LEX. The vertical stabiliser control power is affected by aspect ratio, wing LE sweep, % of wing area LEX, sawtooth or kink, etc.
Fuselage geometry	The aerodynamic control effectiveness varies primarily dependent on the fuselage forebody geometry, and even more pronounced when the aircraft is fuselage dominated. With a short forebody, the forebody vortex is affected by LEX geometry; with a longer forebody there is less interaction between the forebody and LEX vortices but the destabilising effects increase. The forebody shape has powerful effects on zero sideslip aerodynamic forces and moments. The forebody can be the sole source of aircraft directional stability at high angles of attack. In general, any surface ahead of the centre of gravity has adverse effects on stability and secondary effects on control power demands.
Angle of attack	Pitch (LoCE), roll (LaCE), and yaw (DiCE) control effectiveness degrades as angle of attack is increased. The reasons are that pitch (LoCE) and yaw (DiCE) control surfaces become immersed in wing/body wakes. Roll control surfaces (LaCE), being typically placed at outboard wing location become enveloped in low energy air from wing tip stall. Interaction of fuselage forebody and wing vortices create adverse pressure fields in the vicinity of the empenage. Control requirements are typically set by high angle of attack effectiveness.
Angle of sideslip	Pitch (LoCE), roll (LaCE), and yaw (DiCE) control effectiveness degrades as sideslip is increased. The reasons are that symmetric pitch (LoCE) and roll (LaCE) control surfaces become asymmetric with sideslip, thereby creating reduced forces and moments. Sideslip creates sidewash which can affect pitch, roll and yaw forces, but in particular yaw. As sideslip increases, symmetric aircraft become less and less streamlined, inducing difficult to predict non-linear flow effects

(continued)

Table 4.6 (continued)

Effect	Physical explanation
Mach number	Pitch (LoCE), roll (LaCE), and yaw (DiCE) control effectiveness increase in the subsonic speed regime and degrade with increasing supersonic Mach number. At subsonic speeds, any deflection of control surfaces change the circulation in the region of the surface (flow upstream is affected). At supersonic speeds, deflection of any control surface has only an effect on the control surface itself. With respect to Mach effects, control requirements are typically set by high transonic and supersonic aerodynamic effectiveness.
Dynamic pressure (aeroelastic effects)	Airframe flexibility (coupling of structural properties and aerodynamics) is a significant factor in determining control effectiveness. The reasons are that aileron (LaCE) effectiveness is normally reduced due to wing torsion/bending (aileron reversal effect). Rudder (DiCE) effectiveness is reduced due to fuselage bending/torsion. Directional stability is similarly reduced. The effects on horizontal tail (LoCE) effectiveness depend on the hinge position and Mach number.
Interactions/coupling	The effects of aerodynamic flow interactions/couplings on aerodynamic control effectiveness are most difficult to predict. Control coupling/interference effects exist even at zero sideslip angle, $\beta = 0$, due to indirect control forces and moments induced by vortex flow interactions. Examples are lateral-directional stability effects due to tailplane (LoCE) deflection, pitching moments due to angled (non-planar) control surfaces (canted fin, rudder toe-in, V-tail, etc.), or adverse yaw due to roll control surfaces (LaCE) deflection like ailerons, differential aft tail, or differential canard.

Note Data adapted, in part, from Skow [92]

conditions and design parameters, which influence aerodynamic control effectiveness. A more detailed discussion of the aerodynamics of controls is beyond the scope of the current research investigation.

4.3.2.3 Stability Derivatives—Key Performance Indicators

Design-oriented stability and control analysis requires configuration aerodynamics design information in both quantity and quality. At design level, the complete aerodynamic 'thumb-print' of the flight vehicle is commonly defined with a set of stability derivatives.

The analytical representation of an aircraft as a rigid body with six degrees of freedom was first introduced by Prof. Bryan in 1911 [93]. The equations of motion represent aerodynamic forces and moments by means of stability derivatives. This technique assumes that aerodynamic forces and moments can be expressed as a function of the instantaneous values of the perturbation variables. Nelson recapitulates the underlying assumptions in [94]: "*The perturbation variables are the instantaneous changes from the reference conditions of the translational velocities, angular velocities, control deflection, and their derivatives. With this assumption, we can express the aerodynamic forces and moments by means of a Taylor series expansion of the perturbation variables about the reference equilibrium condition.*"

The dimensional or non-dimensional derivatives are either computed analytically, measured in the wind tunnel, or they are obtained from flight test. Kalviste and Eller recall in [95]: "*The advantage of these parameters is that they can be measured and computed throughout the angle-of-attack and sideslip envelope of the aircraft independent of aircraft trim conditions.*" However, the derivative approach (Bryan's hypothesis) provides an approximation only to the aerodynamic forces and moments in the equations of motion. McRuer, Ashkenas, and Graham justify such approach by stating that "*Although none of the aerodynamic coefficients is guaranteed to behave linearly with any of the variables, in most cases, for small perturbations it is reasonably accurate to linearize the coefficients about the operating point* ..." [8].

Obviously, such theory fails when applied to problems where large-amplitude motions are to be expected to result in rapidly changing forces and moments (e.g., spinning, stalled flight). Note that the derivative expressions, dimensional and non-dimensional, are of a particular, commonly-used form. However, little or no distinction in terminology is made amongst them.[10] The definition by McRuer et al. [8] is used throughout this report in that "... *all are referred to as 'stability derivatives,' regardless of the particular form.*"

Ellison and Hoak [96] and Thomas [97] have tried to examine the evolution of the aircraft geometry over the years with the intent, to relate geometric features to stability derivative sensitivity.

1. The commuter type of aircraft represents an engineering extreme at one end of the design scale. This type of aircraft consists largely of independent aerodynamic components (non-integrated design: large aspect ratio, unswept wing and tail surfaces on a long tail arm, etc.). Those primarily independent aerodynamic components make a distinct contribution to each derivative. The 'component build-up' technique of aerodynamic quantities was initially devised for such types of aircraft, culminating in the well-known handbook methods Datcom and ESDU, see Table 4.2.
2. With the introduction of the jet engine, the higher speed transonic type of transport aircraft with moderately swept wings entered the scene. For those aircraft, the coupling between the stability derivatives becomes more pronounced. Non-linear flow phenomena result in non-linear stability derivatives, which require evaluation for each separate flight condition.

[10]The following lists some of the expressions used alone or in combination to define the term 'derivative': coefficient, parameter, static, dynamic, longitudinal, lateral-directional, aerodynamic, stability, control, cross, damping, aeroelastic, static, dynamic, quasi-static, rotary, translational, equivalent, linear, non-linear, rate, time-dependent, …

3. When entering the supersonic speed range, the slender supersonic fighter and transport are characterised by low aspect ratio wings and high sweep angles. The different flow mechanism of those wings and increased fuselage domination results in more incidence dependence of some of the lateral and longitudinal derivatives.
4. The opposite end of the design scale is marked by aerospace lifting-body and wave-rider vehicles. The individual functional parts of the classical subsonic aircraft are blended into a single surface. "*The 'single-body' geometry and its implications regarding derivatives constrains the well known 'component build-up' interpretation. On the other hand, these fully blended vehicles minimise aerodynamic interference effects, so that it can be expected that the estimation of derivatives is simplified*" [97].

The intent to physically characterise and visualise the range of stability derivatives for the spectrum of flying vehicles has been expected to be a major challenge. No single report exists which contains consistent derivative information for the range of conventional and unconventional aircraft from low subsonic to hypersonic speeds. The report by Ross and Thomas [88] is presumably the most complete compilation of control related design information for different aircraft available today. Several researchers have attempted to characterise, and more importantly, have tried to generalise stability derivative information for the conventional TAC only in tabular form (symbol, definition, origin, equation, typical values, variation with Mach number, etc.). Most noteworthy descriptions are those by Nelson [94], Blakelock [98], and Roskam [4].

A generalisation of the derivative information for the range of conventional and unconventional aircraft layouts is a major undertaking, and thus has not been attempted in the present context. Figure 4.20 illustrates the multidimensional design space, wherein the complete aerodynamic representation of the flight vehicle is composed.

Clearly, the aerospace vehicle designer must comprehend the coupling between a particular design decision and its effect on the set of stability derivatives. Such a task becomes a true challenge for aircraft like the asymmetric OFWC, where no classical empirical relations for the estimation of stability derivatives exist at all. This demand has to be addressed twofold. At first, a database of historical derivative information has to be maintained (part of the KBS), capable of illustrating known interrelations between design trends and derivative characteristics. Secondly, a generic aerodynamic estimation method is required, able to consistently estimate a broad range of stability derivatives (a suitable aerodynamic estimation tool is selected in Sect. 4.3.3).

A study of the aerodynamic behaviour of selected case studies (see Fig. 2.9) has been undertaken and documented within the KBS. A dedicated stability derivative database has been constructed and phased into the KBS. As to be expected, the importance of individual stability derivatives on a particular design varies dependent on the choice of aircraft configuration, aircraft concept, and flight condition encountered. Clearly, specific derivative combinations are of significance for

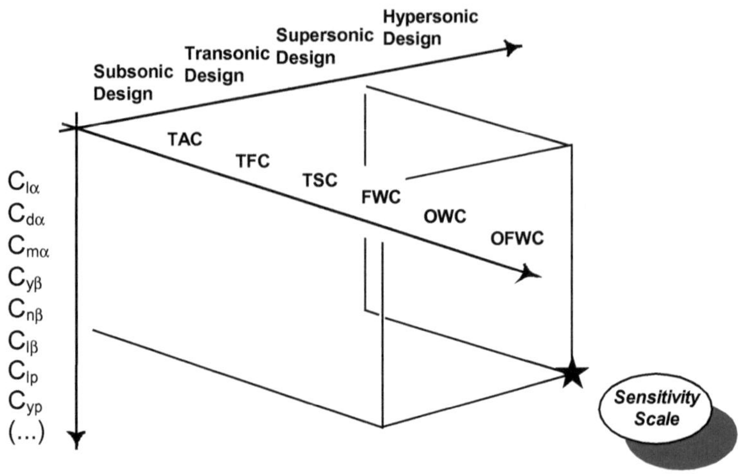

Fig. 4.20 Multi-dimensional dependence of stability derivatives

unconventional aircraft, which, on the other hand, may be neglected for conventional TACs. Figure 4.21 proposes to illustrate the interrelation between design-trends and stability derivatives in the form of a so-called '*Stability Derivative Card*'. Although sufficient derivative information for the range of aircraft configurations and concepts has been phased into the KBS, such analysis/visualisation task is not a subject of the present research undertaking.

When discussing the class of conventional and unconventional aircraft configurations and concepts in symmetric or asymmetric flight, an enlarged range of stability derivatives has to be taken into account. Table 4.7 presents a matrix of translational and rotary stability derivative coefficients thought relevant during conceptual design, without implying the immediate ability to estimate each coefficient. The double-framed coefficients are usually considered during the conceptual design of the conventional TAC. The single-framed coefficients have been additionally considered for the asymmetric AD-1 oblique-wing manned research aircraft, a research programme active from 1979 to 1982 [99]. The table assembles the dependent variables (X, Y, Z, L, M, N), the set of independent static variables (u, v, w, δ_i) and dynamic variables ($u(t)$, $v(t)$, $w(t)$, p, $p(t)$, q, $q(t)$, r, $r(t)$), where the rate (time) dependent variables are of first order only. Control deflections consist of Primary Controls (PC),[11] Secondary Controls (SC),[12] and Configuration Setting (CS).[13]

[11]Primary controls (PC) are: LoCE, DiCE, and LaCE (elevator, elevon, aileron, taileron, rudder, drag rudder, spoiler, canard, body flap, thrustvector, etc.).

[12]Secondary controls (SC) are: trailing-edge flaps, leading-edge flaps, air brakes, etc.

[13]Configuration settings (CS) are: landing gear position, wing tip deflection angle (XB-70), etc.

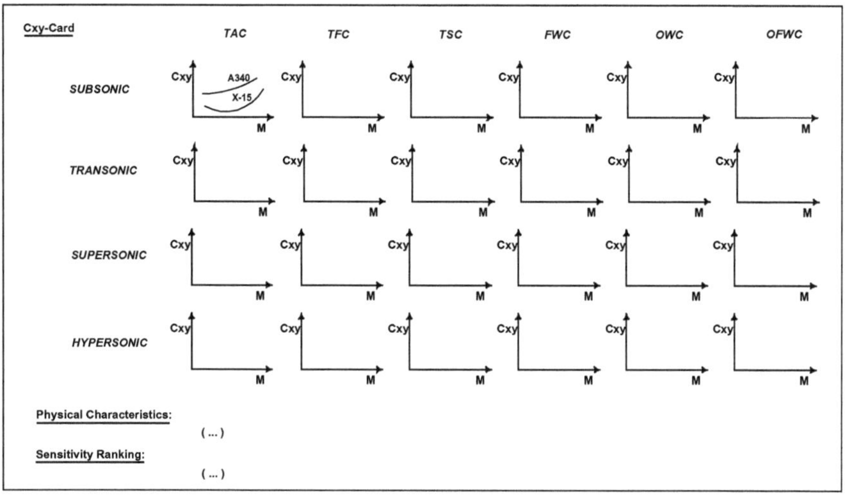

Fig. 4.21 Visualisation proposal of generic stability derivative information: 'Stability Derivative Card'

Overall, Table 4.7 summarises the aerodynamic modelling capability maximum demanded by the generic computer-based estimation tool.

4.3.3 Evaluation of Relevant Aerodynamic Prediction Codes

Although many different types of computer-based aerodynamic analysis methods exist that can predict subsonic-, supersonic-, and hypersonic aerodynamic characteristics, most are not practical for conceptual design. As identified before, potential flow methods are considered most appropriate for conceptual design applications.

4.3.3.1 Survey, Evaluation and Comparison of Potential Flow Methods

A broad survey has been undertaken of subsonic- and supersonic vortex lattice methods (VLM) and panel methods (PM). Figure 4.22 depicts the range of research, industry, and commercially available potential flow methods, which have been investigated along the attribute list as defined in the Figure. The detailed characterisation results of the range of aerodynamic estimation codes has not been included in this report.

Having reviewed the range of vortex lattice methods (VLM) and panel methods (PM) as indicated in Fig. 4.22, the following comments are apt. Enhanced non-linear aerodynamic estimation tools, based on potential flow theory, are most

Table 4.7 Matrix of translational and rotary Stability derivative coefficients

	X[D,T]	Y	Z[L,W]	L	M	N
u	C_{D_u} C_{T_u}	C_{Y_u}	C_{L_u}	C_{l_u}	C_{m_u}	C_{n_u}
v [β]	C_{D_β} C_{T_β}	C_{Y_β}	C_{L_β}	C_{l_β}	C_{m_β}	C_{n_β}
w [α]	C_{D_α} C_{T_α}	C_{Y_α}	C_{L_α}	C_{l_α}	C_{m_α}	C_{n_α}
p	C_{D_p} C_{T_p}	C_{Y_p}	C_{L_p}	C_{l_p}	C_{m_p}	C_{n_p}
q	C_{D_q} C_{T_q}	C_{Y_q}	C_{L_q}	C_{l_q}	C_{m_q}	C_{n_q}
r	C_{D_r} C_{T_r}	C_{Y_r}	C_{L_r}	C_{l_r}	C_{m_r}	C_{n_r}
$\dot{u}$	$C_{D_{\dot{u}}}$ $C_{T_{\dot{u}}}$	$C_{Y_{\dot{u}}}$	$C_{L_{\dot{u}}}$	$C_{l_{\dot{u}}}$	$C_{m_{\dot{u}}}$	$C_{n_{\dot{u}}}$
$\dot{v}$ [$\dot{\beta}$]	$C_{D_{\dot{\beta}}}$ $C_{T_{\dot{\beta}}}$	$C_{Y_{\dot{\beta}}}$	$C_{L_{\dot{\beta}}}$	$C_{l_{\dot{\beta}}}$	$C_{m_{\dot{\beta}}}$	$C_{n_{\dot{\beta}}}$
$\dot{w}$ [$\dot{\alpha}$]	$C_{D_{\dot{\alpha}}}$ $C_{T_{\dot{\alpha}}}$	$C_{Y_{\dot{\alpha}}}$	$C_{L_{\dot{\alpha}}}$	$C_{l_{\dot{\alpha}}}$	$C_{m_{\dot{\alpha}}}$	$C_{n_{\dot{\alpha}}}$
$\dot{p}$	$C_{D_{\dot{p}}}$ $C_{T_{\dot{p}}}$	$C_{Y_{\dot{p}}}$	$C_{L_{\dot{p}}}$	$C_{l_{\dot{p}}}$	$C_{m_{\dot{p}}}$	$C_{n_{\dot{p}}}$
$\dot{q}$	$C_{D_{\dot{q}}}$ $C_{T_{\dot{q}}}$	$C_{Y_{\dot{q}}}$	$C_{L_{\dot{q}}}$	$C_{l_{\dot{q}}}$	$C_{m_{\dot{q}}}$	$C_{n_{\dot{q}}}$
$\dot{r}$	$C_{D_{\dot{r}}}$ $C_{T_{\dot{r}}}$	$C_{Y_{\dot{r}}}$	$C_{L_{\dot{r}}}$	$C_{l_{\dot{r}}}$	$C_{m_{\dot{r}}}$	$C_{n_{\dot{r}}}$
δ_{PC_i}	$C_{D_{\delta_{LoCE_i}}}$ $C_{T_{\delta_{LoCE_i}}}$	$C_{Y_{\delta_{LoCE_i}}}$	$C_{L_{\delta_{LoCE_i}}}$	$C_{l_{\delta_{LoCE_i}}}$	$C_{m_{\delta_{LoCE_i}}}$	$C_{n_{\delta_{LoCE_i}}}$
	$C_{D_{\delta_{DiCE_i}}}$ $C_{T_{\delta_{DiCE_i}}}$	$C_{Y_{\delta_{DiCE_i}}}$	$C_{L_{\delta_{DiCE_i}}}$	$C_{l_{\delta_{DiCE_i}}}$	$C_{m_{\delta_{DiCE_i}}}$	$C_{n_{\delta_{DiCE_i}}}$
	$C_{D_{\delta_{LaCE_i}}}$ $C_{T_{\delta_{LaCE_i}}}$	$C_{Y_{\delta_{LaCE_i}}}$	$C_{L_{\delta_{LaCE_i}}}$	$C_{l_{\delta_{LaCE_i}}}$	$C_{m_{\delta_{LaCE_i}}}$	$C_{n_{\delta_{LaCE_i}}}$
δ_{SC_i}	$C_{D_{\delta_{SC_i}}}$ $C_{T_{\delta_{SC_i}}}$	$C_{Y_{\delta_{SC_i}}}$	$C_{L_{\delta_{SC_i}}}$	$C_{l_{\delta_{SC_i}}}$	$C_{m_{\delta_{SC_i}}}$	$C_{n_{\delta_{SC_i}}}$
δ_{CS_i}	$C_{D_{\delta_{CS_i}}}$ $C_{T_{\delta_{CS_i}}}$	$C_{Y_{\delta_{CS_i}}}$	$C_{L_{\delta_{CS_i}}}$	$C_{l_{\delta_{CS_i}}}$	$C_{m_{\delta_{CS_i}}}$	$C_{n_{\delta_{CS_i}}}$

appropriate for conceptual design application. Clearly, none of the codes evaluated predicts all important parameters. More sophisticated methods do not always give better results. Overall, it is essential to strive for rapid turn-around analyses with the aim '*keep it simple and fast!*'

Vortex lattice methods, like panel methods, are based on solutions to the Laplace (Prandtl-Glauert) equation, thus are subject to the same basic restrictions. Commonalties between both method types are, that the singularities are placed on the surface, the non-penetration condition is satisfied at a number of control points, thus a system of linear algebraic equations needs to be solved to determine the singularity strengths. The VLM is different from the PM, in that it is oriented towards lifting effects and therefore ignores thickness effects in the classical formulation of the VLM. The boundary conditions (non-penetration conditions) are

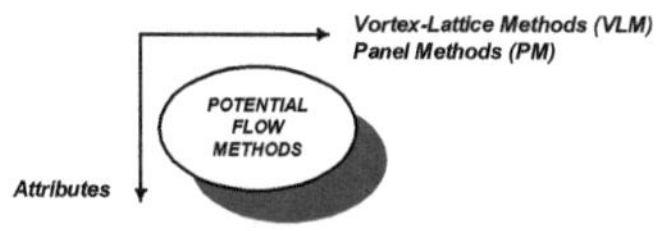

VLM Subsonic: AZTEC [100,2000], VLAERO [101,1996], VLM [102,1976], VLMD [103,1997], VLM4997 [104,1982], NLVLM [105,1990], VORLAT [106,1989], LINAIR PRO [107,1996], SPARROW [108,1997], MASTER [79,1995], TWINS [109,1992], PANSAIL [110,1995], SUB3D [111,1996], HASC95 [112,1996]

VLM Subsonic/Supersonic: VORLAX [113,1977], VORSTAB [114,1993]

PM Subsonic: HESS [115,1985], BOEING-TEA230 [116,1968], MBB-UFE [117,1979], HESS 1 [118,1991], ROBERTS [119,1972], HUNT-SEMPLE [120,1973], HESS 2 [117,1991], MCAERO (MCAIR) [121,1990], SOUSSA [122,1982], VSAERO [123,1997], USAERO [123,1997], LEV [121,1990], NLR [124,1991], SPARV [125,1979], QUADPAN 1 [118,1991], SAAB [126,1984], ECOPAN [127,1993], S-SUB2 [128,1986], PMARC [129,1997], NEWPAN [110,1995], AEROMASTER [130,1997]

PM Subsonic/Supersonic: WOODWARD 1 (WC1) [131,1996], USSAERO (WC2) [132,1997], W12SC3 [133,1983], NASA AMES WING-BODY (AERA RULE) [134,1997], NLRAERO [135,1980], UDP [136,1993], PAN AIR [137,1984], HISSS [127,1993], QUADPAN 2 [115,1985], ADAPT [138,1996]

Note: Each computer-based system quoted is referenced with only one representative source.

CODE NAME: Acronym, Full Name

HISTORY: Approximate Date Introduced, Generation, Responsibility for Implementation, Software Rights

AVAILABILITY: Source Code, Executable Program, Documentation

MATHEMATICAL MODEL: Governing Equation, Evaluation Method, Order, Singularity Type (source-doublet-surface velocity vector), Boundary Condition, Linearity, Steadiness, Viscosity, Heat Transfer, Rotation, Compressibility

PHYSICAL MODEL: Configuration, Surface Model, Panel Geometry, Thickness, Camber, Twist, Forces and Moments, Control Effectors, Aerodynamic Derivatives, Aeroelasticity

NON-LINEARITY: Vacuum Effects/Buffet Onset/Separation, Wake Modelling, Vortex Lift, Unsteady Aerodynamics, Transonic Effects

VALIDATION EFFORT:

COMPUTING REQUIREMENTS: Hardware, Turn-Around Time

REFERENCES:

RECOMMENDED IMPROVEMENTS:

REMARKS:

Fig. 4.22 Survey of potential flow computer-based aerodynamic prediction methods

applied on mean- and not actual surfaces, themselves resembling combinations of thin (planar) lifting surfaces. At last, a specific advantage of the VLM over the advanced PM is that the solution inherently contains the leading edge suction force. The computation of the configuration induced drag is then conducted without resorting to the Trefftz-plane theorem. Overall, vortex lattice methods are "... *very easy to use and are capable of providing remarkable insight into wing aerodynamics and component interaction* ..." [81]. The remarkable accuracy experienced with vortex lattice type methods, its overall simplicity with respect to model set-up (ease of problem description), and its rapid turn-around, qualifies this method superior compared to panel codes for conceptual design application.

The final evaluation sequence and a detailed comparison of three pre-selected vortex lattice methods (LINAIR PRO, HASC95, and VORSTAB) is given in Appendix A.5. All three codes are considered suitable for conceptual design application. The code considered most suitable and finally selected to contribute to the generic stability and control methodology *AeroMech* is VORSTAB, developed by Prof. Lan from Kansas University.

4.3.3.2 Non-linear Vortex Lattice Method VORSTAB

Vortex lattice methods were first formulated in the '30s. Although being simplistic in concept, the purely numerical approach of the method had to await practical application until the early '60s, when adequate computer resources were available.

From then onwards, many different vortex lattice schemes evolved. The following describes the 'standard' implementation of the vortex lattice method (VLM).

Standard Vortex Lattice Aerodynamic Model

The VLM as a potential flow method is based on the idea of the solution of the Prandtl-Glauert equation (Laplace equation in the limit as the freestream Mach number goes to zero),

$$(1 - M_{\infty}^2)\frac{\partial^2 \Phi}{\partial x^2} + \frac{\partial^2 \Phi}{\partial y^2} + \frac{\partial^2 \Phi}{\partial z^2} = 0 \tag{4.4}$$

where Φ is the perturbation velocity potential. The Prandtl-Glauert equation is a linearised form of the full potential equation (assumption of small perturbations), formulated as a linear partial differential equation describing inviscid, irrotational subsonic flow, taking compressibility effects into account.

The logic of the VLM is implemented as follows. The continuous distribution of bound vorticity over the wing surface is modelled by a finite number of discrete horseshoe vortices. The VLM idealises a wing by locating the vortex lattice panels (grid of horseshoe vortices) on the mean camber surface of the wing. A horseshoe vortex is placed in such a way as to position the bound vortex on the c/4 element line of each panel. The lateral panel centroid location is selected for the 3c/4 point of each panel. Lan describes in [139] the underlying implementation concept: "*The VLM is based on the 'three-quarter-chord theorem' which was originally derived by Pistolesi in 1933* [140] *for a flat airfoil. It states that concentrating the discrete vortex at the c/4 point and satisfying the tangency condition at 3c/4 yield the correct circulation for a flat plate.*"

As a consequence, one approach to the solution of the Prandtl-Glauert equation is, to superimpose known solutions, in particular discrete line vortices, representing lift while satisfying the 'three-quarter-chord theorem'. The velocities induced by each horseshoe vortex on a specific control point are calculated by the Biot-Savart law, see Schlichting and Truckenbrodt [141]. A flat wake (linearised theory) is assumed in the 'standard' VLM. A summation over all control points leads to a set of linear algebraic equations that satisfy the boundary condition of no flow through the lifting surface. The vortex strengths Γ_i of each panel are determined by computing the solution of this system of linear equations. Clearly, the vortex strengths are related to the circulation and the pressure differential between lower and upper side of the lifting surface. The pressure differentials may be integrated to yield total forces and moments.

Figure 4.23 illustrates, how vortex filaments are distributed in a horseshoe shape to model a segment of the lifting surface.

The mathematical formulation of the 'standard' VLM is most conveniently described in [81, 139, 142, 143, 144] and will not be reproduced here.

A Revision of the Standard VLM—*VORSTAB*

The primary development aim for VORSTAB has been, to remedy the inadequacy of existing standard-VLMs. This applies in particular to the area of non-linear

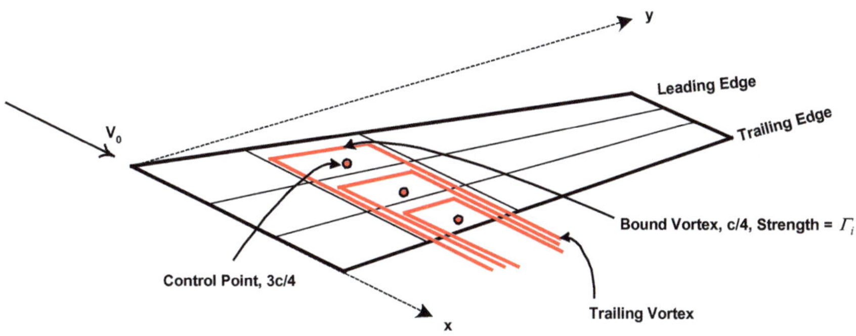

Fig. 4.23 Horseshoe vortex filament implementation of the standard vortex lattice method (VLM)

lateral-directional aerodynamics for vortex dominated configurations at higher angles-of-attack. Mason recalls in [81]: "*Perhaps the most important revision of the vortex lattice method was proposed by Lan, and called the 'quasi vortex lattice method'. In this method Lan uses mathematical methods, rather than the more heuristic arguments … .*"

Clearly, the quasi-VLM VORSTAB, developed by Lan [145], improves the standard-VLM through theoretical considerations[14] and still retains the simplicity of the standard-VLM. In summary, in this highly non-linear VLM are the lifting surfaces modelled with horseshoe vortices, bodies are modelled with vortex multiplets, wing vortex flow is simulated using discrete free vortex filaments emanating from the edges, vortex breakdown is modelled with empirical formulas, and boundary-layer separation effects are taken iteratively into account using sectional non linear data. For more theory detail see Appendix 5 and Refs. [145–148].

The following summarises the computer-based generic aerodynamic prediction capability of VORSTAB:

– Total aerodynamic forces and moments:

$$C_L, C_D, C_M$$

– Subsonic and supersonic longitudinal stability derivative coefficients:

$$C_{L_\alpha}, C_{D_\alpha}, C_{m_\alpha}, C_{L_q}, C_{m_q}$$

– Subsonic and supersonic lateral-directional stability derivative coefficients:

$$C_{Y_\beta}, C_{l_\beta}, C_{n_\beta}, C_{Y_p}, C_{l_p}, C_{n_p}, C_{Y_r}, C_{l_r}, C_{n_r}$$

[14]The following theoretical modifications have been introduced: the wing edge square-root singularity, the logarithmic singularity in the case of flap deflection of the vortex distribution, and the Cauchy singularity in the downwash integral.

- In- and out-ground effects;
- Control effector (CE) deflection;
- Hinge moments;
- Pressure distribution;
- Torsional and bending moment distribution;
- Estimation of aerodynamic coefficients at pre- and post-stall (high angle-of-attack) conditions;
- Asymmetric forebody vortex separation;
- Vortex breakdown effects.

For detail concerning estimation accuracy and validation effort see Chap. 6, Appendix 5, and Refs. [145–148].

4.4 Stability and Control Project Characterisation

Section 3.5 has outlined the significance and shortcomings of design-oriented stability and control analysis during the conceptual design stage. The following is concerned with identifying the parametrics, key to the design of primary control effectors (CE).

4.4.1 Stability and Control Work During Vehicle Synthesis

4.4.1.1 Control Effector Design Rationale

The available CEs, as installed in any aircraft type, have to secure the aircraft's intended flight path by providing adequate stability, control, and trim. Those tasks can only be realised when the complex balance between control power and the inherent airframe stabilities, static-, dynamic-, and manoeuvre stability, has been secured during the conceptual design phase.

Obert reviews in [149] the problem of CE design. " ... *In preliminary design studies they* [the CEs] *often receive no more attention than what is required to estimate horizontal and vertical tailplane volume coefficients. Performance and weight analysts consider tail surfaces as nuisances and pay hardly more attention to them than ascribing a wetted area and a given weight per unit of reference area to them. One would therefore expect that with the advance of technology tail surfaces would diminish in size. The opposite is true however. ... This has been caused by a number of developments which have taken place over the years:*

- *The maximum lift coefficient in take-off and landing has increased.*
- *Available engine power or thrust at low speeds has increased.*

- *The required centre-of-gravity range has increased for many types of aircraft.*
- *Design safety requirements have become much more servere.*
- *Speed, altitude and angle-of-attack limits have greatly expanded.*"

Clearly, the design of CEs results in increased drag and weight for the overall aircraft. The objective to minimise these adverse design implications requires, at first, a review of practiced CE analysis- and design approaches.

4.4.1.2 Design-Oriented Stability and Control Analysis Techniques Reviewed

Table 4.8 reviews standalone and integrated processes for sizing CEs from 1935 until today. The contributions selected belong into *Class II* to *Class IV* aerospace vehicle design synthesis approaches, and each of them discusses methods used for analysing and/or designing CEs. Information provided by typical stability and control textbooks has not been included in the list, since those usually elaborate on the basic ingredients relevant for detailed analysis of aircraft stability and control, but without offering a practical design approach for conceptual design.

One observation evident from the survey in Table 4.8 is, that design-oriented stability and control methods have stagnated in development, although the flying

Table 4.8 Design-Oriented approaches to stability and control analysis

Implementation	Reference, year	Comments
Stand-alone methods		
Root	[150, 1935]	Design contribution to LoCE, based on Gates's dynamic longitudinal stability theory by the use of stability diagrams. Consideration of TAC only.
Silverstein	[151, 1939]	Design of LoCE with emphasis on the evaluation of those design variables, that affect the performance (aerodynamic efficiency) of the CEs. The discussion applies, in part, to the design of DiCE. Consideration of TAC only.
Root	[152, 1939]	Design contribution to LoCE and DiCE based on aerodynamic efficiencies of the CEs supported by analysis of empirical data. Consideration of TAC only.
Morgan et al.	[153, 1945]	Design of LoCE, DiCE, and LaCE based primarily on the analysis of empirical data compared to the theoretical approach. Consideration of TAC only.
Wimpenny	[154, 1954]	Design of LoCE, DiCE, and LaCE based on volume coefficients, empirical data, and stability and control requirements. Consideration of TAC and FWC.

(continued)

Table 4.8 (continued)

Implementation	Reference, year	Comments
Lee	[155, 1961]	Design of LoCE, DiCE, and LaCE with modern non-aerodynamic stability methods by fully integrating power control systems, auto stabilisation, and fly-by-wire as contrasted by the classical (aerodynamic) means of airframe design.
Wood	[156, 1963]	Design of LoCE, DiCE, and LaCE based on empirical data and analysis of stability and controllability. Discussion of design parameters. Consideration of TAC only.
Burns	[157, 1972]	Design of LoCE, DiCE, and LaCE based on static and dynamic stability and control design criteria for satisfactory handling characteristics. Detailed discussion of design parameters. Aircraft configuration independent discussion.
Torenbeek	[158, 1990]	Design of LoCE, DiCE, and LaCE based on static and dynamic stability and control design criteria, empirical data, volume coefficients, and design-critical flight conditions. Detailed discussion of design parameters. Consideration of TAC only.
Nicolai	[159, 1984]	Design of LoCE, DiCE, and LaCE based on static only stability and control design criteria, empirical data, volume coefficients, and design-critical flight conditions. Detailed discussion of design parameters. Consideration of TAC, TFC, and FWC.
Hünecke	[160, 1987]	Design of LoCE, DiCE, and LaCE based on static and dynamic stability and control design criteria, empirical data, volume coefficients, and design-critical flight conditions. Detailed discussion of design parameters. Consideration of TAC only.
Whitford	[161, 1987]	Design of LoCE, DiCE, and LaCE based on static and dynamic stability and control design criteria, empirical data, volume coefficients, and design-critical flight conditions. Detailed discussion of design parameters. Consideration of TAC, TFC, and FWC.

(continued)

Table 4.8 (continued)

Implementation	Reference, year	Comments
Stinton	[162, 1991]	Design of LoCE, DiCE, and LaCE based on static and dynamic stability and control design criteria, empirical data, volume coefficients, and design-critical flight conditions. Detailed discussion of design parameters. Consideration of TAC and FWC.
Raymer	[163, 1992]	Design of LoCE, DiCE, and LaCE based on static and dynamic stability and control design criteria, empirical data, volume coefficients, and design-critical flight conditions. Detailed discussion of design parameters. Consideration of TAC, TFC, and FWC.
Heinemann	[164, 1997]	Design of LoCE, DiCE, and LaCE based on static and dynamic stability and control design criteria and design-critical flight conditions. Detailed discussion of design parameters. Consideration of TAC only.
Hünecke	[165, 1998]	Design of LoCE, DiCE, and LaCE based on static and dynamic stability and control design criteria, empirical data, volume coefficients, and design-critical flight conditions. Detailed discussion of design parameters. Consideration of TAC only.
Stinton	[166, 1998]	Design of LoCE, DiCE, and LaCE based on static and dynamic stability and control design criteria, empirical data, volume coefficients, and design-critical flight conditions. Detailed discussion of design parameters. Consideration of TAC and FWC.
Anderson	[167, 1999]	Design of LoCE, DiCE, and LaCE based on empirical data and volume coefficients. Consideration of TAC only. Descriptive character.
Jenkinson et al.	[168, 1999]	Design of LoCE, DiCE, and LaCE based on empirical data and volume coefficients. Consideration of TAC only. Descriptive character.
Scholz	[169, 1999]	Design of LoCE, DiCE, LaCE based on the volume coefficient, static and dynamic stability and control design criteria, empirical data, volume coefficients, and design-critical flight conditions. Discussion of design parameters. Consideration of TAC only.

(continued)

Table 4.8 (continued)

Implementation	Reference, year	Comments
Howe	[170, 2000]	Design of LoCE, DiCE, LaCE based on volume coefficient, static and dynamic stability, control design criteria, empirical data, volume coefficients, and design-critical flight conditions. Discussion of design parameters. Consideration of TAC, TFC, and FWC.
Integrated into synthesis environment		
Oman	[171, 1977]	Design of LoCE and DiCE with tail volume coefficients and computed tail arms. Consideration of TAC only.
Thorbeck	[172, 1984]	Design of LoCE via evaluation of controllability and stability criteria. Statistical data for the design of DiCE and LaCE. Consideration of TAC only.
Alsina	[173, 1987]	Design of LoCE and DiCE for design-critical flight conditions and with the use of statistical data.
Bil	[174, 1988]	Design of LoCE, DiCE and LaCE with statistical data and volume coefficients. A follow-on design sequence designs the LoCE and the DiCE via evaluation of controllability and stability criteria. Consideration of TAC only.
Kay	[175, 1993]	Design of LoCE, DiCE, and LaCE based on static and dynamic stability and control design criteria and design-critical flight conditions. Detailed discussion of design parameters. Consideration of TAC, TFC, and FWC.
Heinze	[176, 1994]	Design of LoCE, DiCE, and LaCE with tail volume coefficients and design-critical flight conditions. Consideration of TAC and FWC.
Roskam	[4, 1995]	Design of LoCE, DiCE, and LaCE based on static and dynamic stability and control design criteria, empirical data, volume coefficients, and design-critical flight conditions. Detailed discussion of design parameters. Consideration of TAC only.
Nunes	[177, 1995]	Design of LoCE for design-critical flight conditions. Consideration of TAC, TFC, and TSC.

(continued)

Table 4.8 (continued)

Implementation	Reference, year	Comments
MacMillin	[178, 1996]	Design of LoCE, DiCE, and LaCE based on static and dynamic stability and control design criteria and design-critical flight conditions. Detailed discussion of design parameters. Consideration of TAC only.
Pohl	[179, 1997]	Design of LoCE, DiCE, and LaCE based on static only stability and control design criteria, empirical data, volume coefficients, and design-critical flight conditions. Detailed discussion of design parameters. Consideration of TAC only.
Lee et al.	[9, 1998]	Design of LoCE, DiCE, and LaCE based on static and dynamic stability and control design criteria taking a FCS and design-critical flight conditions into account. Detailed discussion of design parameters. Consideration of FWC (X-33) only.
Nicolai	[180, 1999]	Design of LoCE, DiCE, and LaCE with a generic VLM (VORLAX) and a 3d PM (QUADPAN). Consideration of a range of conventional and unconventional aircraft configurations.

vehicle continues to evolve. The classical CE design approach centres on using statistical data to determine volume coefficients for the LoCE, DiCE, and LaCE. If practiced at all at conceptual design, static stability, dynamic stability, and selected quasi-static design-critical flight conditions are evaluated in follow-on steps, to cross-check the initial tail volume assumptions.

Overall, the 'classical' CE sizing procedures have evolved for the conventional TAC only, thus are based on well-known design principles. The resulting CE hardware design proposals are intended to be conservative, primarily aiming to cover all design-critical operational eventualities, subsequently leading to over-engineered individual CEs. However, it regularly happens that CEs are under-dimensioned, due to having overlooked a particular design-critical flight condition with unfortunate design-parameter couplings (see Concorde accident). Clearly, the 'classical' CE sizing procedure is the currently available and therefore accepted approach to design CEs of TAC only. The approach is generally limited, in that it is not capable of providing alternative tail design solutions even for the TAC, as required in the case of relaxed static stability aircraft.

The following summarises and discusses four noteworthy approaches to the design of controls.

CE Sizing Procedure by Nicolai [159]

The CE sizing procedure by Nicolai represents the 'classical' approach. At first, the size of the DiCE, LoCE, and LaCE are determined with historical trends of the *volume coefficient*. The second step calculates the *static stability* criterion $C_{m\alpha}$ via differentiation of the longitudinal one-degree-of-freedom equation of motion (EOM) with assumptions valid for the TAC only. The component build-up approach is used to estimate the lateral-directional stability derivative coefficients $C_{n\beta}$ and $C_{l\beta}$. The investigation of dynamic stability[15] is thought to be not feasible due to lack of aerodynamic and inertial data. The third step performes *trim* calculations to determine the longitudinal trim drag by estimating the location of the neutral point (n.p.) and subsequently specifying the static margin (SM). The deflection angle of the LoCE is selected to be a direct measure for the trim drag. The LoCE deflection required to trim the aircraft in $n = 1$ flight is calculated with the longitudinal EOM, showing the dependence of trim drag on the centre of gravity (c.g.) location. The CE deflections required to *manouver* in a pull-up and in a level turn are estimated. The follow-on fourth step analyses and designs the CEs for the three axes by using a more complete set of sizing criteria while considering civil and/or military certification requirements. The LoCE is sized for stability by defining an appropriate $C_{m\alpha}$, while it is sized for control by investigating the Design-Constraining Flight Conditions (DCFCs): (a) high g manoeuverability, (b) take-off rotation, (c) high angle-of-attack low-speed flight and associated trim drag. The largest of the CE areas required for stability or control is finally selected. The DiCE is sized for stability by selecting a CE area to satisfy the desired $C_{n\beta}$. Spin resistance is checked with the dynamic directional stability parameter $C_{n\beta dyn}$ and with the so-called lateral control spin parameter, LCSP. The DiCE is sized for control by investigating the DCFCs: (a) crosswind landing, (b) asymmetric power, and (c) adverse yaw. The largest of the CE areas is finally selected. The LaCE is primarily sized for adequate roll power but no procedure is suggested.

Nicolai's approach is the typical manual-style (stand-alone) design/analysis sequence of CEs. Although he attempts to discuss the TFC and the FWC, the approach proposed can not be considered to be of generic character. Although relaxed stability is mentioned, this important design variable has not been integrated into the CE sizing process.

CE Sizing Procedure by Pohl [179]

The work by Pohl intends, to advance the 'classical' sizing procedure for the LoCE only.[16] In particular, the method aims to be flexible to changes of aircraft geometry, aerodynamics, and test conditions. Furthermore, it intends to establish 'unconventional' CE sizing criteria and solutions, it should be able to take into account a more refined aerodynamic data set, should identify relevant airworthiness requirements, and be able to calculate the longitudinal trim control surface deflection. The single-degree of freedom longitudinal EOM is used to determine trim, control, and

[15]The dynamic analysis is expected to fine-tune the configuration.

[16]The method is primarily developed for Airbus Industrie type transonic transport aircraft.

stability. The c.g. range is determined in accordance with pertinent certification requirements and design guidelines. The input data required are primarily geometry and aerodynamics. The aerodynamic data set is generated using the multi-method approach, consisting of a panel method (PM), ESDU data sheets, engineering methods, and existing data. The trim control surface deflection for a given configuration and c.g. position is estimated. The analysis sequence estimates the velocity, CE deflection, CE stall margin, angle-of-attack, and the stability for key DCFCs. The DCFCs to *trim* are (a) low speed performance with forward c.g., high-lift devices deflected, under-wing engines idle, (b) $0g$ manoeuvre with the aircraft in trim. The DCFCs to evaluate *control* are (a) trimmed wing stall at $1.1\ V_{s1g}$, (b) the $\pm 0.4g$ pull-out/push-over at all velocities down to $1.1\ V_{s1g}$, (c) the take-off rotation manoeuvre, (d) hold the nose gear on the ground at aft c.g. before the rotation velocity, and finally (e) speed recovery. The requirement for evaluating *stability* is to provide 5% manoeuvre margin, a typical design guideline for transonic transport aircraft. Either trim, control, or the stability requirement determines the available c.g. range.

Pohl's approach is operational as a stand-alone sequence. The assumptions implicit in the single-degree-of-freedom longitudinal EOM naturally limit the potential of the method to the determination of LoCEs of TAC-types only. No provision is made to implement a logic which deals with relaxed stability vehicles. The non-generic multi-method aerodynamic data estimation task risks data-inconsistencies while representing the main effort in executing the CE sizing procedure.

CE Sizing Procedure by Lee et al. [9]

Lee et al. employ a consistent approach to the design of the LoCE, DiCE, and the LaCE.[17] Key design/analysis parameters and -requirements are stability, control coupling, departure resistance, control power, and manoeuverability. CE types considered are aerodynamic surfaces, engine thrust vector control (TVC), and reaction control system (RCS) thrusters. The *static stability* characteristics are assessed using the stability derivative coefficients: C_m, C_n, C_l, and C_{ndyn}. *Dynamic stability* is assessed along the modal characteristics of the short-period mode and the Dutch-roll mode. Static and dynamic stability are explained in terms of the undamped natural frequencies. *Roll control departure susceptibility* is assessed via the parameter known as LCDP (Lateral Control Departure Parameter) and C_{ndyn}. LaCE-to-DiCE *interconnect ratios*, needed to achieve decoupled roll and yaw response, are estimated. *Control power* requirements for static trim and stabilisation are estimated at pre-selected DCFC throughout the flight envelope. Control power required for stability augmentation in the longitudinal and lateral-directional axis is estimated using closed-loop FCS gains. *Manoeuverability* is assessed based on the vehicle rotational rate and rotational acceleration capability.

Lee's approach is highly flexible and it can be assumed that it is integrated into a multidisciplinary synthesis environment, see Nicolai [180]. The high degree of

[17]The approach is used at Lockheed Martin Skunk Works and it is only an assumption that it is integrated into a multidisciplinary synthesis environment, see Nikolai [180].

versatility is, however, limited to a specific class of symmetric flying vehicles only, since a range of assumptions is embedded into the formulation of the analytical CE sizing requirements. As indicated by Nicolai, the aerodynamic data set is estimated using a non-linear VLM. Static and dynamic stability are evaluated, a logic is integrated to assess the effects of relaxed static stability. Overall, the conception of Lee's approach is a step towards a generic stability and control method.

CE Sizing Procedure by Kay et al. [175]

The CE analysis/design approach by Kay et al. is set up to operate in the stand-alone mode.[18] To apply the method, the following *information* regarding the candidate concept is needed: (a) vehicle layout of major components and CEs, (b) mass properties, (c) extreme performance objectives (quantified design and certification requirements). In a second step, *critical combinations of the flight condition variables* are identified for each DCFC. In the third step, a linear VLM is employed to determine the *aerodynamic data set*. The fourth step determines the *control power* required to satisfy the range of DCFCs. The DCFC describing the equilibrium/performance cases are: (a) classical 1 g trim during normal trimmed flight and control allocation, (b) longitudinal manoeuvring flight, (c) steady sideslip, and (d) engine-out trim. Dynamic DCFCs are (a) take-off and landing rotation, (b) time to bank, (c) inertia coupling (pitch due to roll and yaw due to loaded roll), (d) coordinated velocity axis roll, (e) short period and CAP (Control Anticipation Parameter) requirements, and (f) high angle-of-attack departure. The measure for control power available/required is the deflection of the CEs for the individual DCFCs. A logic is implemented to optimally allocate individual LoCEs in the case of an over-estimated configuration like the TSC for minimum trim drag.

Kay's approach has not been integrated into a multidisciplinary synthesis environment. As a consequence, any design feedback (iteration) has to be performed manually. Although the aerodynamic tool employed is of generic character, the calculation routines used to estimate the CE deflection angles to trim, to stabilise, and to control, are clearly of non-generic character. The method is setup to function predominantly for fighter aircraft of TAC layout. The control allocation logic schedules longitudinal controls only. Although longitudinal dynamic stability is estimated, the lateral-directional Dutch roll mode is neglected. Only aerodynamic CEs are considered at low angle-of-attack subsonic flight conditions.

The approaches by Kay and Lee are considered the state-of-the-art in designing vehicle controls at conceptual design level. Clearly, both methods have been the starting point for developing the generic stability and control method subject of the present research investigation. It is becoming obvious, that the modern aerospace vehicle designer is very ready to employ non-aerodynamic design solutions to stability and control problems, if operational advantages can be achieved. The recent paper by Sauvinet [181] confirms such thinking and points the direction.

[18]The approach is primarily developed for fighter type configurations and the High-Speed Commercial Transport (HSCT).

The method developed in the present context is defined, from the outset, to operate in a multidisciplinary synthesis environment. Applicable design guidelines and certification requirements have to be taken into account. The aerodynamics tool employed must be of generic and non-linear character, since CEs are usually sized in the non-linear corners of the flight envelope. A common set of DCFCs, valid for the range of aircraft configurations and concepts, has to be identified to permit evaluation of only design critical areas in the flight envelope. The DCFCs have to represent the link between conceptual design and flight test. The analytical expressions used to analyse trim, static and dynamic stability, and control have to be of generic character. A logic has to be integrated to take the design variables longitudinal and lateral-directional relaxed static stability into account. The primary parameter of concern in the context of analysing/designing CEs is 'control power'. The control power available/required for the range of aerodynamic CEs has to be quantified for each DCFC, by specifying the volume coefficient (geometry), stability derivative coefficients (aerodynamics), and the CE deflection angle (operation) required. In modern aerospace vehicle design, adequate stability-, control-, and trim characteristics are ensured with provision of sufficient control power.

4.4.2 *Concepts and Technologies*

4.4.2.1 Concept 'Stability Derivative Coefficients'

The estimation of a flight vehicle's stability and control characteristics is commonly performed via two different sets of parameters. The Equations of Motion (EOM) represent aerodynamic forces and moments by means of aerodynamic stability derivatives. The derivative approach (Bryan's hypothesis) provides an approximation only to the aerodynamic forces and moments. In contrast, the dynamic parameter approach as proposed by Etkin [182], with parameters like damping ratio, natural frequency and time constant, represents the aerodynamic forces and moments with suitably defined aerodynamic transfer functions. The dynamic parameter approach has to be regarded as a generalisation of the derivative approach,[19] able to overcome the shortcomings of the derivative approach. However, it is most convenient to use the derivative approach at conceptual design, whereas the dynamic parameter approach is used during detailed stability and control analysis. The primary benefit of the derivative approach for conceptual design is the physical transparency it offers to the engineer, see Thomas [183].

[19]The dynamic parameter approach is based on eigenvalues of the aircraft equations of motion, linearised about a specific trimmed or steady-state flight condition.

4.4.2.2 Concept 'Volume Coefficient'

The volume coefficient allows statistical comparisons of designs, to guide the designer with the first approximations of workable stabiliser and control areas. The aerodynamic CE stabiliser- and control volume coefficient is classically defined as

$$V_{CE} = \frac{l_{CE}}{\bar{c}} \frac{S_{CE}}{S} \tag{4.5}$$

representing " ... *the ratio of two volumes characteristic of the airplane's geometry* ..." [184], where l_{CE} is the moment arm measured from the CE aerodynamic centre (a.c.) to the aircraft c.g. Clearly, the volume coefficient is a geometric ratio representing the effectiveness of aerodynamic CEs relative to the wing. Since the c.g. position is not necessarily fixed, the volume coefficient V_{CE} is not a constant. It is therefore more convenient to choose a fixed reference point like the wing-body mean aerodynamic centre instead of the c.g., leading to a constant volume coefficient.

Overall, the volume coefficient is a relative measure of the ability of the CE, to provide both stability and control. The question arises, if it is useful to distinguish between a volume coefficient for control (e.g., the rudder) and one for stability (e.g., vertical fin). It must be remembered, that the control-volume contributes to stability, whereas the stability-volume is an integral ingredient required for the generation of aerodynamic control forces in the case of hinged control surfaces.

With reference to the above statements, it is necessary to discuss the coupling between stability and control. It is a valid statement that stability and control are inseparable when assessed against a handling quality scale. Clearly, the pilot is only concerned to experience an aircraft with well-balanced stability and control characteristics, measured against a well defined handling quality rating scale. However, what can be influenced by the designer are the means to deliver this balance between control power and inherent stabilities. Modern technology enables the designer to detach the stability issue to a certain degree from the actual design. The stability characteristics can be tailored, to a large degree, hardware-independent by a FCS, resulting into an additional design degree-of-freedom for the airframe. Having substituted, in part, the classical aerodynamic stability constraints against artificial stability constraints given by a certifiable FCS, the control power issue takes the precedence in airframe design. However, physical control forces can not be substituted by artificial means, thus modern relaxed static stability (RSS) vehicles are constrained by control power and associated FCS issues (actuator loads, rate, bandwidth, etc.).[20]

[20]To recall, in classical aircraft design, design for stability and control follows a predefined schedule. First, a c.g. range is pre-defined as an operational requirement. Consequently, the range of stability is given and must be provided by hardware design decisions. With this pre-defined stability scenario, control is evaluated in a second step throughout the flight envelope. In the case of deficient control authority, modifications are unavoidable which in turn effect stability as well.

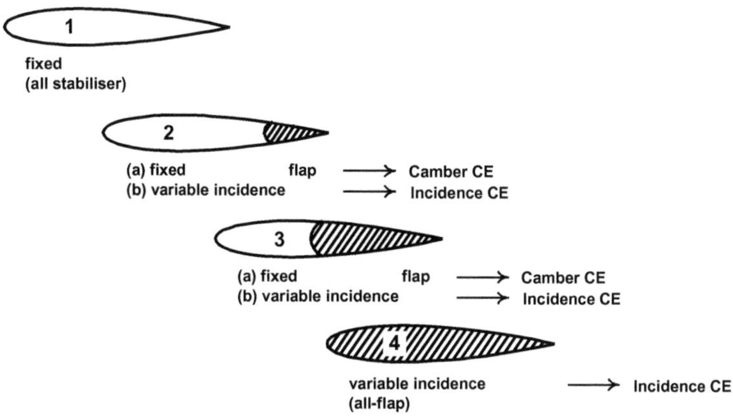

Fig. 4.24 Aerodynamic control effector (CE) family

Figure 4.24 illustrates the difficulty in separating the total volume coefficient into a stability-volume coefficient and a control-volume coefficient.

The fixed CE (case 1) may be idealised with a stability-volume coefficient only and no control-volume coefficient contribution. The other engineering extreme is the variable incidence (all-flap) CE (case 4). Here, the entire CE may be seen, at first, to contribute to the control-volume coefficient only. However, any aerodynamic lifting surface has a definite effect on stability. Case 4 illustrates the problem, in that it is difficult to separate, thus distinguish between the stability-volume coefficient and the control-volume coefficient. Longitudinally, the activity of deflecting a lifting surface does not modify the n.p.-position and consequently aircraft stability.[21] For the sake of simplicity and with respect to generality it has been decided, to avoid transforming the total volume coefficient V_{CE} into the stability-volume coefficient and control-volume coefficient, since no new information is gained by doing so.

Table 4.8 illustrates, that the volume coefficient is a standard indicator for control power available during conceptual design. However, the difference between the individual members of the CE-family lies in their aerodynamic efficiency. The total efficiency of an aerodynamic CE is dependent on its composition as camber control and/or incidence control, see Fig. 4.24. It is a well-known fact that the incidence CE (case 4) has the highest aerodynamic efficiency. Thus, the CE-efficiencies of cases 2 to 4 are significantly different, a detail not implied in the classical formulation of the volume coefficient. Clearly, the information contained

This hardware design coupling of stability and control can not be avoided for unaugmented aircraft types and poses design trade-off constraints on behalf of the designer.

[21]The concept of the free-floating canard must be regarded as an exceptional case. The n.p.-position of the aircraft is, however, not influenced by permanent deflection of the canard surface. For more detail see Middel [185].

in the volume coefficient alone is not sufficient to determine control power. It has been decided that the combined information given with (a) the control volume describing geometry-relations, (b) the stability derivative coefficients describing the aerodynamic control function, and (c) the CE deflection angle describing its operational status, are together sufficient to quantify control power for stability and control evaluations. As a consequence, the parameters V_{CE}, C_{xyz}, and δ_{CE} are interrelated for any aerodynamic CE. As an example, if the calculated CE deflection (δ_{CE}) for a specific DCFC is too large, then modifying the geometry description (V_{CE}) and/or the aerodynamic design (C_{xyz}) of the CE may fix the control power deficiency.

4.4.2.3 Concept 'Tailplane Sizing Diagram'

The classical CE sizing diagram (called tailplane sizing diagram or scissors sizing plot) is a visualisation means to relate the LoCE size (tail volume coefficient or S_{LoCE}) to the forward and aft position of the c.g. The optimal c.g. range is obtained when the c.g. range is simultaneously in contact with the forward and aft c.g. limits. The sizing diagram is utilised as a convenient medium, to physically visualise and harmonise a variety of design contraints. The forward (control) and aft (stability and control) c.g. design criteria are determined by constraints like stability margin, manoeuvre margin, loading range, nose up and down control (take-off rotation, mistrim, go-around, V_{min} recovery, nose gear tip up, stall recovery, etc.), $C_{L\ min/max\ CE}$, and the stability augmentation system (SAS) saturation limit. A representative schematic for a CE sizing diagram is reproduced from Hofmann and Clement [189] in Fig. 4.25 without any further comments. Illustrative case-study examples for LoCE sizing plots are given for the F-4E (CCV) in [187], the B-52 in [188, 189], and the A3XX in [39].

It has been initially intended to construct LoCE-, DiCE-, and LaCE sizing diagrams for the range of conventional and unconventional aircraft configurations and concepts (TAC, TFC, TSC, FWC, OWC, OFWC, etc.). The following indicates the constraints and implications faced.

Some comments are apt to distinguish the design of the LoCE from the design of the DiCE and the LaCE for the symmetric class of aircraft. Overall, the LoCE is more demanding in the flight-physical context compared to the DiCE and LaCE types, due to the non-symmetric flight condition and inertial force-effects involved. However, the LoCE may be easier to size. Conversely, the lateral-directional CEs are less demanding in the flight-physical context (reduced complexity due to symmetric cruising flight conditions and no involvement of inertial forces), but they are more difficult to size due to control coupling effects. Clearly, the designer's task to aim for a satisfactory level of lateral-directional stability and control is a more difficult task compared to the longitudinal case, because there is no single measurable end product equivalent to the c.g. position, against which to evaluate the lateral-directional characteristics.

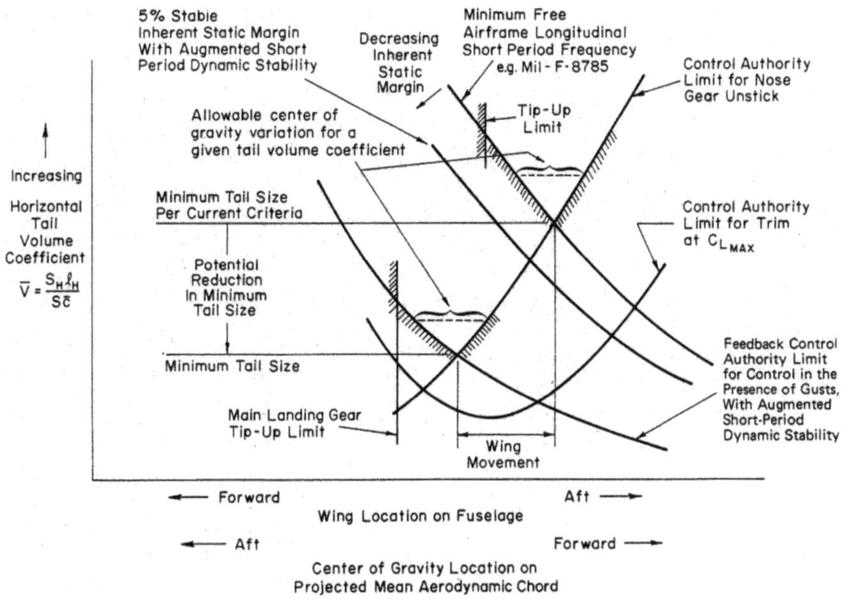

Fig. 4.25 Classical LoCE sizing diagram with design criteria for the TAC-type aircraft configuration [186]

Mueller proposes in [190], to visualise design constraints for the DiCE in a diagram spanning the weathercock stability, $C_{n\beta}$, on the abscissae, and the vertical tail area S_{DiCE} on the ordinate. In contrast, Hofmann and Clement [186] present DiCE sizing criteria in a diagram, showing the c.g. location and moment arm on the abscissae, and the area ratio S_{DiCE}/S on the ordinate. Overall, the c.g. position has not an analogous meaning for the lateral-directional case compared to the longitudinal situation, since no lateral-directional c.g./n.p.-concept exists. In addition, fore- and aft wing location on the fuselage in case of the TAC, has not the aerodynamic influence on DiCE sizing that it has on LoCE sizing, because unswept and even swept wings generate little asymmetric side force and yawing moment in a steady sideslip, when compared with the DiCE. The c.g. position does, however, affect the moment arm and the yawing radius of gyration. As a consequence, it is not meaningful nor practical to transform the LoCE sizing diagram into a DiCE sizing diagram by having the same abscissae. No sizing diagram is known to be published for the LaCE.

Apart from the physical visualisation aspect, the CE sizing diagram has no real design relevance. Its existance is clearly not required to advance the design, since it is based on information which is available even before its construction. In the context of generic method development, the CE sizing diagram is not a convenient means to present design constraints in a consistent style, since the physical

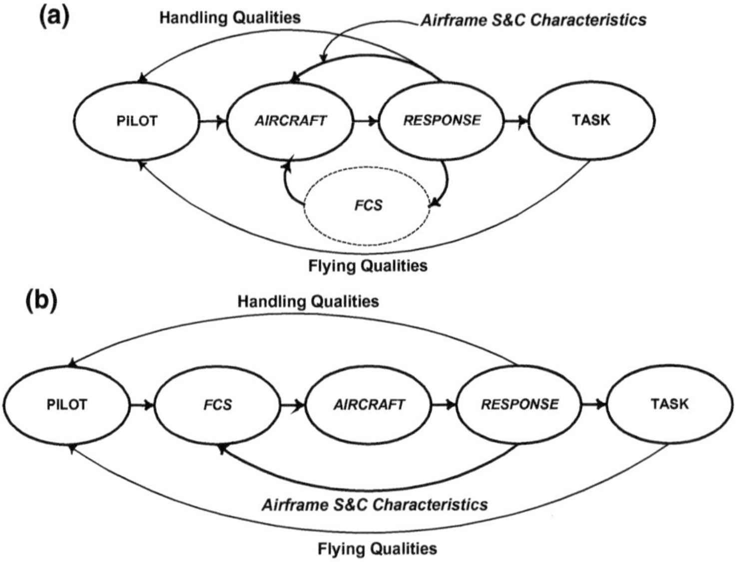

Fig. 4.26 Flying qualities, handling qualities, and airframe stability and control characteristics of: **a** the conventional aircraft, and **b** the FBW aircraft. Data adapted, in part, from Cook [191]

character implicit with the range of symmetrical and asymmetrical aircraft configurations and CE-families is too different.[22] Thus, it has been decided that the information provided by the diagram does not justify the effort to construct it. The design constraints encountered during a CE optimisation cycle are most conveniently presented in a format offered by the modern optimiser environment.

4.4.2.4 Flying Qualities, Handling Qualities, and Airframe S&C Characteristics

It needs to be distinguished between aircraft *flying qualities*, *handling qualities*, and *airframe stability and control characteristics* as shown in Fig. 4.26.

Cook defines in [191]: "*The pilot's perception of flying qualities is considered to comprise a qualitative description of how well the aeroplane carries out the commanded task. On the other hand, the pilot's perception of handling qualities is considered a qualitative description of the adequacy of the short term dynamic response to controls in the execution of the flight task.*" The present research undertaking is primarily concerned with airframe stability and control

[22]The construction of the sizing diagram for the CEs of a OFWC represents a real challenge. An elevon functions as a LoCE, LaCE and eventually as well as a DiCE. Design aspects like control allocation schemes need to be considered, leading to sizing diagrams which consequently will have lost physical transparency and simplicity.

characteristics, excluding the overall mission task (e.g., passenger comfort, re-entry cross-range, gun pointing task) and the pilot in the loop.

Gibson [192] replies to the question whether or not to consider handling qualities during the conceptual design phase: "... *Handling is, to a large extent, dependent on FBW* [Fly-By Wire]. *The configuration itself has little effect on it so long as the controls have enough power to do what they have to do in extreme conditions to* α_{max}, *pitch down after stall, etc. These subjects are not a function of a control system but control power. But the handling within these limits, the handling will always lie in those limits, is very much a function of the control system and control laws. In that sense, the handling qualities have less information of the configuration and this will be the case for a long time, even if you do not rely completely on the control system for normal handling qualities, even if the control system fails completely. What you really need to get the aircraft down is a minimum stability* ..." For more detail see Gibson [193, 194], Hodgkinson [195], and McRuer [196].

In conclusion, the current research undertaking concentrates on sizing the flight vehicle's control effectors for sufficient control power. The *flying quality* task clearly belongs to mission simulation, and is not subject of the current research undertaking. The *handling quality* task has to be seen, at first, as a configuration fine-tuning activity after the airframe controls have been designed for adequate control power. However, the strength of the affiliation between handling qualities and trim drag is dependent on the choice of aircraft configuration. Therefore, there is a clear research and development incentive for integrating control system design and handling qualities into conceptual design. Most noteworthy developments are by Morris [83] and Mavris et al. [197].

4.4.2.5 Control Configured Vehicle (CCV)—Stability Augmentation

As indicated earlier in this chapter, the CCV design approach capitalises on the potential of considering advanced flight control concepts during the conceptual design phase. The report by Holloway [33] outlines the scope of analytical studies performed from the mid 1970s onwards, indicating most significant performance improvements achievable from six control functions: (1) augmented stability, (2) gust load alleviation, (3) manoeuvre load control, (4) fatigue reduction, (5) ride control, and (6) flutter mode control. The current research investigation considers only the CCV-function of relaxing stability. Overall, the relaxed static stability (RSS) function has, compared to the other CCV-functions, the most significant design impact on the flight vehicle.[23]

It has to be recalled that birds fly statically unstable in all three axes, having a magnificent flight control system on board! Man-made technology is slowly following this role-model towards RSS-vehicles, one obstacle being the ability to certify the FCS. It is a fact that the currently operational FCSs will gain maturity in

[23]The term RSS implies relaxed stable and indifferent, but as well unstable airframes.

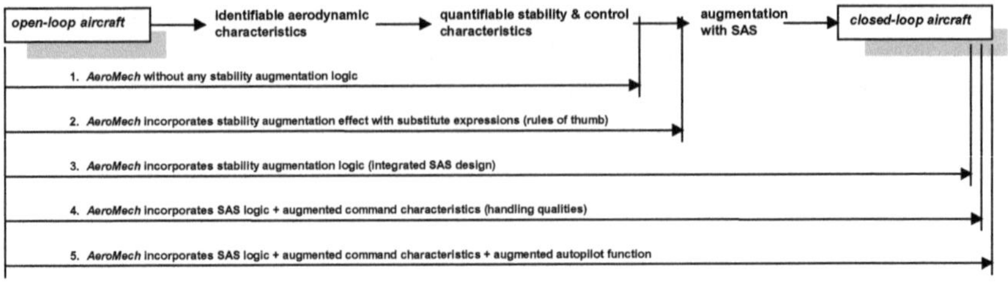

Fig. 4.27 *AeroMech* FCS options shown qualitatively along the open-loop and closed-loop aircraft chain

the future, leading to routinely certifiable aircraft with performance-optimal relaxed stability characteristics. However, the impact of RSS-enabling technologies is rarely taken into account in today's aircraft synthesis procedures. To be precise, it is current practice to relax longitudinal stability as an aftermath activity for a given design, using design rules to reduce trim drag,[24] see for example Graeber [198]. Anderson and Mason [199] comment: "*... there appears to be no current systematic method through which the configuration can be optimized within the constraints of control system structure and control power.*"

Clearly, the design parameters describing relaxed stability need to be integrated as design variables into the aircraft synthesis process. Such logic not only reduces trim drag throughout the flight envelope, but it enables the reduction of the gross-vehicle dimensions, thereby significantly affecting vehicle weight, drag, etc. The assignments by Ashkenas and Klyde [34], Morris [83], Anderson and Mason [199], and Beaufrere [200] deliver leading-edge thinking towards CCV-vehicles, in particular RSS-vehicles. A historical review of the development of stability augmentation systems (SAS) is provided by Abzug and Larrabee in [12].

Figure 4.27 summarises the FCS options considered during the definition phase of the generic method *AeroMech*.

The following remarks relate to each option shown:

1. Abandoned—neglects effects of the FCS on aircraft sizing.
2. Abandoned—enables only incomplete (non-generic) assessment of the effects of a FCS on aircraft sizing.
3. **Selected**—enables complete assessment of the effects of a SAS on aircraft sizing.
4. Abandoned—no immediate relevance for aircraft sizing.
5. Abandoned—no immediate relevance for aircraft sizing.

It must be noted that option 2 has been considered as an alternative approach for application in the *AeroMech* methodology. However, various complications

[24]Relaxing stability has been traditionally an add-on performance improvement measure for commercial transport aircraft, with little but usually no effect on the overall aircraft layout; for fighters, the implementation of RSS has been dictated by manoeuverability demands.

invalidate the simplicity gains offered by this solution. The main problem areas are: (a) how to assess an inherently unstable airframe without SAS; (b) how to assess the design-dependency on a fictitious FCS; and (c) rules of thumb have a very narrow range of approximate validity, thus are not compatible with a generic analysis/design environment. Consequently, the integration of a simplistic SAS (option 3) has been selected, since it is required to map the interdependency between control power and inherent and/or artificial stability.

AeroMech has to be capable of handling the following three design philosophies. **A**: Open-loop aircraft with sufficient inherent stability characteristics. Design for (i) control power and (ii) inherent stability characteristics. **B**: Closed-loop aircraft with rest stability. Design for (i) control power and (ii) rest stability (inherent airframe stability). This is the current design philosophy for today's commercial transonic transport aircraft. The SAS has only a modest influence on aircraft sizing, thus on flight performance. **C**: Closed-loop aircraft without safety-relevant rest stability. Design for (i) control power and (ii) artificial stability to enforce a performance-optimal balance. This is the design philosophy for most of the advanced aircraft configurations, in particular the modern FWC, OFWC, etc. A fully integrated SAS is obligatory for such aircraft types. It is expected, that design option (C) will be applied to commercial transport type aircraft in the near future.

Emulation of the FCS

In general, adequate stability is required for all flight phases. To achieve the desired static and dynamic stability levels with high performance aircraft, control augmentation is required to minimise overall configuration changes with adverse performance repercussions on weight and drag. Usually, classical control theory covers the analysis and synthesis of sophisticated augmentations systems. However, a different (simplistic) approach is required for conceptual design. The present research investigation realises the stability augmentation function with the concept of Equivalent Stability Derivatives (ESD).

The idea of ESD has been first published at NACA by Imlay [201]. The approach is used in the present context to avoid servomechanism analysis. For a modern treatment of the ESD subject, see Etkin [202] and Roskam [4]. A fundamental assumption to the ESD-approach is, that control surface actuators, feedback control computers, and the required sensors operate infinitely fast (no-lag assumption). ESDs are the normal controls-fixed values, to which have been added increments consisting of gain factors times control derivatives. Thus, the ESD is the sum of the inherent stability derivative of the aircraft and the contribution to that derivative by the idealised SAS, see Eq. 4.6.

$$\underbrace{C_{x_{y_{SAS}}}}_{\textit{Equivalent Stability Derivative}} = \underbrace{C_{x_{y_{Airframe}}}}_{\textit{Inherent Stability Derivative}} + \underbrace{C_{x_{\delta}}}_{\textit{Control Derivative}} \underbrace{K}_{\textit{Gain}} \tag{4.6}$$

The SAS and the airframe together comprise the augmented aircraft. The SAS operates through the feedback path in a closed-loop system. The FCS is emulated with a simple negative feedback, using a single variable to augment selectively the stability characteristics of the aircraft. With the selection of a suitable value of the *feedback gain*, the stability may be augmented. In general, the two main tasks for the design of the SAS are: *First* it is necessary to survey the sensitivities of all individual feedback options for all aircraft configurations of interest. The *second* task is the estimation of appropriate feedback gains. In general, the greater the required change in the stability characteristics, the greater the feedback gains needed to effect the change. Having included the gain as a design variable into the CE sizing logic, the designer is able to retrieve information related to required actuator performance (bandwidth, actuation power, etc.), see Roskam [203] and Morris [84].

The choice of feedback variables is important in determining the nature of the change of damping and stiffness, since each variable results in a unique combination of changes. It must be remembered, that however complex the functional structure of the FCS in the application, the basic augmentation-effect of each feedback variable does not change. Therefore, the FCS emulation with the ESD-approach simulates the augmentation-effect of the most complicated FCS by virtue of considering the primary augmentation drivers.[25] The present context is primarily concerned with the integration of a simplistic stability augmentation system (SAS), to restore pitch- and yaw stiffness and damping characteristics. The SAS has to function for the stable vehicle (p, q, r feedback for damping restoration), and the indifferent to unstable vehicle (α, β feedback for stiffness restoration, and p, q, r feedback for damping restoration). For more analysis detail see Chap. 5. Obviously, the CE-sizing rules are building up in complexity, since the SAS is offering additional degrees of design-freedom. The emulation of the FCS with the ESD-approach represents, however, the minimum-complexity solution.[26]

4.4.2.6 Control Allocation

In classical aerospace vehicle design, the CEs are single moment-generating controllers for each rotational degree of freedom: LoCE for pitch, DiCE for yaw, and LaCE for roll. The performance of CEs may be constrained, in general, by design-specific and operational limitations. As a consequence, modern high performance aircraft have multiple moment- and force generating controllers for each axis. The allocation, or blending, of a sub-set of CEs to achieve specific objectives (sufficient

---

[25]The emulation of a FCS using the simplified control law (ESD-approach) has certain limitations. It is impossible to emulate a FCS representation valid for generic conceptual design. The typical pre-selected feedback variables for conventional aircraft might be misleading for novel aircraft applications. Although the ESD-approach has an overall generic character, the selection of the feedback variables might be case-specific. Follow-on studies have to determine the most suitable choice of feedback variables for the range of aircraft configurations and concepts. The following assumes the classical feedback variables.

[26]Although the design of a simplified SAS appears not too difficult, the main challenge, however, arises in off-design conditions.

control power throughout the flight envelope, minimisation of trim drag in cruise, etc.) is the control allocation problem. Examples are the TSC, which is statically indeterminate in longitudinal trim, and the modern BWB FWC, which is usually statically indeterminate in longitudinal and lateral trim.

There is an unlimited number of load distributions, each capable to satisfy the trim constraints of aircraft with redundant CEs while representing an additional degree of freedom on which the designer can capitalise. The control allocation problem is a classical mathematical optimisation task, usually handled with servomechanism analysis, as illustrated by Durham [204], Bordignon. and Durham [205], Durham and Bordignon [206], Buffington [207], Cameron and Princen [23], Page and Steinberg [208], and Ikeda and Hood [209]. However, a strategy has to be devised to sidestep servomechanism analysis, thus to integrate a simplistic control allocation logic into *AeroMech*. Two control allocation concepts have been identified suitable for integration.

The ad hoc approach pre-defines the control momentum (e.g., deflection angles) for all redundant (e.g., aerodynamic) CEs of a flight vehicle, thereby reducing the statically indeterminate system to a statically determined system. Either a pre-defined set of CE allocation concepts is sequentially evaluated, or a single schedule is manually selected.[27] Sensible information on CE operating schedules (e.g., setting angles) of redundant (e.g., aerodynamic) controls can be retrieved from engineering experience (statistics). The KBS, see Sect. 2.5, describes CE deflection schedules for a range of statically indeterminate aircraft case studies. Case study examples are given in Appendix A.6. Overall, the ad hoc control allocation approach must be regarded as a generic but manual only 'optimisation' process for getting started.

An optimisation approach to the longitudinal control allocation problem of symmetric aircraft has been developed by Goodrich et al. [210]. "*LOTS* [linear optimum trim solution] *enables the rapid calculation of the longitudinal load distribution resulting in the minimum trim drag in level, steady state flight for airplanes with a mixture of three or more aerodynamic surfaces and propulsive control effectors.*" The LOTS algorithm is of closed form (analytical solution to the trim problem), is computationally efficient, thus suitable for conceptual design application. The method considers any number of aerodynamic CEs and a single jet-exhaust nozzle. The trim equations are linearised and a Lagrange formulation is used to minimise the drag function while satisfying trim constraints. Overall, significant trim drag reductions have been demonstrated with LOTS, see Goodrich et al. [210], Kay et al. [175], and Rakowitz [144]. The feasibility of developing an analogous approach for laterally staggered CEs has not been evaluated.

Clearly, the manual ad hoc approach and the non-generic optimum trim solution (LOTS) are methods to determine a first allocation schedule for redundant CEs. It has been considered not sensible to advance LOTS towards handling both, the

[27]An ad hoc distribution frequently advocated for the TSC is to carry no load on the tail. An alternative is to carry equal but opposite loads on the canard and tail.

Table 4.9 Non-generic reduced-order longitudinal and lateral-directional dynamic mode approximations

Implementation	References	Mode	Degrees of freedom	Calculation parameter	Comments
Kolk	[211]	SP, P	SP (Z, M), P (X, Z)	SP ($\omega_{n\ SP}$, ζ_{SP}, $T_{1/2}$), P ($\omega_{n\ P}$, ζ_P, T_P)	SP anomaly at aft cg; P: assumption of constant α
		DR, S	DR (Y, L, N), S (L, N)	(−)	DR, S: detailed discussion of design variables and parameters
Babister	[212]	SP, P	SP (Z, M), P (X, M)	SP ($T_{1/2}$, T), P (ζ_P, T_P)	SP, P: discussion of TAC and FWC design effects at subsonic and supersonic speeds
		$DR_{1,2,3}$, S, R	DR_1 (N), DR_2 (Y, N), DR_3 (Y, L), S (L, N), R (L)	$DR_{1,2}$ ($T_{1/2}$, T), DR_3 ($\omega_{n\ DR}$, ζ_{DR}), S ($T_{1/2}$, T_2), R ($T_{1/2}$)	Y analogous to SP; L_2 is damped oscillation; discussion of design variables and parameters
Woodcock and Drake	[213]	SP, P	SP (Z, M), P (X, Z)	SP ($\omega_{n\ SP}$, ζ_{SP})	SP: inclusion of manoeuvre point; P approximation inadequate at supersonic and hypersonic speeds
		DR, S, R	DR (Y, L, N), S (L, N), R (L, N)	DR ($\omega_{n\ DR}$, ζ_{DR}), S ($1/T_S$), R ($1/T_R$)	Extensive discussion of the validity of constant-coefficient assumptions
McRuer, Ashkenas, and Graham	[8]	SP, P	SP (Z, M), P_1 (Z, M), P_2 (X, Z, M)	SP ($\omega_{n\ SP}$, ζ_{SP}), P_2 ($\omega_{n\ P}$, ζ_P)	SP: high accuracy compared with complete EOM; discussion of the 2DOF and 3DOF P motion
		DR, S, R	DR_1 (Y, N), DR_2 (Y, L, N), S (L, N), R (L)	$DR_{1,2}$ ($\omega_{n\ DR}$, ζ_{DR})	DR: comparison of 2DOF and 3DOF approximations; discuss coupling effects
Hofmann and Clement	[186]	SP	SP (Z, M, δ)	SP ($\omega_{n\ SP}$, ζ_{SP})	SP formulation with a **feedback logic**; relation of augmented SP to design parameters
		DR	(−)	DR ($\omega_{n\ DR}$, ζ_{DR})	Discussion of DR formulation with a **feedback logic**
McLean	[214]	SP, P	SP (Z, M), P_1 (Z, M), P_2 (X, Z, M)	SP ($\omega_{n\ SP}$, ζ_{SP}, T_2), $P_{1,2}$ ($\omega_{n\ P}$, ζ_P)	P: $M_u = 0$ (Lanchester); discussion of divergent tuck mode (both assumptions not valid for DC-8)
		DR, S, R	(−)	(−)	Discussion of EOM and transfer functions; DR-approximation rarely used

(continued)

Table 4.9 (continued)

Implementation	References	Mode	Degrees of freedom	Calculation parameter	Comments
Blakelock	[98,]	SP, P	SP (Z, M), P (X, Z)	SP ($\omega_{n\ SP}$, ζ_{SP}), P ($\omega_{n\ P}$, $\zeta_{n\ P}$)	SP: dependency on density; effects of stability derivatives; P: comparison of approx./exact results
		DR, S, R	DR (Y, N), S (L, N), R (L)	DR ($\omega_{n\ DR}$, ζ_{DR}), S (τ_S), P (τ_R)	DR: discussion of 1DOF assumption; S/R: discuss design effects
Stevens and Lewis	[215]	SP, P	SP (X, Z, M), P (X, Z, M)	SP ($\omega_{n\ SP}$, ζ_{SP}), P ($\omega_{n\ P}$, ζ_P)	SP expressions more complicated than P expressions; valid for TAC only
		DR, S, R	DR (Y, L, N), S/R (L, N)	DR ($\omega_{n\ DR}$, ζ_{DR}), S (τ_S), R (1/τ_R)	DR: ω- more accurate than ζ-expression; S: τ accurately predicted when mode instable
Brockhaus	[216]	SP, P	SP (Z, M), P (X, Z)	SP ($\omega_{n\ SP}$, ζ_{SP}), P ($\omega_{n\ P}$, ζ_P)	SP, P: simplest expressions; accuracy demands required for control law design
		S, R	S (L, N), R (L)	S (1/T_S), R (1/T_R)	S/R: discuss design effects
Roskam	[4]	SP, P	SP (Z, M), P (X, Z)	SP ($\omega_{n\ SP}$, ζ_{SP}), P ($\omega_{n\ P}$, ζ_P)	SP, P: approximations valid for inherently stable aircraft
		DR, S, R	DR (Y, N), S (L, N), R (L)	DR ($\omega_{n\ DR}$, ζ_{DR}), S (τ_S), R (1/τ_R)	DR accuracy: frequency compares well, damping is poor
Hancock	[217]	SP, P	SP (X, Z, M), P (X, Z)	SP ($\omega_{n\ SP}$, ζ_{SP}), P ($\omega_{n\ P}$, ζ_P)	SP, P: design effects, subsonic and supersonic speed
		DR, S, R	DR (W, L, N), S(L, N), R (L)	DR ($\omega_{n\ DR}$, ζ_{DR}, $T_{1/2\ DR}$), S ($T_{1/2\ R}$), R ($T_{1/2\ R}$)	DR: inclusion of drag forces; description of design effects
Russell	[218]	SP, P	SP (Z, M)	SP (T_{SP})	P: three derivations
		DR, S, R	DR_1 (N), DR_2 (N, L), DR_3 (Y, N, L), DR_4 (L), S (Y, R)	DR_2 ($T_{1/2}$, T, $\omega_{n\ DR}$), DR_4 ($T_{1/2}$, T), S (derivative ratio), R (L_p)	DR: various types of extreme DR mode; DR_4 valid for inertially slender aircraft; changing DR character

(continued)

Table 4.9 (continued)

Implementation	References	Mode	Degrees of freedom	Calculation parameter	Comments
Etkin and Reid	[184]	SP, P	SP (Z, M), P_1 (X, Z), P_2 (X, Z, M)	SP (characteristic equation), P_1 (T), P_2 ($\omega_{n\ P}$, ζ_P)	SP: good for wide range of vehicles and flight conditions; P: Lanchester model
		DR, S, R	DR (Y, N), S/R (Y, L, N)	DR (transfer function), S/R (transfer function)	DR: use with caution; high accuracy for S and R approximations
Cook	[191]	SP, P	SP (Z, M), P_1 (X, Z), P_2 (X, Z, M)	SP ($\omega_{n\ SP}$, ζ_{SP}), P_1 ($\omega_{n\ P}$, ζ_P=0), P_2 ($\omega_{n\ P}$, $\zeta_{n\ P}$)	SP: comparison to mass-spring-damper system (defined for TAC); P_1 (Lanchester)
		DR, S, R	DR (Y, N), S (L, N), R (L)	DR ($\omega_{n\ DR}$, ζ_{DR}), S (T_S), R (T_R)	DR: rather poor approximation but gaining of physical insight; S/R: derivative ratio
Nelson	[94]	SP, P	SP (Z, M), P (X, Z)	SP ($\omega_{n\ SP}$, ζ_{SP}), P ($\omega_{n\ P}$, ζ_P=0)	SP approximation in general more accurate than the P approximation; no compressibility effects
		DR, S, R	DR (Y, N), S (L, N), R (L)	DR ($\omega_{n\ DR}$, ζ_{DR}), S (characteristic root), R (T_R)	The DR is truly a 3DOF motion with strong coupling between the equations; derivative ratio
Schmidt	[219]	SP, P	SP_1 (M), SP_2 (Z, M), P (X, Z)	SP_2 ($\omega_{n\ SP}$, ζ_{SP}), P ($\omega_{n\ P}$, ζ_P, T_P)	SP: drag force has no strong effect; SP, P: evaluation of accuracy
		DR, R	DR_1 (N), DR2 (Y, N), R (L)	$DR_{1,2}$ ($\omega_{n\ DR}$, ζ_{DR}), S (characteristic root), R (T_R)	DR: results are airframe specific
Hodgkinson	[195]	SP, P	SP (M), P (X, Z)	SP ($\omega_{n\ SP}$, ζ_{SP}), P ($\omega_{n\ P}$, ζ_P)	P: Lanchester's classical expression for the damping (energy derivation)
		DR, R	DR (Y, N), R (L)	DR (sideslip response to rudder), R (roll rate response to aileron)	DR: design contributions and certification requirements; coupled roll-spiral
Phillips	[220]	P	P (Z, M)	P ($\omega_{n\ P}$, ζ_P)	P: detailed discussion of design effects
Phillips	[221]	DR	DR (Y, N, L)	DR ($\omega_{n\ DR}$, ζ_{DR})	DR: detailed discussion of design effects

longitudinal and lateral control allocation problem. Both approaches are intended to provide a statically determined starting configuration only. However, integration of *AeroMech* into a multidisciplinary synthesis environment enables fine-tuning the pre-selected control allocation scheme for the range of existing CEs (de-coupled and coupled), while satisfying a nominated objective function.

4.4.2.7 Reduced Order Models—Physical Visibility

Classical stability and control analysis is concerned with the exact description of aircraft stability and response characteristics by solving the small perturbation equations of motion (EOM). Cook remarks in [191], that although this is usually the object of a flight dynamics investigation, "... *it is difficult, if not impossible, to establish the relationships between the stability characteristics and their aerodynamic drivers.*" He continues that "... *these disadvantages can be avoided by seeking approximate solutions, which can also provide considerable insight into the physical phenomena governing the dynamic behaviour of the aircraft.*"

The inherent problem with the dynamic EOM is the lack of physical transparency they offer. Minimisation of this deficiency to an acceptable level is vital for the practitioner working in the conceptual design area. It must be remembered that the experience-database for advanced aircraft configurations is limited, thus the designer is 'low on the learning curve' when discussing advanced sizing ideas. As a consequence, it is of benefit to calculate the dynamic response of an aircraft with reduced order models (approximate solutions focusing on highest-of-importance), to qualitatively visualise the mode's dependence on stability derivatives and flight conditions, while solving in parallel the dynamic EOM to obtain high-quality quantitative results. Such undertaking assists building understanding for the range of aircraft configurations and concepts, since some physical transparency is gained. Clearly, calculated results from the reduced order models have no direct design-impact with the availability of the better quality results by the EOM.

The following investigates the suitability of rigid body algebraic expressions (reduced order models) for use in the generic stability and control design environment. Table 4.9 summarises reduced order dynamic mode analysis approaches and their application to aerospace vehicle design. Abbreviations used in the column describing the *Mode* are: SP (Short-Period mode), P (Phugoid mode), DR (Dutch Roll mode), S (Spiral divergence), R (Roll subsidence); in the column describing *Degrees of Freedom*: X, Y, Z (x-, y-, and z-force equation), L, M, N (rolling-, pitching-, and yawing moment equation), δ (control deflection).

The magnitudes of undamped natural frequency, damping ratio, and time constant are intimately tied to acceptable or unacceptable stability and control behavior of the aircraft. For that reason, it is important for the aerospace vehicle designer to understand which design factors are the 'design drivers' to affect the dynamic stability parameters. Understanding the modes and their dependence on certain design drivers requires analytical solutions, which are, however, not available for the full system of equations (EOM). When longitudinal-lateral decoupling occurs, as in the case for symmetric aircraft configurations, it becomes feasible to manipulate the aircraft transformed state equations algebraically. Unlike the longitudinal EOM, it is more difficult to solve the lateral-directional EOM approximately. Because of the motion coupling present to a greater or lesser extent for the lateral-directional dynamics (DR, S, R), the modes are not so distinct and simplifying assumptions are less relevant with the consequent loss of accuracy. Overall, the longitudinal and lateral-directional resulting analytical solutions are approximate (Table 4.9), but they provide some insight into the design drivers influencing the dynamic modes for the symmetric aircraft type only.

It has to be acknowledged that the closed-form approximations discussed have been derived based on a foreknowledge of the modal characteristics of the aircraft (predominantly the TAC), to arrive at the approximate equations of lower order than the exact ones. In either case, an approximation is required to arrive at reasonably compact and usable expressions that delineate dominant, as opposed to trivial, effects. Such effects usually vary among aircraft configurations and concepts, so it is to be expected that literal approximate factors, which apply to all vehicles for all flight conditions, will be an exception rather the rule. Clearly, no simple and generic analytical approximation is available to give accurate results for the dynamics of the TAC and of the other aircraft configurations under all conditions. It is expected that the reduced-order expressions in Table 4.9 are applicable, in part, to the range of stable symmetric unconventional aircraft configurations (e.g., TFC, TSC, FWC). The longitudinally and lateral-directionally coupled asymmetric aircraft types (e.g., OWC, OFWC) have, unfortunately, to discard the methods presented in Table 4.9. Summarising, the physical transparency gained by using the classical reduced-order methods warrant their integration into the stability and control method for the symmetric aircraft types only. The stability augmentation logic proposed by Hofmann and Clement [186] enables the consideration of relaxed stable to unstable airframes. No further effort has been invested in advancing reduced-order models describing the dynamics of aircraft.

4.5 Flight Evaluation Characterisation

4.5.1 Flight Evaluation Work During Vehicle Synthesis

The stability and control characteristics of any flight vehicle have to comply with design- and certification requirements throughout the flight envelope. In that sense, the CEs are responsible to ensure a flyable and safe vehicle, despite their adverse effect on flight performance. The obstacle at the heart of this undesirable situation is the significant information requirement (in both quantity and quality) of design-related stability and control analysis, since the early conceptual design has to guarantee satisfactory flight characteristics which will be confirmed or censured during the flight test phase at the end of the design chain. Clearly, there is a strong incentive to establish a competent link between the first (conceptual design) and the final (flight test) vehicle development stage.

The connection between conceptual design and flight test is established, when evaluating the vehicle's stability and control characteristics using flight simulation techniques. Techniques proposed feasible during the conceptual design stage are, first to identify and then to evaluate design-constraining flight conditions (DCFC) modelled with the coupled static Six-Degree-Of-Freedom Equations Of Motion (static 6-DOF EOM),[28] as a second step to solve the coupled dynamic 6-DOF EOM, and finally to utilise a dedicated conceptual design engineering simulator. Clearly, the static 6-DOF EOM are only concerned with individual flight conditions (DCFCs), whereby the dynamic 6-DOF EOM and the conceptual design engineering simulator are both able to simulate the entire flight envelope, thereby demanding an increased amount of information.

The present research undertaking concentrates solely on the identification and formulation of a generic set of DCFCs including the analytical framework (static 6-DOF EOM), and the formulation of the dynamic 6-DOF EOM, both used for the initial design of controls. The engineering simulator developed by Burdun [222] is, however, considered an ideal follow-on step to systematically interconnect conceptual design with flight test. The capability of the system was put to test when reconstructing the Air France Concorde accident in July 2000 by using minimum input information available within a short time frame, typical for a vehicle simulation scenario at the conceptual design stage, see Burdun [223] and Chudoba and Burdun [224]. Summarising, such approach to flight simulation enables the conceptual design engineer, to understand operational limits of new and existing vehicle designs along FAR Part 25 based test- and certification scenarios during the conceptual design phase.

4.5.2 *Design-Constraining Flight Conditions (DCFCs)*

The simplicity requirement central to initial conceptual design work[29] can be fulfilled, when considering only those flight case scenarios, which have a primary and interdisciplinary effect on CE hardware sizing.[30] A minimum set of so-called *Design-Constraining Flight Conditions* (DCFC) defines a CE-feasibility space for the aircraft configuration and concept under investigation, while taking reference to quantified design guidelines and certification requirements.

[28]The coupled static 6-DOF EOM are called trim EOM.

[29]Simplicity has highest priority during conceptual design evaluations due to permanent design data shortage and computing time limitations.

[30]In case of no interdisciplinary coupling effect of a design parameter at conceptual design level, its investigation can be done with more freedom and accuracy at a more detailed analysis level.

DEFINITION: *Design-Constraining Flight Conditions (DCFC) are flight conditions with an overall governing effect on aircraft hardware sizing.*

Consequently, only DCFCs are relevant for the initial sizing of CEs, while taking flight test expertise into account embedded in the set-up of the DCFCs themselves and in quantified design- and certification requirements (see Chaps. 5 and 6). Clearly, the quality of the process, to identify and derive a generic set of DCFCs, is considered central to the success of the generic stability and control methodology. This process requires an understanding (a) of the design-evolution of conventional and unconventional aircraft configurations, (b) of design- and certification requirements, and (c) of the flight test processes involved. As a result, the success or failure of an individual aircraft configuration and concept selected during a design cycle, and its overall engineering and management persuasive power, both depend on the engineering capability available to significantly reduce design-risks ahead of detail design and flight test.

A measure of control power is given with (i) the volume coefficient V_i (geometry), (ii) stability derivatives C_{xyz} (aerodynamics), and (iii) the individual CE deflection angles δ_i (operation). Those parameters are dependent on the choice of aircraft configuration and concept, and in particular on the range of DCFCs, consisting of control-power demanding combinations of flight condition parameters:

$$Control\,Power\,(V_i, C_{xyz}, \delta_i) = f\left\{\begin{array}{l} Aircraft\,Configuration\,\&\,Concept; \\ Design - Constraining\,Flight\,Conditions\,(DCFC): \\ (i)\,Configuration\,Setting\,(CS), \\ (ii)\,Flight\,Condition\,Variable\,(FCV); \\ (iii)\,Failure\,Condition\,(FC) \end{array}\right\}$$

Note, the control power available for an individual DCFC (e.g. take-off rotation) depends largely on the individual conditioning (CS, FCV, FC) of the aircraft. There may be more than one particular combination of CS, FCV, and FC belonging to a single DCFC, relevant to size an individual CE. Therefore, it is the objective of *AeroMech* to be able to identify the sensitivities (weighting and ranking) of individual DCFCs with their design-critical permutations of CS, FCV, and FC throughout the flight envelope (see Fig. 4.28).

4.5.2.1 Selection Process of DCFCs

Considerable effort has been invested in identifying and selecting stability and control DCFCs. The problem at the heart of this activity is the difficulty to conclusively identify and automate the selection process of relevant DCFCs. Reasons for this undesirable situation are: (a) the flight conditions relevant for aircraft certification are usually based on certification requirements, which themselves are of rather non-quantitative character, and (b) the flying characteristics and design sensitivities of unconventional aircraft layouts are not known in advance throughout

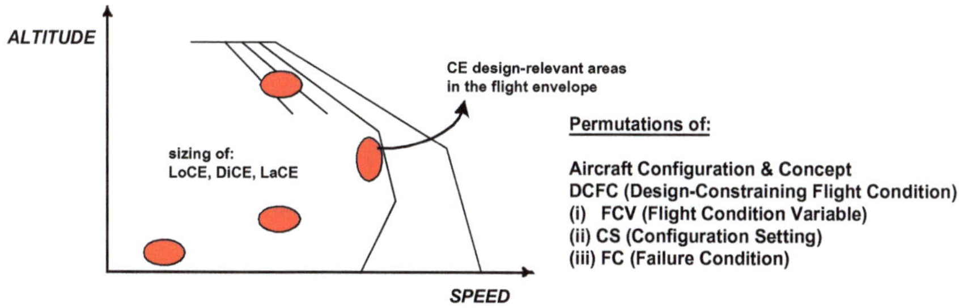

Fig. 4.28 Control effector design regions qualitatively in the flight envelope

the mission profile during the conceptual design stage. As a consequence, the following question requires answering: To what extent does the set of DCFCs belonging to a TAC differ from the set belonging to a FWC, a TSC, etc.?

A review of pertinent literature has revealed a remarkable range of DCFCs thought relevant for sizing of CEs. Leyman describes in [225] DCFCs relevant to size the CEs of Concorde. Nicholls compiles in [226] a matrix of critical flight cases for the design of ESCT[31] CEs. Le Tron presents in [227] relevant JAR/FAR 25 certification requirements and Airbus design rules for sizing of CEs for A3XX. Burns [157] and Kay et al. [175] are primarily concerned with CE sizing rules specific to combat aircraft. Clearly, the variance observed between the sets of DCFCs is a result of the multitude of permutations possible between CS, FCV, and FC for individual DCFCs, the broad range of relevant DCFCs to be checked, the consideration of static only and/or dynamic effects, and finally the dependence on the choice of aircraft configuration and concept. The inconsistency observed vividly confirms, that individual sets of DCFCs apply to specific aircraft configurations and concepts. Furthermore it illustrates the lack of a consistent approach to size CEs, implying the risk of missing critical flight cases as aviation history repeatedly confirms.

The foundation for the selection of the generic set of DCFCs has been experience gained with the KBS (see Sect. 2.5), the consultation of stability and control experts, e.g. [228–240], a one-day Concorde simulator session enabled by British Airways in Bristol/Filton UK, see Chudoba [241], and finally the compilation of a synoptic table called "*Stability & Control Design and Test Condition Matrix*", see Chudoba [242].

Concorde Full-Motion Simulator Observation

The two existing Concorde full-motion simulators[32] assemble the best analytical and numerical representation of Concorde available, since the underlying algorithms have been painstakingly derived through the development process

[31] ESCT stands for European Supersonic Commercial Transport.

[32] The simulator in Toulouse/France is owned by Air France, the simulator in Bristol/United Kingdom is owned by British Airways.

underlying Concorde.[33] For the present research undertaking, a flight test programme has been set up to investigate the inherent stability and control characteristics of the aircraft, using the British Airways simulator [241]. The primary objective of the undertaking has been, to evaluate DCFCs of an unconventional supersonic FWC throughout the flight envelope. Clearly, the Concorde simulator has been flown through extreme flight conditions, which have not been investigated at all before in real flight or using a simulator.

A description of the test flight schedules, their significance on design, and the post-simulation interpretation have not been included in the present report. In summary, much design-related experience has been gained with respect to slender-body type aircraft, in particular the characteristic relationship between the a. c. and the c.g. position, the stability augmentation function for each axis activated or de-activated, the operation and failure conditions of the fuel transfer system throughout the speed range, etc. The understanding attained from the Concorde full-motion simulator session has contributed to the final selection of the generic set of DCFCs. The subject of each test procedure card is summarised in Appendix A.7.

Stability and Control Matrix

The aim of the document '*Stability & Control Aerospace* Vehicle Design and Test Condition Matrix*vehicle design*' [242] has been, to accumulate, to introduce, and to provide configuration setting detail (FCV, CS, FC) of the variety of flight conditions (DCFC), which must be considered during the design, flight test, and certification process of large subsonic and supersonic commercial transport aircraft with respect to stability and control. This condensation is the basis for the final identification process of a generic set of DCFCs, thus providing the opportunity not to by-pass a design-critical flight condition.

The primary requirement for the set up of the stability and control matrix has been, that the flight cases accumulated must be relevant for the design of airframe stability and control characteristics. As has been discussed before, the correct choice of the airworthiness code influences, in part, the selection of DCFCs. Airbus aircraft are certified with reference to JAR-25 (subpart B-Flight) [243]. The specific control philosophy engaged with Airbus aircraft[34] requires so-called Special Conditions (SC), which are an annexation to the standard JAR-25 document for certification of advanced technology features. Consequently, the s&c matrix has to refer to JAR-25 including SCs, which result in additional design criteria and flight cases for modern transonic aircraft (see Sect. 3.2). The airworthiness requirements used for supersonic transport aircraft are the TSS Standards set up for Concorde [244]. These requirements supply additional design relevant flight cases for slender body type aircraft of advanced configuration layout. Further design flight cases are implied in the military certification requirements. The code selected is the MIL-F-8785C (Flying Qualities of Piloted Airplanes) [245] relevant for open-loop

---

[33]The underlying aerodynamic database includes $\alpha-$ and β-sweeps well beyond standard operational limits [233].

[34]FBW system incorporating C* law.

aircraft, in favour of the newer MIL-STD-1797 [246] relevant for military FBW aircraft.

Familiarisation studies have been conducted to relate aircraft stability and control design- and flight test routines to JAR-25, TSS-Standards, and MIL-Specs. Original flight test schedules have been incorporated into the stability and control matrix, to give realistic and detailed information about the variety of flight tests to be performed. A340 test flight schedules have been selected to represent the modern multi-engined transonic commercial transport aircraft. Concorde test flight schedules have been selected to provide the best knowledge available for a commercial supersonic transport aircraft of advanced configuration layout.

The flight conditions listed in the main body of the stability and control matrix are subdivided into stability- and control related characteristics rather than flight phases. The matrix contains in total 324 flight cases pertinent to stability and control with the following subheadings: No, Item, Test, Reference, Specific Test Condition, Test Description, Results and Conclusion. The table of contents of this document is given in Appendix A.8. The document '*Stability & Control Design and Test Condition Matrix*' is not a generic database. It is based on specific aircraft configurations, which provide, however, the best knowledge available representative for modern subsonic-, transonic-, and supersonic civil transport aircraft. Although similarities exist between sets of DCFCs for the variety of aircraft applications, the differences depend on the peculiarity of the individual application. In this context, the stability and control matrix is a fairly complete compendium of DCFCs, which supports the identification process of the generic set of DCFCs. The generic set of DCFCs significantly reduces the risk of missing a CE-relevant design-critical flight condition.

JAR/FAR 25 Stability and Control Certification Requirements

Before introducing the proposed generic set of DCFCs, it is necessary to enlist the current stability and control requirements for sizing of LoCEs, DiCEs, and LaCEs of any modern conventional or unconventional aircraft. Passenger transport aircraft have to comply with JAR/FAR 25 certification requirements and foreseeable evolutions of it. It is of special interest to note, that today's commercial transport aircraft still have to demonstrate acceptable open-loop stability and control characteristics.

Tables 4.10 and 4.11 consider civil certification requirements (JAR/FAR 25) only, which quantify the requirements to a lesser degree but are more flexible for design-interpretation compared to the military equivalent. The certification requirements selected for this overview are those applicable for Airbus widebody aircraft (A330 and A340), see [243]. Several JAR/FAR paragraphs are replaced or deleted by special conditions (SC), dependent on the technology features integrated.

No attempt has been made to discuss company design guidelines in the present context.[35] Obviously, design rules for transonic transport aircraft are tailored to

[35]Design guidelines for the variety of aircraft configurations and concepts are embedded in the KBS, see Sect. 2.5.

Table 4.10 JAR/FAR 25 certification requirements for the design of directional and lateral CEs

CE	Certification requirement	Paragraph	Title	Comments
DiCE, LaCE	Manoeuvrability	JAR 25.143	*Controllability and Manoeuvrability*	Predominantly qualitative requirement.
		JAR 25.147	*Directional and lateral control*	Directional and lateral control requirements with and without all engines operating.
	Stability	JAR 25.177	*Static directional and Lateral stability*	Static directional stability must be positive; the static lateral stability may not be negative; JAR 25.177 (C) is replaced by SC F-4 [Static Directional and Lateral Stability]; requirements for straight, steady sideslips.
		JAR 25.181 (b)	*Dynamic stability*	The Dutch roll mode must be controllable with normal use of the primary controls.
	Engine failure handling	JAR 25.161 (b) (d)	*Trim*	The aircraft must be trimmable with one and two engines failed.
		JAR 25.147 (a) (1)(b)(1) (C)(d)	*Directional and lateral control*	Requirement to hold the wings level and make sudden changes in heading; lateral control with one engine inoperative.
		JAR 25.149	*Minimum control speed*	Minimum control speeds are taken into account for performance calculations (take-off and landing); design speeds are related to fin and rudder characteristics.
	Roll capability	JAR 25.147 (C) (2)(e)	*Directional and lateral control*	Required peak roll rate with critical engine inoperative throughout the flight profile; availability of sufficient excess lateral control in sideslips, for the recovery from upsets produced by gusts and evasive manoeuvres, and to provide a peak roll rate for safety.
		JAR 25.253 (a) (3)	*High-speed characteristics*	Adequate roll capability to assure prompt recovery from lateral upset condition.
		JAR 25.149 (h) (2)	*Minimum control speed*	Requirement for two critical engines inoperative.
	Cross wind capability	JAR 25.237	*Wind velocities*	Specification of a cross component of wind velocity for take-off and landing.

Note The foundation to this table is the Airbus Industrie A330/A340 joint certification basis from 1994, see [243]

Table 4.11 JAR/FAR 25 certification requirements for the design of longitudinal CEs

CE	Certification requirement	Paragraph	Title	Comments
LoCE	Stall demonstration and recovery	JAR 25.103	Stall speed	Replaced by SC F-1 § 3 [Minimum steady flight speed and 1g stall speed]; definition of V_{s1g} at fwd cg.
		JAR 25.201	Stall demonstration	Replaced by SC F-1 § 6.1 [High incidence handling demonstrations]; defines flight demonstrations to be performed.
		JAR 25.203	Stall characteristics	Replaced by SC F-1 § 6.2 [Characteristics in high incidence manoeuvres]; IM F-1 § 3,7 [Minimum steady flight speed and 1g stall speed entry rate, position of deceleration devices during handling to high incidence]; defines conditions for stall recovery.
	Speed recovery	JAR 25.145 (a)	Longitudinal control	Pitch the nose downward so that acceleration to the selected trim speed is prompt; a special case is the 'CEV-manoeuvre', a mis-trim flight case with high deceleration rate to investigate tailplane stall characteristics.
	Push-over	JAR 25.255	Out-of-trim characteristics	Requirement of satisfactory manoeuvre stability and control in both nose-up and nose-down direction.
		JAR 25.1419	Ice protection	Investigation of flight in icing conditions.
	Trim ability	JAR 25.161 (c)	Trim	Longitudinal and lateral-directional possibility to trim while still having enough control power throughout the operational flight envelope.
	Stability	JAR 25.173	Static longitudinal stability	Deleted by SC F-3 [Static Longitudinal Stability]; the longitudinal control laws provide a neutral stability within the normal flight envelope].
		JAR 25.175	Demonstration of static longitudinal stability	Deleted by SC F-3 [Static Longitudinal Stability]; the longitudinal control laws provide a neutral stability within the normal flight envelope].
		JAR 25.181	Dynamic stability	Any short period oscillation must be heavily damped.

(continued)

Table 4.11 (continued)

CE	Certification requirement	Paragraph	Title	Comments
	Manoeuvrability	JAR 25.143	Controllability and manoeuvrability	Predominantly qualitative requirement.
	High-speed characteristics	JAR 25.253(a)	High-speed characteristics	[Speed increase and recovery characteristics]; no control reversal about any axis at any speed up to V_{DF}/M_{DF} and V_D/M_D; sufficient lateral control power for a prompt recovery from a laterally upset condition.
		JAR 25.255	Out-of-trim characteristics	With the aircraft trimmed up to V_D/M_D, a capability of a specified ±g has to be shown.
		JAR 25.335 (b) (1)	Design airspeeds	Replaced by SC A-4 [Design Dive Speed]; the high speed protection system limits nose down pilot authority at speeds above V_C/M_C.
		JAR 25.107 (e) (4)	Take-off speeds	Consideration of ground handling: normal take-off is required even with out-of-trim conditions.

Note The foundation to this table is the Airbus industrie A330/A340 joint certification basis from 1994, see [243]

meet relevant JAR/FAR requirements, as to obtain a well-behaved aircraft. The design rules evolved for the transonic-type aircraft apply predominantly to the state-of-the-art TAC only (distinct under-wing podded engines, -horizontal and vertical tailplane, -fuselage, etc.). Clearly, the success of future advanced aircraft, depends, to a large degree, on the inevitable modification of company-specific design rules and current certification requirements.

4.5.2.2 Generic Set of DCFCs

The stability and control methodology *AeroMech* evaluates DCFCs through two successive complexity levels, using the same consistent set of calculation methods for both levels.[36] The first set of generic DCFCs aims to define the CE design space, thus results in rather conservative CE sizing. The second and more refined set of generic DCFCs delivers, in contrast, a first competitive CE design proposal.

The design of CEs in two successive steps naturally creates an inconsistent physical coverage of control power and stability at the first step. Inconsistency means, that for example the dynamic modes may not be assessed in the first loop, since the modes usually have secondary effects on CE-sizing. Furthermore, the poor analytical representation of the modes with first-level conceptual data warrants their investigation at the more detailed (second) design level. Thus, the second CE-sizing loop evaluates the more complete set of DCFCs by using the same calculation routines, finally resulting in the consistent physical coverage of control power and stability.

4.6 1st-Level and 2nd-Level DCFCs

The generic sets of 1st-level and 2nd-level DCFCs, given in Tables 4.12, 4.13 and 4.14, have been identified relative to the design of LoCEs, DiCEs, and LaCEs during conceptual design. The '*Stability and Control Design and Test Condition Matrix*' [246] has been the basis for the selection activity, taking JAR/FAR 25, MIL Specs, TSS certification requirements, and flight test schedules into account. The following briefly indicates design implications of the individual flight conditions on LoCE-, DiCE-, and LaCE sizing, without providing configuration-specific detail about test procedures, about CS & FCV & FC, and about the certification requirements applicable, since this information is compiled in [242]. The numbers in brackets (see column CS & FCV & FC) refere to specific DCFCs in this reference.

---

[36]The calculation methods are discussed in Chap. 5.

Table 4.12 Generic 1st-level and 2nd-level DCFCs for the conceptual design of LoCEs

LoCE–DCFC	CS & FCV & FC	Level
Static		
1 g trim stability	Longitudinal trim (161)	1
	Trim curves (169)	1
	Aft C.G. clearance (182)	1
	Double hydraulic failure (295–315)	1
	Approach trim	1
	Minimum control speed, take-off climb (V_{MCA}) (144)	1
	Minimum speed at high incidence (V_{MIN}, $V_{\alpha\ max\ 1g}$) (149)	1
	Go-around on 4 engines without ground effect (90)	2
	Cruise with trim jam (170)	2
	Trim tank failure (168,127)	2
	C.G. shifting speed with fuel transfer system	2
	Emergency descent with partial loss of forward C.G. transfer facility (127)	2
	Emergency descent with reverse thrust operating, partial loss of C.G. transfer facility (131)	2
	Slats/flaps failures (320–324)	2
	Foreplane runaway, failing to a fixed position (control allocation)	2
Manoeuvre		
Rotation capability	Rotation on take-off/nosewheel lift-off (11)	1
	Rotation on landing with ground effect	1
Load factor capability	Dive recovery/pull up	2
	Speed recovery/push over (5)	1
	Load factor capability (2, 12, 20, 104, 114)	1
Dynamic		
Mode	Short period oscillation (203)	2
	Phugoid oscillation (203)	2
Transient response	Nose wheel load at break release	2
	Power application (8, 13, 24)	2

Longitudinal Control Effector (LoCE)—DCFCs

Longitudinal Trim: Proof that the LoCE (in particular the trimmable stabiliser surface) remains inside its defined range. Demonstration of $\delta_{min/max}$ throughout the flight envelope with forward and aft c.g. position.

Trim Curves: Estimation of the trim curves for the most demanding DCFCs.

Aft C.G. Clearance: Investigation of c.g. changes at constant speed and height.

Table 4.13 Generic 1st-level and 2nd-level DCFCs for the conceptual design of DiCEs

DiCE–DCFC	CS & FCV & FC	Level
Static		
β-Trim stability	Minimum control speed, approach and landing, 2 engines out ($V_{MCL\text{-}2}$) (148)	1
	Engine failure during take-off	1
	Trim 2 engines inoperative (164)	1
	Demonstration of max. cross wind on landing (one critical engine failed) (248)	1
	Straight sideslips (46)	1
	Adverse yaw	1
	Landing from approach slope 4° with 2 critical engines failed (81)	2
Manoeuvre		
Yaw control capability	Directional control, 2 engines inoperative (33)	1
	Time to yaw	2
	Double hydraulic failure (295–315)	2
Dynamic		
Mode	Dutch roll oscillation (216)	2
	Roll subsidence (216)	2
	Spiral divergence (216)	2
Spin recovery		
		2
Inertia coupling		
	Pitch due to velocity axis roll	2
	Yaw due to loaded roll	2

Double Hydraulic Failure: Hydraulic system failures must be compensated with CE-redundancy, to satisfy emergency (minimum) stability and control requirements. A typical design rule is the availability of 50% rest CE-efficiency. CE-redundancy is coupled with CE surface area, since too many CEs are too expensive.

Approach Trim: This flight case is flown with configuration ‘Full’, all high-lift devices deployed. The aerodynamic pitching moment is largest, thus highly demanding for the LoCE.

Minimum Control Speed, TO-Climb (V_{MCA}): Estimation generally in configuration ‘Full’ at low speed.

Minimum Speed at High Incidence (V_{min}, $V_{Alpha\ max\ 1g}$): Flight case considers s&c characteristics in the α-protect range. It is required to control α_{max}: select α_{max} in configuration ‘Full’ at low speed and lowest weight, aircraft must be controllable with a TBD-reserve.

Table 4.14 Generic 1st-level and 2nd-level DCFCs for the conceptual design of LaCEs

LaCE–DCFC	CS & FCV & FC	Level
Static		
β-Trim stability		
	Minimum control speed, approach and landing, 2 engines out ($V_{MCL\text{-}2}$) (148)	1
	Engine failure during take-off	1
	Landing from approach slope 4° with 2 critical engines failed (81)	1
	Trim 2 engines inoperative (164)	1
	Demonstration of max. cross wind on landing (one critical engine failed) (248)	1
	Straight sideslips (46)	1
	Demonstration of max. cross wind on landing (248)	2
Manoeuvre		
Roll control capability		
	Directional control, 2 engines inoperative (33)	1
	Time to roll	1
	Double hydraulic failure (295–315)	2
Dynamic		
Mode		
	Dutch roll oscillation (216)	2
	Roll subsidence (216)	2
	Spiral divergence (216)	2
Spin recovery		
		2
Inertia coupling		
	Pitch due to velocity axis roll	2
	Yaw due to loaded roll	2

Go-Around on 4 Engines Without Ground Effect: Flight condition considers the trimmed case in configuration 'Full' and sudden climb thrust setting.

Cruise With Trim Jam: This flight case relates to a system failure (e.g., actuator failure). For a specific c.g. position one receives $\delta_{trim} = 0$, for c.g.-deviations one receives $\delta_{trim} \neq 0$. It is required to estimate the allowable c.g. deviation from the ideal position. First, the c.g. position for $\delta_{trim} = 0$ must be determined. Any deviation of the c.g. from this point reduces the remaining control power left for controlling and stabilising. A trim jam results in mis-trim conditions with reduced control authority and performance penalties (trim drag).

Trim Tank Failure: Fuel transfer failure relates to a system failure (e.g., sensor failure). In analogy to 'Cruise With Trim Jam', the present case results in a mis-trim condition. The hinge moments, surface areas, and deflection angles of aerodynamic CE must be large enough, to compensate the Δx_{cg} due to trim tank failure. After specifying the maximum allowable c.g.-deviation for this case, the CEs and actuators can be determined.

C.G. Shifting Speed With Fuel Transfer System: This flight case balances the acceleration from subsonic to supersonic speeds and vice versa and the accompanying shift of the a.c. with the corrective action of the fuel transfer system, to keep the lag between the n.p. and the c.g. to a minimum.

Emergency Descent With Partial Loss of Forward c.g. Transfer Facility: (see case before).

Emergency Descent With Reverse Thrust Operating, Partial Loss of C.G. Transfer Facility: Analogous to mis-trim cases before, but with reverse thust effects (hazardous analysis).

Slat/Flap Failure: Each slat/flap configuration is tuned to a different α_{limit} for the aircraft. In case of slat and/or flap failure, the reference flight conditions are changing and the flight control system inevitably requires this information to modify the flight control laws. Slat failure is critical since it results in modifying α_{limit} (change in attitude) with the danger of stall. Flap failure is less critical; it primarily results in a change of trim lift, thus trim speed. Summarising, slat failure modifies the reference flight condition (attitude), whereby flap failure retains the original reference condition with a modified altitude. Consequently, slat failure must be less probable than flap failure. It must be prevented that both failure cases happen together.

LoCE Runaway—Failing to a Fixed Position (Control Allocation): This flight case relates to a system failure (e.g., actuator failure). This failure case is the primary reason to design for redundant CEs. It is particularly critical for the TFC, since a deflected and failed to a fixed position canard reduces static longitudinal stability significantly. Sufficient control power must be available to overpower a predefined LoCE runaway occurence.

Rotation on Take-Off/Nosewheel Lift-Off: This is a classical flight case for sizing the LoCE. The criticality of this condition depends on the sign of the thrust arm. The LoCE of aircraft configurations with a positive thrust arm (engine thrust line below the c.g.) is usually not constrained by this flight case.

Rotation on Landing With Ground Effect: It is required to check the de-rotation capability of the aircraft and the nose-wheel loads on the ground in configuration 'Full', low speed, low power setting, and with ground effect. This condition becomes more important on aircraft with elaborate high-lift systems, especially those with powered lift, where downwash at the tail is large. Furthermore, it might be critical for configurations with high c.g. positioning, like the B747 with Shuttle Orbiter or the Beluga.

Dive Recovery/Pullup: This flight case evaluates the load factor capability to recover from a dive. It implicitly checks the tendency of the aircraft to tuck, since the C_{mo} of the aircraft tends to change in the negative direction with increasing Mach number. The availability of sufficient longitudinal control power is fundamental.

Speed Recovery/Push Over: The aircraft must be able to pitch nose downward, so that the acceleration to this manoeuvre is 'prompt', away from the aerodynamic stall-attitude. A special variation of this flight condition is the 'CEV' manoeuvre,[37] specifically CE stall-demanding for FBW TAC aircraft with the thrust line below the c.g. and a trimmable stabiliser (highly non-linear flight case). The aircraft is flown at low speed at α_{limit}. With Airbus aircraft, the trimmable stabiliser is adjusted for high attitude flight at this condition, thus elevator (LoCE) push-authority is required (most demanding at rearward c.g.) to fly the push over manoeuvre. The more control authority that exists in the push direction, the more critical is the tail stall. The 'power-off' test is rather uncritical, whereby the 'power-on' case, elevator full down, and stabiliser full up is critical for negative tail stall. This LoCE sizing case defines the LoCE design in Normal Law (Airbus), or it alternatively defines the aft c.g. position or max thrust setting programmed in the control laws.[38] This maneouvre is one of the most demanding mis-trim flight cases, since low dynamic pressure prevails. Note that the F-22 pitch trim was sized by the speed recovery/ push over maneuver.

Load Factor Capability: This flight condition demonstrates load factor capability throughout the flight envelope. It is particularly critical for sizing of CE-systems when investigating certain system failures (e.g., hydraulic system failure resulting in reduced rest-control authority). The flight case primarily determines the maximum hingemoment requirements for the CE-actuators. Note that the flight cases dive recovery and speed recovery may already be included in the category load factor capability.

Short Period Oscillation/Phugoid Oscillation: With modern FBW aircraft, testing dynamic stability serves primarily the evaluation of the control laws raher than the open-loop airframe stability and control characteristics. However, it is mandatory during design to properly design for balanced dynamic characteristics, but to fine-tune only the dynamic characteristics with the FCS to avoid excessive system demands.

Nose Wheel Load at Break Release: Aircraft with a trycycle gear arrangement require sufficient loads on the nose gear to guarantee adequate ground handling characteristics. During the acceleration process of the aircraft (the worst condition is high thrust setting, low thrust line, aft and high c.g. position) sufficient nose gear loads have to be demonstrated, supported with the LoCE, but primarily adjusted

[37]CEV (Centre d'Essais en Vol) – French Flight Test Centre (Certification Authority—DGAC).

[38]On Airbus aircraft, the problem has been resolved with the Attitude-Protection system, which reduces the elevator-pull authority; the system prevents the dangerous exceedance of α_{max}.

with the c.g. position, main gear position, wing position, etc. Note that aerodynamic controls have usually no effect at brake release.

Power Application: The flight case intends to evaluate control authority margins throughout the flight envelope, dependent on rapid changes to thrust (max. thrust variations). The primary design parameters are the thrust moment and the magnitude of thrust variation.

Directional Control Effector (DiCE) and Lateral Control Effector (LaCE)—DCFCs

Minimum Control Speed, Approach and Landing, 2 Engines Out ($V_{MCL\text{-}2}$) (DiCE, LaCE) : For this test, the aircraft is trimmed for approach with two critical engines failed. It is required to maintain heading while demonstrating certain roll authority. The test is critical since it is a low dynamic pressure flight case with configuration setting for approach.

Engine Failure During Take-Off (DiCE): Dependent on the phase into the take-off, the aircraft has to remain controllable on the ground, secured by tyre steering and aerodynamic control.

Trim 2 Engines Inoperative (DiCE, LaCE): Demonstrate aircraft ability to maintain trim with two critical engines inoperative for the critical corners in the flight envelope.

Demonstration of Max. Cross Wind on Landing (One Critical Engine Failed) (DiCE, LaCE): Demonstrate aircraft ability to maintain trim with one critical engine failed. The cross-wind problem is not particularly critical for a BWB-type FWC, since this configuration has minimum side-area.

Straight Sideslips (DiCE, LaCE): Primarily a system design flight case, to avoid actuator stall limitations due to excessive hingemoments.

Adverse Yaw (DiCE, LaCE): The deflection of LaCEs generates adverse yaw, primarily at high angles of attack and low dynamic pressure, a moment to be compensated by the DiCE. The magnitude of adverse yaw depends on the choice of LaCE-type. This flight condition is seldom the critical condition for sizing the DiCE of transport aircraft, but it reduces the remaining excess control power of the DiCE. Fighter aircraft DiCEs (rudders) are usually sized by adverse yaw control in rapid roll maneuvers (very large roll rates required).

Landing from Approach Slope 4° with 2 Critical Engines Failed (DiCE, LaCE): Low dynamic pressure go-over flight case. Having two critical engines failed is not relevant for take-off, but for climb, cruise, descent, and landing.

Directional Control, 2 Engines Inoperative (DiCE, LaCE): Low dynamic pressure flight case linked to 'time to yaw' and 'time to roll'.

Time to Yaw (DiCE): This flight case must be seen in analogy to the 'time to roll' flight case. It might be a DiCE-critical flight condition, depending on the operational requirements (e.g., de-crab manoeuvre, heading control).

Time to Roll (LaCE): Flight case represents the classical condition for sizing the LaCE on military aircraft. This is because MIL-F-8785 has severe time-to-roll requirements which always design roll control capability. The MIL-F-8785 roll requirements had to be relaxed for the C-5 aircraft because it was calculated that the C-5 could not meet the original time-to-roll requirements even if one wing were severed instantaneously at the root. This is not the case for civil transports.

Double Hydraulic Failure (DiCE, LaCE): Hydraulic system failures must be compensated with CE redundancy, to satisfy emergency (minimum) stability and control requirements. A typical design rule is the availability of 50% rest CE-efficiency. CE-redundancy is coupled with CE surface area, since too many CEs are too expensive.

Dutch Roll Oscillation, Roll Subsidence, Spiral Divergence (DiCE, LaCE): With modern FBW aircraft, testing dynamic stability serves primarily the evaluation of the control laws raher than the open-loop airframe stability and control characteristics. However, it is mandatory during conceptual design to properly design for balanced dynamic characteristics, but to fine-tune the dynamic characteristics with the FCS only to avoid excessive system demands. If at all, the size of the DiCE is chosen to support Dutch roll damping.

Spin Recovery (DiCE): The need to test for spin and departure depends on whether the aircraft configuration is spin prone or not. It is a well-known fact that the modern transonic transport TAC is not spin prone, whereby the modern BWB FWC has to be investigated for spin and departure characteristics during the initial design stages. The most efficient design measure to prevent spin departure is having effective CEs available.

Pitch Due to Velocity Axis Roll, Yaw Due to Loaded Roll (LoCE, DiCE, LaCE): Highly manoeuvrable aircraft require feedback loops that minimise inertial coupling effects. Those feedback commands make the aircraft roll about the velocity vector rather than about the longitudinal axis, thereby preventing angle-of-attack from being converted into sideslip angle at high angle-of-attack. This DCFC identifies the magnitude of control power required for roll co-ordination.

4.7 Summary of Results

The ability to define the problem solving capabilities of *AeroMech* naturally depends on knowing the true impact of stability and control on aerospace vehicle design in the first place. With this intention in mind, the current chapter has presented an attempt to identify, isolate, and interpret relevant design parameters required for the development of the generic stability and control methodology *AeroMech*. The parameter reduction process performed vividly has demonstrated shortcomings of the traditional CE design-approach, a reality most appropriately expressed by Anderson: "*One of the most empirical and least precise aspects of the airplane design process is the sizing of the tail.*" [167]

Technology does not exist in a vacuum, and having only the final data of a successful program tells little about how that accomplishment might be replicated. Consequently, the strategy to initially invest in the aircraft conceptual design Data-Base System (DBS) and the Knowledge-Based System (KBS), has enabled a multi-disciplinary parameter reduction process of generic character. The problem with developing methods for conceptual design application manifests in the fact, that the identification of gross design parameters requires, at first, a fairly detailed understanding of the problem itself. From this point of view, the capability and potential of the DBS and KBS has not been even demonstrated in the present context.

The generic characterisation of geometry and mass has been supported by an attempt, to consistently define flight vehicle configurations and concepts. The design guide parametrics evolved has been set up with the intention, to keep things 'as simple as possible, but no simpler'. However, the generic design parametrics presented must be considered a first attempt only, since it is felt that a generic representation requires a constant and enduring evolution, a true challenge reaching beyond the current research undertaking.

Having identified a generic set of gross aerodynamic parameters, a suitable configuration-aerodynamics estimation technique has been selected. Clearly, the design implication of those parameters hinges, in part, on the capability of the computational aerodynamic prediction code selected. It is an unfortunate situation, that the advancement of aerodynamic estimation methods suitable for conceptual design application have been virtually forgotten by code developers. However, the capability and potential of the aerodynamic method selected has been judged a key component to realise the generic capability of *AeroMech*.

CE design techniques in use during the conceptual design phase have been reviewed. It has been shown, that the methods in use have stagnated in development, although the flight vehicle has continued to evolve. A variety of design concepts and techniques related to the design of CEs have been discussed. The stability and control properties of many modern aircraft depend greatly on the characteristics of the FCS employed, but as well on the basic aerodynamic and inertial characteristics of the aircraft. Thus, a new approach to stability and control is presented in this report, since unstable designs, servo characteristics, and others, do not fit well into the classical frameworks available for designing controls during conceptual design.

Considerable effort has been invested in closing the loop between aircraft conceptual design and flight test. JAR/FAR 25 certification-relevant formulations have been reviewed, which inevitably guide the design of CEs. A generic set of DCFCs has been defined with the support of A340 and Concorde flight test schedules, finally taking JAR/FAR, MIL Specs, and TSS certification requirements into account. This generic set of DCFCs has been grouped into two successive calculation phases, both representing conceptual design fidelity. The proposed follow-on utilisation of a dedicated conceptual design engineering flight simulator is not a subject of the current investigation.

References

1. Küchemann, D. and Bagley, J.A., "Twenty Years' Progress in Aerodynamics and the Changing Shape of Aeroplanes," *Interavia*, 1966, pp. 487–489.
2. Försching, H.W., "Grundlagen der Aeroelastik," First Edition, *Springer-Verlag*, 1974.
3. Nickel, K. and Wohlfahrt, M., "Tailless Aircraft – In Theory and Practice," First Edition, *Edward Arnold*, Translator E.M. Brown, 1994.
4. Roskam, J., "Airplane Flight Dynamics and Automatic Flight Controls—Part I," Third Edition, *DARcorporation*, 1995.
5. Goldsmith, H.A., "Stability and Control of Supersonic Aircraft at Low Speeds," ICAS Paper 64.588, *4th International Council of the Aeronautical Sciences*, Paris, France, 24-28 August 1964.
6. Sim, A.G., "A Correlation Between Flight-Determined Derivatives and Wind-Tunnel Data for the X-24B Research Aircraft," NASA TM 113084, NASA, August 1997.
7. Pinsker, W.J.G., "The Lateral Motion of Aircraft, and in Particular of Inertially Slender Configurations," ARC R.&M. No. 3334, 1963.
8. McRuer, D., Ashkenas, I., and Graham, D., "Aircraft Dynamics and Automatic Control," *Princeton University Press*, 1973.
9. Lee, H.P., Chang, M., and Kaiser, M.K., "Flight Dynamics and Stability and Control Characteristics of the X-33 Vehicle," AIAA Paper 98-4410, *AIAA Guidance, Navigation, and Control Conference and Exhibit*, Boston, MA, 10-12 August 1998.
10. Nickel, K. and Wohlfahrt, M., "Tailless Aircraft – In Theory and Practice," Translated by Brown, E.M., *Edward Arnold*, 1994.
11. Sim, A.G. and Curry, R.E., "Flight Characteristics of the AD-1 Oblique-Wing Research Aircraft," NASA TP 2223, NASA, March 1987.
12. Abzug, M.J. and Larrabee, E.E., "Airplane Stability and Control – A History of the Technologies That Made Aviation Possible," First Edition, Cambridge Aerospace Series No. 6, *Cambridge University Press*, 1997.
13. Miller, J., "The X-Planes X-1 to X-29," First Edition, *Speciality Press*, 1983.
14. Sternfiel, L., "Some Considerations of the Lateral Stability of High-Speed Aircraft," NACA TN 1282, NASA, 1947.
15. Hallion, R.P., "Test Pilots: The Frontiersmen of Flight," *Doubleday*, 1981.
16. Ford, D., "Glen Edwards – The Diary of a Bomber Pilot," First Edition, *Smithsonian Institution Press*, 1998.
17. Pape, G.R. and Campbell, J.M., "Northrop Flying Wings – A History of Jack Northrop's Visionary Aircraft," First Edition, *Schiffer Military*, Aviation History, 1995.
18. Northrop, J.K., "The Development of All-Wing Aircraft," 35th Wilbur Wright Memorial Lecture, The Royal Aeronautical Society, London, 1947.
19. Wilson, J.R., "New Blend For An Old Wing Design," Aerospace America, April 2000, pp. 28–35.
20. Phillips, E.H., "NASA To Fly Sub-Scale Blended Wing Body," Aviation Week & Space Technology, 07 February 2000, pp. 48–49.
21. Rech, J. and Leyman, C.S., "A Case Study By Aerospatiale and British Aerospace on the Concorde," AIAA Professional Study Series, AIAA, 01 April 2003.

22. Sachs, G., "Minimum Trimmed Drag and Optimum C.G. Position," Vol. 15, No. 8, *AIAA Journal of Aircraft*, August 1978, pp. 456–459.
23. Cameron, D. and Princen, N., "Control Allocation Challenges and Requirements for the Blended Wing Body," AIAA Paper 2000-4539, *AIAA Guidance, Navigation, and Control Conference and Exhibit*, 14-17 August 2000.
24. Anon., "Concorde Flying Manual – Vol. 1," ATP. No. E.8021, Serial No. 4, Holder no. 2304, British Airways, January 1986.
25. Anon., "A300-600ST: Technical Description – Vol. 1 – General Characteristics," SATIC/TE DN 0268/95-24/3/1995, Satic, 1995.
26. Jenkins, D.R., "The History of Developing the National Space Transportation System – The Beginning Through STS-50," Second Edition, *Motorbooks Inernational*, 1993.
27. Weightman, G.D., "Vertical Centre of Gravity Flight Issues," *Very Large Transport Aeroplane Conference*, Leeuwenhorst Congres, The Netherlands, 13–16 October 1998.
28. Heidmann, H., "Trimmtank-System zum Erreichen widerstands-optimaler Schwerpunktslagen," Jahrestagung der DGLR, October 1982.
29. Anon., "Introduction to Supersonics," BA/SST/47/A, British Aerospace Airbus, August 1986.
30. Jenkins, D.R., "B-1 Lancer – The Most Complicated Warplane Ever Developed," Vol. 2, Military Aircraft Series, *McGraw-Hill*, 1999.
31. Moon, H., "Soviet SST – The Technopolitics of the Tupolev-144," First Edition, *Orion Books*, 1989.
32. Ross, J.W. and Rogerson, D.B., "XB-70 Technology Advancements," A83-1048, Rockwell International Corporation, 1983.
33. Holloway, R.B., "Introduction of CCV Technology Into Airplane Design," AGARD CP-147, Vol. 1, October 1973.
34. Ashkenas, I.L. and Klyde, D.H., "Tailless Aircraft Performance Improvements With Relaxed Static Stability," NASA CR 181806, NASA, March 1989.
35. Sanders, K.L., "Simpler Wing Location for a Specified Longitudinal Stability," *Aerospace Engineering*, March 1960, pp. 67–72.
36. Torenbeek, E., "Synthesis of Subsonic Airplane Design," Delft University Press, Kluwer Academic Publishers, 1990.
37. Hoey, R.G., "Testing Lifting Bodies at Edwards," *A PAT Projects, Inc. Publication*, July 1997.
38. Erickson, B.A., "Flight Characteristics of the B58 Mach 2 Bomber," Vol. 66, Nr. 623, *Journal of the Royal Aeronautical Society*, November 1962, pp. 665–671.
39. Greff, E., "Aerodynamic Design and Technology Concepts For a New Ultra-High Capacity Aircraft," ICAS Paper 96-4.6.3, *20th Congress of the International Council of the Aeronautical Sciences*, 08-13 September 1996.
40. Sacco, G., "P.180: Reasons and Evolution of an Unconventional Aerospace Vehicle Design," The Michigan State University, October 1989.
41. Wakayama, S. and Kroo, I., "The Challenge and Promise of Blended-Wing-Body Optimization," AIAA Paper 98-4736, *7th AIAA/USAF/NASA/ISSMO Symposium on Multidisciplinary Analysis and Optimization*, St. Louis, MO, 02-04 September 1998.
42. Morris, S.J. and Tigner, B., "Flight Tests of an Oblique Flying Wing Small Scale Demonstrator," AIAA Paper 95-3327-CP, *Guidance, Navigation, and Control Conference*, Baltimore, MD, 07-10 August 1995.
43. Davies, P.E. and Thornborough, T., "Boeing B-52 Stratofortress," First Edition, Crowood Aviation Series, *The Crowood Press*, 1998.
44. Yaros, S.F., Sexstone, M.G., *et al*, "Synergistic Airframe-Propulsion Interactions and Integrations," NASA TM-1998-207644, NASA, March 1998.
45. Myhra, D., "The Horten Brothers and Their All-Wing Aircraft," First Edition, Schiffer Military Aviation History, *Schiffer Publishing Ltd.*, 1998.
46. Grellmann, H.W., "B-2 Aerodynamic Design," AIAA Paper 90-1802, *Aerospace Engineering Conference and Show*, Los Angeles, CA, 13-15 February 1990.

47. Crickmore, P.F., "Lockheed SR-71 – The Secret Missions Exposed," First Edition, *Osprey Aerospace*, 1993.
48. McCarty, C.A., Feather, J.B., *et al*, "Design and Analysis Issues of Integrated Control Systems for High-Speed Civil Transports," NASA CR 186022, NASA, May 1992.
49. Stillwell, W.H., "X-15 – Research Results With A Selected Bibliography," NASA SP-60, NASA, November 1964.
50. Robinson, M.R. and Herbst, W.B., "The X-31A and Advanced Highly Maneuverable Aircraft," ICAS Paper 90-0.4, *17th Congress of the International Council of the Aeronautical Sciences*, Stockholm, Sweden, 09-14 September 1990.
51. Moore, M. and Frei, D., "X-29 Forward Swept Wing Aerodynamic Overview," AIAA Paper 83-1834, *Applied Aerodynamics Conference*, Danvers, MA, 13-15 July 1983.
52. Stollery, J., "Aerodynamics, Past, Present and Future," The Sydney Goldstein Memorial Lecture, College of Aeronautics, Cranfield University, 1 November 1995.
53. Prem, H., "Cooperation Know-How in High-Tech Products," Binational Conference, University Hohenheim, Stuttgart, 16–17 October 1986, in "Forschung und Entwicklung – Technisch-Wissenschaftliche Veröffentlichungen 1956-1987—Ein Rück- und Ausblick," Jubiläumsausgabe anläßlich des 75. Geburtstag von Dipl.-Ing. Dr.-Ing. E.h. Ludwig Bölkow, MBB, 1987.
54. Prandtl, L., "Applications of Modern Hydrodynamics to Aeronautics," NACA TR 116, NASA, 1921.
55. Busemann, A., "Aerodynamischer Auftrieb bei Überschallgeschwindigkeit," *5th Volta Congress*, Rome, Italy, 30 September-06 October1935, pp. 328–360.
56. Weissinger, J., "The Lift Distribution of Swept-Back Wings," NACA TM 1120, NASA, 1942.
57. Jones, R.T., "Properties of Low-Aspect Ratio Pointed Wings at Speeds Below and Above the Speed of Sound," NACA Report 835, NACA, 1946.
58. Multhopp, H., "Methods for Calculating the Lift Distribution of Wings (Subsonic Lifting Surface Theory)," Aeronautical Research Council R&M 2884, 1950.
59. Küchemann, D., "A Simple Method of Calculating the Span and Chordwise Loading on Straight and Swept Wings Of Any Given Aspect Ratio at Subsonic Speeds," Aeronautical Research Council R&M 2935, 1952.
60. Anon., "Design of Body-Wing Junctions for High Subsonic M, for Swept Back Wings and Symmetrical Bodies," RAE R Aero 2336, 1949.
61. Hoerner, S.F., "Aerodynamic Drag," First Edition, Published by the Author, May 1951.
62. Hoerner, S.F., "Fluid-Dynamic Drag," Third Edition, Published by the Author, 1965.
63. Hoerner, S.F. and Borst, H.V., "Fluid-Dynamic Lift," Second Edition, Published by L.H. Hoerner, April 1985.
64. Hoak, D.E., Finck, R.D., *et al*, "USAF Stability and Control Datcom," Flight Control Division, Air Force Flight Dynamics Laboratory, Wright-Patterson Air Force Base, 1978.
65. Anon., "Method for Predicting the Pressure Distribution on Swept Wings With Subsonic Attached Flow," ESDU Transonic Data Memo 6312, 1963.
66. Schemensky, R.T., "Development of an Empirically Based Computer Program to Predict the Aerodynamic Characteristics of Aircraft – Volume I Empirical Methods," TR AFFDL-TR-73-144, Air Force Flight Dynamics Laboratory, Wright-Patterson AFB, October 1973.
67. Vukelich, S.R., "Development Feasibility of Missile Datcom," AFWAL-TR-81-3130, Air Force Flight Dynamics Laboratory, Wright-Patterson AFB, October 1981.
68. Falkner, V.M., "The Calculation of Aerodynamic Loading on Surfaces of Any Shape," ARC R&M 1910, Aeronautical Research Council, National Physical Laboratory, August 1943.
69. Hess, J.L. and Smith, A.M.O., "Calculation of Nonlifting Potential flow About Arbitrary Three-Dimensional Bodies," Douglas Report ES40622, Douglas Aircraft Company, 1962.
70. Adam, Y., "A Hermitian Finite Difference Method for the Solution of Parabolic Equations," No. 1, Comp. Math. Applications, 1975, pp. 393–406.

71. Chung, T.J., "Finite Element Analysis in Fluid Dynamics," First Edition, *McGraw-Hill*, 1978.
72. Rizzi, A.W. and Inouye, M., "Time Split Finite Volume Method for Three-Dimensional Blunt-Body Flows," Vol. 11, No. 11, *AIAA Journal*, November 1973, pp. 1478–85.
73. Gottlieb, D. and Orszag, S.A., "Numerical Analysis of Spectral Methods: Theory and Applications," SIAM, Philadelphia, 1977.
74. Snyder, J.R., "CFD Needs in Conceptual Design," AIAA Paper 90-3209, *Aircraft Design, Systems and Operations Conference*, Dayton, OH, 17-19 September 1990.
75. Shevell, R.S., "Aerodynamic Bugs: Can CFD Spray Them Away?," AIAA Paper 85-4067, *3rd Applied Aerodynamics Conference*, Colorado Springs, CO, 14-16 October 1985.
76. Sinclair, J. (Editor in Chief), "BBC English Dictionary," First Edition, *BBC English and HarperCollins Publishers Ltd.*, 1992.
77. Blake, W.B., "Prediction of Fighter Aircraft Dynamic Derivatives Using Digital Datcom," AIAA Paper 85-4070, *3rd Applied Aerodynamics Conference*, Colorado Springs, CO, 14-16 October 1985.
78. Blake, W.B. and Simon, J.M., "Tools for Rapid Analysis of Aircraft and Missile Aerodynamics," AIAA Paper 98-2793, *16th AIAA Applied Aerodynamics Conference*, Albuquerque, NM, 15-18 June 1998.
79. Razgonyaev, V. and Mason, W.H., "An Evaluation of Aerodynamic Prediction Methods Applied to the XB-70 for Use in High Speed Aircraft Stability and Control System Design," AIAA Paper 95-0759, *33rd Aerospace Sciences Meeting and Exhibit*, Reno, NV, 09-12 January 1995.
80. Rubbert, P.E. and Tinoco, E.N., "Impact of Computational Methods on Aerospace Vehicle Design," AIAA Paper 83-2060, August 1983.
81. Mason, W.H., "Applied Computational Aerodynamics," Class Notes for AOE 4114, Department of Aerospace and Ocean Engineering, Virginia Polytechnic Institute and State University, 1995.
82. Nicolai, L.M. and Carty, A., "Role of the Aerodynamicist in a Concurrent Multi-Disciplinary Design Process," Paper ADP010502, *Symposium of the RTO (Research and Technology Organization) Applied Vehicle Technology Panel (AVT)*, Ottawa, Canada, 18–21 October 1999.
83. Morris, S.J., "Integrated Aerodynamics and Control System Design for Tailless Aircraft," AIAA Paper 92-4604, *Astrodynamics Conference*, Hilton Head Island, SC, 10-12 August 1992.
84. Morris, S.J., "Integrated Aerodynamic and Control System Design of Oblique Wing Aircraft," Ph.D. Thesis, SUDAAR #620, Department of Aeronautics and Astronautics, Stanford University, January 1990.
85. Dorsett, K.M. and Peters, S.E., "Needs Description," NASA CP 10138, *Proceedings of the Non-Linear Aero Prediction Requirements Workshop*, NASA, March 1994.
86. Hancock, G.J., "Dynamic Effects of Controls," AGARD R-711, March 1983.
87. Young, A.D., "Introductory Remarks and Review of 1979 Symposium," AGARD R-711, March 1983.
88. Ross, A.J. and Thomas, H.H.B.M., "A Survey of Experimental Data on the Aerodynamics of Controls, in the Light of Future Needs," AGARD CP-262, September 1979.
89. Sweetman, B., "Northrop B-2 Stealth Bomber – The Complete History, Technology, and Operational Development of the Stealth Bomber," First Edition, Mil-Tech Series, *Motorbooks International*, 1992.
90. Moul, T.M., Fears, S.P., Ross, H.M., and Foster, J.V., "Low-Speed wind-Tunnel Investigation of the Stability and Control Characteristics of a Series of Flying Wings With Sweep Angles of 60°," NASA TM 4649, NASA, August 1995.
91. Thomas, H.H.B.M., "The Aerodynamics of Aircraft Control – A General Survey in the Context of Active Control Technology," AGARD Report R-711, March 1983.
92. Skow, A.M., "Control of Advanced Fighter Aircraft," AGARD Report R-711, March 1983.

93. Bryan, G.H., "Stability in Aviation," *MacMillan*, London, 1911.
94. Nelson, R.C., "Flight Stability and Automatic Control," Second Edition, *WCB/McGraw-Hill*, 1998.
95. Kalviste, J. and Eller, B., "Coupled Static and Dynamic Stability Parameters," AIAA Paper 89-3362-CP, *16th Atmospheric Flight Mechanics Conference*, Boston, MA, 14-16 August 1989.
96. Ellison, D.E. and Hoak, D.E., "Stability Derivative Estimation at Subsonic Speeds," Vol. 2, No. 154, *Annals of the New York Academy of Sciences*, 1968, pp. 367–396.
97. Thomas, H.H.B.M., "Estimation of Stability Derivatives (State of the Art)," ARC CP No. 664, 1963.
98. Blakelock, J.H., "Automatic Control of Aircraft and Missiles," Second Edition, *John Wiley*, 1991.
99. Sim, A.G. and Curry, R.E., "Flight-Determined Aerodynamic Derivatives of the AD-1 Oblique-Wing Research Airplane," NASA TP-2222, NASA, 1984.
101. Anon., "VLAERO," Prospectus, Analytical Methods, Inc., Redmond, Washington, October 1996.
102. Lamar, J.E., "A Vortex Lattice Method for the Mean Camber Shapes of Trimmed Noncoplanar Planforms with Minimum Vortex Drag," NASA TN D-8090, NASA, 1976.
103. Carmichael, R., "Public Domain Aeronautical Software - A Collection of Public Domain Software from NASA and USAF for the PC," Internet, http://pdas.com/, June 1997.
104. Lamar, J.E. and Herbert, H.E., "Production Version of the Extended NASA-Langley Vortex Lattice FORTRAN Computer Code," Volume 1, User's Guide, NASA TM 83303, NASA, April 1982.
105. Rom, J., Melamed, B., and Almosnino, D., "Comparison of Experimental Results with the Non-Linear Vortex Lattice Method Calculations for Various Wing-Canard Configurations," ICAS-90-3.3.4, *17th Congress of the International Council of the Aeronautical Sciences*, Stockholm, Sweden, 09-14 September 1990.
106. Anon., "Program VORLAT," http://cac.psu.edu/~lnl/497/mpi/vorlat.f , with reference to Bertin, J.J. and Smith, M.L., "Aerodynamics for Engineers," Second Edition, *Prentice Hall*, 1989.
107. Anon., "LinAir Pro Users Guide," Desktop Aeronautics, Inc., 1996.
108. Gallman, J.W., Kaul, R.W., Chandrasekharan, R.M. and Hinson, M.L., "Optimization of an Advanced Business Jet," Vol. 34, No. 3, *AIAA Journal of Aircraft*, May-June 1997.
109. Buresti, G., Lombardi, G. and Petagna, P., "Wing Pressure Loads in Canard Configurations: A Comparison Between Numerical Results and Experimental Data," Vol. 96, Issue 957, *Aeronautical Journal*, August-September 1992, pp. 271-279.
110. Gaydon, J.H., "Improved Panel Methods for the Calculation of Low-Speed Flows Around High-Lift Configurations," Ph.D. Thesis, Bristol University, September 1995.
111. Anon., "3-D Subsonic Aerodynamics Software - Sub3D for Windows – Version 1.0 User's Guide," User's Guide Rev. A, SoftwAeronautics, Inc., 1996.
112. Albright, A.E., Dixon, C.J., and Hegedus, M.C., "Modification and Validation of Conceptual Design Aerodynamic Prediction Method HASC95 With VTXCHN," NASA CR-4712, NASA, March 1996.
113. Miranda, L.R., Elliott, R.D., and Baker, W.M., "A Generalized Vortex Lattice Method for Subsonic and Supersonic Flow Applications," NASA CR-2865, NASA, December 1977.
114. Lan, C.E., "Methods of Analysis in The VORSTAB Code (Version 3.1)," The University of Kansas, May 1993.
115. Margason, R.J., Kjelgaard, S.O., Sellers, W.L., et. al., "Subsonic Panel Methods - A Comparison of Several Production Codes," AIAA 85-0280, *23rd Aerospace Sciences Meeting*, Reno, NV, 14-17 January 1985.
116. Rubbert, P.E. and Saaris, G.R., "A General Three-Dimensional Potential Flow Method Applied to V/STOL Aerodynamics," SAE Paper 68004, *National Air Transportation Meeting*, February 1968.

117. Petrie, J.A.H., "Development of an Efficient and Versatile Panel Method for Aerodynamic Problems," Ph.D. Thesis, University of Leeds, March 1979.
118. Katz, J. and Plotkin, A., "Low-Speed Aerodynamics - From Wing Theory to Panel Methods," *McGraw-Hill, Inc.*, 1991.
119. Roberts, A. and Rundle, K., "Computation of Incompressible Flow About Bodies and Thick Wings Using the Spline-Mode System," BAC (CAD) Report Aero Ma 19, 1972.
120. Hunt, B. and Semple, W.G., "The BAC (MAD) Program to Solve the 3-D Lifting Subsonic Neumann Problem Using the Plane Panel Method," Report ARG 97 BAC (MAD), 1973.
121. Erickson, L.L., "Panel Methods - An Introduction," NASA-TP-2995, NASA Ames Research Center, 01 December 1990.
122. Lamar, J.E., "The Use of Linearized-Aerodynamics and Vortex-Flow Methods in Aerospace Vehicle Design (Invited Paper)," AIAA Paper 82-1384, August 1982.
123. Anon., "AMI - Specialists in Computational Fluid Dynamics Services and Software Products," Prospectus, Analytical Methods, Inc., Februar 1997.
124. Hoeijmakers, H.W.M., "Panel Methods for Aerodynamic Analysis and Design," AGARD Report 783, May 1991.
125. Petrie, J.A.H., "Development of an Efficient and Versatile Panel Method for Aerodynamic Problems," Ph.D. Thesis, University of Leeds, March 1979.
126. Lötstedt, P., "A Three-Dimensional Higher-Order Panel Method for Subsonic Flow Problems - Description and Applications," SAAB-SCANIA Rep. L-0-1 R100, 1984.
127. Roggero, F. and Larguier, R., "Aerodynamic Calculation of Complex Three-Dimensional Configurations," Vol. 30, No. 5, *AIAA Journal of Aircraft*, Sept.-Oct. 1993.
128. Baston, A., Lucchesini, M., Manfriani, L., *et. al.*, "Evaluation of Pressure Distributions on an Aircraft by two Different Panel Methods and Comparison with Experimental Measurements," ICAS-86-1.5.3, *15th Congress of the International Council of the Aeronautical Sciences*, London, England, 07-12 September 1986.
129. Anon., "PMARC_12 - Panel Method Ames Research Center, Version 12," Internet, http://www.cosmic.uga.edu./abstracts/arc-13362, July 1997.
130. Anon., "AeroMaster," http://www.ssmotion.com/aeromstr.htm, July 1997.
131. Manning, V.M., "A Comparison of the Woodward-Carmichael Code and PAN AIR," Internet, http://www-leland.stanford.edu/~valman/research/WCMST/WCMST.html, 8 October 1996.
132. Anon., "USSAERO - Unified Subsonic Supersonic Aerodynamic Analysis Program," Internet, http://cognac.cosmic.ug…bstracts/lar-11305.html, July 1997.
133. Mason, W.H. and Rosen, B.S., "The COREL and W12SC3 Computer Programs for Supersonic Wing Design and Analysis," NASA CR-3676, NASA Langley Research Center, December 1983.
134. Carmichael, R., "Public Domain Aeronautical Software - A Collection of Public Domain Software from NASA and USAF for the PC," Internet, http://pdas.com/, June 1997.
135. Hoeijmakers, H.W.M., "A Panel Method for the Determination of the Aerodynamic Characteristics of Complex Configurations in Linearized Subsonic or Supersonic Flow," Report NLR TR 80124, 1980.
136. Maughmer, M., Ozoroski, L., Straussfogel, D., and Long, L., "Validation of Engineering Methods for Predicting Hypersonic Vehicle Control Forces and Moments," Vol. 16, No. 4, *AIAA Journal of Guidance, Control, and Dynamics*, July-August 1993.
137. Siclari, M., Visich, M., Cenko, A., Rosen, B., and Mason, W., "Evaluation of NCOREL, PAN AIR, and W12SC3 for Supersonic Wing Pressures," Vol. 21, No. 10, *AIAA Journal of Aircraft*, October 1984, pp. 816–822.
138. Anon., "Introducing ADAPT," Version 1.0, Synaps, Inc., 1996.
139. Lan, C.E., "Applied Airfoil and Wing Theory," First Edition, *Cheng Chung Book Company*, 1988.
140. Pistolesi, E., "Betrachtungen über die gegenseitige Beeinflussung von Tragflügelsystemen," *Collected Presentations of the 1937 Lilienthal-Gesellschaft Meeting*, 1937, pp.214-219.

141. Schlichting, H. and Truckenbrodt, E., "Aerodynamik des Flugzeuges – Erster Band – Grundlagen aus der Strömungsmechanik – Aerodynamik des Tragflügels (Teil I)," Second Edition, *Springer*, 1967.
142. Bertin, J.J. and Smith, M.L., "Aerodynamics For Engineers," Second Edition, *Prentice Hall*, 1989.
143. Moran, J., "An Introduction to Theoretical and Computational Aerodynamics," First Edition, *John Wiley & Sons*, 1984.
144. Rakowitz, M.E., "Evaluation of LinAir as a Design Tool for the Lift Distribution of a Three-Surface Aircraft," M.Sc. Thesis, College of Aeronautics, Cranfield University, September 1997.
145. Lan, C.E., "A Quasi-Vortex-Lattice Method in Thin Wing Theory," Vol. 11, No. 9, *AIAA Journal of Aircraft*, September 1974, pp. 518–527.
146. Lan, C.E., Emdad, H. *et al*, "Calculation of High Angle-Of-Attack Aerodynamics of Fighter Configurations," AIAA Paper 89-2188-CP, *7th Applied Aerodynamics Conference*, Seattle, WA, 31 July-02 August 1989.
147. Lan, C.E., "VORSTAB – A Computer Program For Calculating Lateral-Directional Stability Derivatives With Vortex Flow Effect," NASA CR 172501, NASA, January 1985.
148. Lan, C.E., "Methods of Analysis in The VORSTAB Code (Version 3.1)," Department of Aerospace Engineering, The University of Kansas, May 1993.
149. Obert, E., "Tail Design," Report No. H-0-93, Issue No. 1, Fokker Aircraft B.V., Lecture Notes to the ECATA Postgraduate Aerospace Vehicle Design Course, 22 March 1992.
150. Root, L.E., "Dynamic Longitudinal Stability Charts for Design Use," Vol. 2, No. 3 *Journal of Aeronautical Sciences* (JAS), May 1935, pp. 101–108.
151. Silverstein, A., "Towards a Rational Method of Tail-Plane Design," Journal of the Aeronautical Sciences, 1939.
152. Root, L.E., "Empennage Design With Single and Multiple Vertical Surfaces," Vol. 6, No. 9, *Journal of the Aeronautical Sciences*, July 1939, pp. 353–360.
153. Morgan, M.B. and Thomas, H.H.B.M., "Control Surface Design in Theory and Practice," Vol. 49, Issue 416, *The Aeronautical Journal*, The Royal Aeronautical Society, August 1945, pp. 431–510.
154. Wimpenny, J.C., "Stability and Control in Aerospace Vehicle Design," Vol. 58, Journal of the Royal Aeronautical Society, May 1954, pp. 329–360.
155. Lee, G.H., "The Aeroplane Designer's Approach to Stability and Control," AGARD Report 334, April 1961.
156. Wood, K.D., "Aerospace Vehicle Design – Volume I – Aerospace Vehicle Design," Eleventh Edition, *Johnson Publishing Company*, 1963.
157. Burns, B.R.A., "Design Considerations for the Satisfactory Stability and Control of Military Combat Aeroplanes," AGARD CP 119, 1972.
158. Torenbeek, E., "Synthesis of Subsonic Airplane Design," 6th Printing, Delft University Press, *Kluwer Academic Publishers*, 1990.
159. Nicolai, L.M., "Fundamentals of Aerospace Vehicle Design," Second Edition, *METS, Inc.*, 1984.
160. Hünecke, K., "Modern Combat Aerospace Vehicle Design," Second Edition, *Airlife Publishing Ltd.*, 1987.
161. Whitford, R., "Design for Air Combat," First Edition, *Jane's Publishing Company Limited*, 1987.
162. Stinton, D., "The Design of the Aeroplane," *BSP Professional Books*, 1991.
163. Raymer, D.P., "Aerospace Vehicle Design: A Conceptual Approach," Second Edition, *AIAA Education Series*, AIAA, 1992.
164. Heinemann, E., "Aerospace Vehicle Design," First Edition, *Aerospace Vehicle Designs Inc.*, 1997.
165. Hünecke, K., "Die Technik des modernen Verkehrsflugzeuges," First Edition, *Motorbuchverlag*, 1998.
166. Stinton, D., "The Anatomy of the Aeroplane," Second Edition, *Blackwell Science*, 1998.

167. Anderson, J.D., "Aircraft Performance and Design," First Edition, *WCB/McGraw-Hill*, 1999.
168. Jenkinson, L.R., Simpkin, P., and Rhodes, D., "Civil Jet Aerospace Vehicle Design," First Edition, *Arnold*, 1999.
169. Scholz, D., "Skript zur Vorlesung Flugzeugentwurf," First Edition, Fachhochschule Hamburg, Fachbereich Fahrzeugtechnik, Sommersemester, 1999.
170. Howe, D., "Aircraft Conceptual Design Synthesis," First Edition, *Professional Engineering Publishing*, October 2000.
171. Oman, B.H., "Vehicle Design Evaluation Program (VDEP)," NASA CR 145070, NASA, 01 January 1977.
172. Thorbeck, J., "Ein Beitrag zum Rechnergestützten Entwurf von Verkehrsflugzeugen," Ph.D. Thesis, Technical University Berlin, 1984.
173. Alsina, J., "Development of an Aerospace vehicle design Expert System," Ph.D. Thesis, Cranfield Institute of Technology, College of Aeronautics, October 1987.
174. Bil, C., "Development and Application of a Computer-Based System for Conceptual Aerospace Vehicle Design," Ph.D. Thesis, Delft University, Delft University Press, 1988.
175. Kay, J., Mason, W.H., *et al*, "Control Authority Issues in Aircraft Conceptual Design: Critical Conditions, Estimation Methodology, Spreadsheet Assessment, Trim and Bibliography," VPI-Aero-200, Virginia Polytechnic Institute and State University, Department of Aerospace and Ocean Engineering, November 1993.
176. Heinze, W., "Ein Beitrag zur quantitativen Analyse der technischen und wirtschaftlichen Auslegungsgrenzen verschiedener Flugzeugkonzepte für den Transport großer Nutzlasten," Ph.D. Thesis, ZLR-Forschungsbericht 94.01, Institut für Flugzeugbau und Leichtbau, Technical University Braunschweig, 1994.
177. Nunes, J.M.B., "Aerospace Vehicle Design Optimisation – Conceptual Evaluation of a Three-Lifting Surface Turbo-Fan Airliner," Ph.D. Thesis, College of Aeronautics, Cranfield University, February 1995.
178. MacMillin, P.E., "Trim, Control, and Performance Effects in Variable-Complexity High-Speed Civil Transport Design," M.Sc. Thesis, Department of Aerospace and Ocean Engineering, Virginia Polytechnic Institute and State University, May 1996.
179. Pohl, T., "Improvement and Extension of Tail Sizing Procedures for Preliminary Design Purposes," M.Eng. Thesis, Department of Aeronautics, Imperial College of Science, June 1997.
180. Nicolai, L.M. and Carty, A., "Role of the Aerodynamicist in a Concurrent Multi-Disciplinary Design Process," Paper RTO MP-35, *RTO AVT Symposium on 'Aerodynamic Design and Optimization of Flight Vehicles in a Concurrent Multi-Disciplinary Environment'*, Research and Technology Organization, Applied Vehicle Technology Panel (AVT), Ottawa, Canada, 18–21 October 1999.
181. Sauvinet, F., "Longitudinal Active Stability: Key Issues for Future Large Transport Aircraft," ICAS Paper 4101, *22nd Congress of International Council of the Aeronautical Sciences*, Harrogate, UK, 28 August-01 September 2000.
182. Etkin, B., "Transfer Functions: Improvement on Stability Derivatives for Unsteady Flight," UTIA Report 42, 1958.
183. Thomas, H.H.B.M., "State of the Art of Estimation of Derivatives," AGARD Report 339, April 1961.
184. Etkin, B. and Reid, L.D., "Dynamics of Flight – Stability and Control," Third Edition, *John Wiley & Sons, Inc.*, 1996.
185. Middel, J., "Development of a Computer Assisted Toolbox for Aerodynamic Design of Aircraft at Subcritical Conditions with Application to Three-Surface and Canard Aircraft," Ph.D. Thesis, Delft University Press, 1992.
186. Hofmann, L.G. and Clement, W.F., "Vehicle Design Considerations for Active Control Application to Subsonic Transport Aircraft," NASA CR-2408, August 1974.

187. Jenny, R.B., Krachmalnick, F.M., and LaFavor, S.A., "Air Superiority with Controlled Configured Fighters," AIAA Paper 71-764, *3rd Aircraft Design and Operations Meeting*, Seattle, WA, 12-14 July 1971.
188. Pasley, L.H. and Kass, G.J., "Improved Airplane Performance Through Advanced Flight Control System Design," AIAA Paper 70-875, July 1970.
189. Pasley, L.H., Rohling, W.J., and Wattman, W.J., "Compatibility of Maneuver Load Control and Relaxed Static Stability," AIAA Paper 73-791, *5th Aircraft Design,Flight Test and Operations Meeting*, St. Louis, MO, 06-08 August 1973.
190. Mueller, L.J., "Pilot and Aircraft Augmentation on the C-5," Vol. 7, No. 6, *AIAA Journal of Aircraft*, November-December 1970, pp. 550–553.
191. Cook, M.V., "Flight Dynamics Principles," First Edition, *Arnold*, 1997.
192. Gibson, J.C. (Handling Qualities Expert, BAe Military) and Chudoba, B., Personal Communication, 6 February 1997.
193. Gibson, J.C., "The Definition, Understanding and Design of Aircraft Handling Qualities," Report LR-756, Delft University of Technology, February 1995.
194. Gibson, J.C., "Development of a Methodology for Excellence in Handling Qualities Design for Fly By Wire Aircraft," Ph.D. Thesis, Published in Series 03, Control and Simulation 06, Delft University Press, 1999.
195. Hodgkinson, J., "Aircraft Handling Qualities," First Edition, *Blackwell Science Ltd*, 1999.
196. McRuer, D.T., "Aviation Safety and Pilot Control – Understanding and Preventing Unvavorable Pilot-Vehicle Interactions," National Research Council, *National Academy Press*, 1997.
197. Mavris, D.N., DeLaurentis, D.A., and Soban, D.S., "Probabilistic Assessment of Handling Qualities Characteristics in Preliminary Aerospace Vehicle Design," AIAA Paper 98-0492, *36th AIAA Aerospace Sciences Meeting and Exhibit*, Reno, NV, 12-15 January 1998.
198. Graeber, U.P., "Realisation of Relaxed Static Stability on a Commercial Transport," AGARD Paper CP-384, October 1984.
199. Anderson, M.R. and Mason, W.H., "An MDO Approach to Control-Configured-Vehicle Design," AIAA Paper 96-4058, *6th Symposium on Multidisciplinary Analysis and Optimization*, Bellevue, WA, 04-06 September 1996.
200. Beaufrere, H., "Integrated Flight Control System Design for Fighter Aircraft Agility," AIAA Paper 88-4503, Aircraft Design, Systems and Operations Conference, Atlanta, GA, 07-09 September 1988.
201. Imlay, F.H., "A Theoretical Study of Lateral Stability with an Automatic Pilot," NACA Report 693, 1940.
202. Etkin, B., "Dynamics of Atmospheric Flight," Second Edition, *John Wiley & Sons, Inc.*, 1972.
203. Roskam, J., "Airplane Flight Dynamics and Automatic Flight Controls – Part II," First Edition, *DARcorporation* (Design, Analysis and Research Corporation), 1995.
204. Durham, W.C., "Constrained Control Allocation," Vol. 16, No. 4, *AIAA Journal of Guidance, Control, and Dynamics*, July-August 1993, pp. 717–725.
205. Bordignon, K.A. and Durham, W.C., "Null-Space Augmented Solutions to Constrained Control Allocation Problems," AIAA Paper 95-3209-C, *Guidance, Navigation, and Control Conference*, Baltimore, MD, 07-10 August 1995.
206. Durham, W.C. and Bordignon, K.A., "Multiple Control Effector Rate Limiting," AIAA Paper 95-3208-CP, Guidance, *Navigation, and Control Conference*, Baltimore, MD, 07-10 August 1995.
207. Buffington, J.M., "Tailless Aircraft Control Allocation," AIAA Paper 97-3605, Guidance, Navigation, and Control Conference, New Orleans, LA, 11-13 August 1997.
208. Page, A.B. and Steinberg, M.L., "A Closed-Loop Comparison of Control Allocation Methods," AIAA Paper 2000-4538, *AIAA Guidance, Navigation, and Control Conference and Exhibit*, Denver, CO, 14-17 August 2000.

209. Ikeda, Y. and Hood, M., "An Application of L1 Optimization to Control Allocation," AIAA Paper 2000-4566, *AIAA Guidance, Navigation, and Control Conference and Exhibit*, Denver, CO, 14-17 August 2000.
210. Goodrich, K.H., Sliwa, S.M., and Lallman, F.J., "A Closed-Form Trim Solution Yielding Minimum Trim Drag for Airplanes With Multiple Longitudinal-Control Effectors," NASA TP 2907, May 1989.
211. Kolk, W.R., "Modern Flight Dynamics," *Prentice-Hall*, 1961.
212. Babister, A.W., "Aircraft Stability and Control," *Pergamon Press*, 1961.
213. Woodcock, R.J. and Drake, D.E., "Estimation of Flying Qualities of Piloted Airplanes," AFFDL-TR-65-218, April 1966.
214. McLean, D., "Automatic Flight Control Systems," First Edition, Prentice Hall International Series in Systems and Control Engineering, *Prentice Hall International Ltd.*, 1990.
215. Stevens, B.L. and Lewis, F.L., "Aircraft Control and Simulation," First Edition, *John Wiley & Sons, Inc.*, 1992.
216. Brockhaus, R., "Flugregelung," First Edition, *Springer-Verlag*, 1994.
217. Hancock, G.J., "An Introduction to the Flight Dynamics of Rigid Aeroplanes," First Edition, Ellis Horwood Series in Mechanical Engineering, *Ellis Horwood Limited*, 1995.
218. Russell, J.B., "Performance & Stability of Aircraft," First Edition, *Arnold*, 1996.
219. Schmidt, L.V., "Introduction to Aircraft Flight Dynamics," First Edition, *AIAA Education Series*, AIAA, 1998.
220. Phillips, W.F., "Phugoid Approximation for Conventional Airplanes," Vol. 37, No. 1, *AIAA Journal of Aircraft*, January-February 2000, pp. 30–36.
221. Phillips, W.F., "Improved Closed-Form Approximation for Dutch Roll," Vol. 37, No. 3, *AIAA Journal of Aircraft*, May-June 2000, pp. 484.490.
222. Burdun, I.Y. and Parfentyev, O.M., "Analysis of Aerobatic Flight Safety Using Autonomous Modeling and Simulation," SAE Paper 2000-01-2100, *2000 Advances In Aviation Safety Conference & Exposition*, April 2000.
223. Burdun, I.Y., "Virtual Test and Evaluation of Air France Concorde Flight No. AF4590," Preliminary Case Study, Atlanta, 26 July 2000.
224. Chudoba, B. and Burdun, I.Y, "Virtual Test and Evaluation of Air France Concorde Flight No. AF4590," Presentation at Fairchild Dornier, Oberpfaffenhofen, 27 July 2000.
225. Leyman, C.S., "Concorde Flight Mechanics/Aircraft Sizing," Presentation at the Future Projects Office, British Aerospace Airbus, Filton, 4 February 1997.
226. Nicholls, K., "Critical Flight Cases for Handling Qualities," Memorandum B57M/SST/KPN/11890, Future Projects, British Aerospace Airbus, 2 May 1996.
227. Le Tron, X., "Handling Qualities Requirements," Memo 822.008/97, AI/LE-D, Airbus Industrie, 25 April 1997.
228. Bennett, F., Priestly, and Chudoba, B., Personal Communication, Concorde Training Centre, British Aerospace Airbus, Bristol/Filton, United Kingdom, 29 May 1996.
229. Hammer, J., Cros, T., and Chudoba, B., Personal Communication, Flight Test Division, Airbus Industrie, Toulouse, France, 11 June 1996.
230. Morton, R.F. and Chudoba, B., Personal Communication, Concorde Simulator, British Aerospace Airbus, Bristol/Filton, United Kingdom, 26 June 1996.
231. Graeber, U. and Chudoba, B., Personal Communication, College of Aeronautics, Cranfield University, 20–21 August 1996.
232. Pacull, M., Hugo, F., Druot, T., Irvoas, J., Smith, H., and Chudoba, B., Personal Communication, Aérospatiale Aéronautique Airbus, Toulouse, 10 September 1996.
233. Green, P., Miller, A., Reid, S., Hyde, L., Smith, H., and Chudoba, B., Personal Communication, Future Projects Office, British Aerospace Airbus, Bristol/Filton, United Kingdom, 19 September 1996.
234. Leyman, C.S., Hyde, L., Haddrell, A., Nicholls, K., and Chudoba, B., Personal Communication, Future Projects Office, British Aerospace Airbus, Bristol/Filton, United Kingdom, 4 February 1997.

235. Khaski, E., Irvoas, J., and Chudoba, B., Personal Communication, Aérospatiale Aéronautique Airbus, Toulouse, 11 February 1997.
236. Morton, C., Britton, D., and Chudoba, B., Personal Communication, Concorde Simulator, British Aerospace Airbus, Bristol/Filton, United Kingdom, 21 February 1997.
237. Bailey, R., Morton, C., Gaudrey, J., Graeber, U., and Chudoba, B., Personal Communication, Concorde Flight Simulator Session, British Aerospace Airbus, Bristol/ Filton, United Kingdom, 27 February 1997.
238. Green, P., Morton, C., and Chudoba, B., Personal Communication, Concorde Simulator Logic, Concorde Simulator, British Aerospace Airbus, Bristol/Filton, United Kingdom, 6 March 1997.
239. Perrin, K.M., Reid, J., and Chudoba, B., Personal Communication, Civil Aviation Authority (CAA), Gatwick Airport, United Kingdom, 24 March 1997.
240. Rauscher, E., Smith, B., Hammer, J., and Chudoba, B., Personal Communication, SATIC-Beluga, Toulouse, 2 April 1997.
241. Chudoba, B., "Investigation of Inherent Slender-Body Characteristics Using the CONCORDE Simulator," CoA Report NFP0104, Department of Aerospace Technology, College of Aeronautics, Cranfield University, 27 February 1997.
242. Chudoba, B., "Stability & Control Aerospace Vehicle Design and Test Condition Matrix," Technical Report EF-039/96, Daimler-Benz Aerospace Airbus, September 1996.
243. Regis, Y., "A330/A340 Joint Certification Basis," AI/EA-A 414.000/89, Issue 4, Airbus Industrie, July 1994.
244. Anon., "Concorde TSS Standards," Avion de Transport Supersonique, Supersonic Transport Aircraft, Part 3 - Issue 4 - Flying Qualities, Part 7-3 - Issue 3 - Flying Controls, The Air Registration Board, 1969–1976.
245. Anon., "Flying Qualities of Piloted Airplanes – Military Specification," MIL-F-8785C, 1980.
246. Anon., "Flying Qualities of Piloted Vehicles – Military Standards," MIL-STD-1797A, 1990.

Chapter 5
'*AeroMech*'—Conception of a Generic Stability and Control Methodology

5.1 Introduction

Having constructed the foundation of the research undertaking as documented throughout Chaps. 1–4, the generic stability and control methodology *AeroMech* and key algorithms are presented in this chapter. It has been the primary development target throughout the research undertaking, to strive towards a feasible framework of a generic methodology. As a consequence, it has been expected that the methodology developed will represent a 'sparse matrix',[1] but capable of demonstrating its overall logic, functionality, flexibility, and potential to cope with state-of-the-art and future applications.

5.2 Methodology Concept

The overall objective of the current research undertaking has been the development of a generic stability and control methodology with the capability, to evaluate control power design issues (CE-sizing) of fixed wing aircraft. The range of aircraft configurations to be considered includes subsonic to hypersonic designs of symmetric layout (tail-aft configuration [TAC], tail-first configuration [TFC], three-surface configuration [TSC], flying-wing configuration [FWC]), and of asymmetric layout (oblique-wing configuration [OWC], and oblique flying-wing configuration [OFWC]). In fact, the methodology is not limited to the above mentioned vehicle configurations and concepts.

AeroMech has to be functional in two different calculation modes. Its main capability and strength is accessible when integrated into a multidisciplinary design environment. Furthermore, it must be possible to execute *AeroMech* in the

[1]The transformation of the *AeroMech* methodology conception into an executable software has been beyond the current research undertaking.

B. Chudoba, *Stability and Control of Conventional and Unconventional Aerospace Vehicle Configurations*, Springer Aerospace Technology,
https://doi.org/10.1007/978-3-030-16856-8_5

stand-alone mode, independent from the synthesis environment, which is the principal mode for validation and calibration of the calculation routines.

The method shall be able to evaluate stability and control aspects through two successive complexity levels, both residing at conceptual design. The first level requires a minimum of input information to determine the design space available, the second step subsequently is capable of delivering a first competitive design proposal at a more refined design resolution. It is absolutely vital that a consistent analytical approach is used for both steps, to avoid the implications of method switching.

5.2.1 AeroMech *Logic—Flowchart*

Any computer-based analysis functions according to the sequence *input*, *analysis*, *output* (IAO). As a consequence, the *AeroMech* flowchart is presented along the sections *Input-File*, *Analysis—Aerodynamics*, *Analysis—Stability and Control*, and *Output File*.

5.2.1.1 Input File

The input file, see Fig. 5.1, contains the following information:

1. *Design-Constraining Flight Conditions* (*DCFC*): The generic set of DCFCs is prepared, when the individual DCFCs are defined with selection of CE-critical combinations of Configuration Setting (CS), Flight Condition Variable (FCV), and Failure Case (FC)[2]. Reference [1] is set up to be a reference guide for the selection of flight test relevant settings of these variables.
2. *Constraints*: Design guidelines and certification requirements have to be defined, to serve as design constraints. Quantified design guidelines are manufacture-specific design rules (e.g., limitation on tailplane aspect ratio to avoid tail stall), which often reflect a specific design philosophy. Quantified certification requirements are based on either the JAR/FAR and/or the MIL Specs. Although not easily quantifiable, it is however clear which limitations have to be numerically defined (static-, dynamic-, and manoeuvre stability boundaries, etc.).
3. *Control Allocation*: It is necessary to select a control allocation logic for aircraft with redundant CEs. In case of such redundance, a variety of strategies exist for stabilising, controlling, and trimming the aircraft.

[2]Any flight condition requires a certain configuration setting (CS) like the position of the landing gear, setting of the high-lift devices, etc. Definition of flight-condition variables (FCV) includes parameters like Mach number, altitude, cross-wind component, etc. The category failure conditions (FC), specifies: engine failure, hydraulic failure, etc.

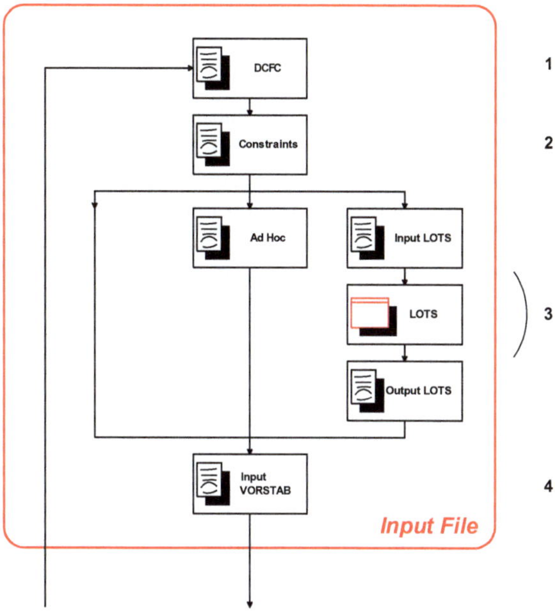

Fig. 5.1 *AeroMech* flowchart —input file definition

(i) *None*: It is assumed that a single CE is responsible to generate forces and moments for one aircraft axis only. The control allocation (CA) problem does not exist.

(ii) *Ad Hoc*: Ad hoc control allocation settings for the LoCE, DiCE, and LaCE are pre-selected to reduce the static undetermined system to a determined system. These settings are either based on experience (statistics) or are alternatively based on some theoretical foundation.

(iii) *LOTS*: The *Linear Optimum Trim Solution* (LOTS) developed by Goodrich et al. [2] is selected to automatically allocate longitudinal controls for minimum trim drag (the first iteration loop has to assume a CE setting).

The information to be obtained with the control allocation (CA) logic consists of start values for the deflection angles of individual CEs (LoCE, DiCE, and LaCE). The automatic CA sequence LOTS is, however, only concerned with LoCEs. No automatic and generic CA-sequence has been developed within this research undertaking for allocating the DiCEs and LaCEs.

4. *VORSTAB Model Setup*: The input file belonging to the aerodynamic estimation code VORSTAB has to be prepared. It specifies the flight vehicle geometry for the determination of the aerodynamic influence coefficients. Two input-options have to be considered. In case of a pre-defined configuration (geometry, positioning, and hinge lines), no further geometry modelling activity is required. If the CE geometry has not yet been specified, a CE start layout has to be selected. In the multidisciplinary context, this information requires a re-evaluation of the structural concept, to adjust weight and subsequently the c.g. position and inertias.

5.2.1.2 Analysis—Aerodynamics

The computational aerodynamic code has to predict the aerodynamic influence coefficient matrix [AIC] for subsonic and supersonic speeds.[3] Overall, a specific sequence is employed to obtain a trimmed aerodynamic data set, and to finally predict stability derivative information.

The logic developed for *AeroMech*, see Fig. 5.2, is dependent on VORSTAB's internal programming structure (code development history). An investigation of the source code has revealed inconsistencies related to the internal stability derivative estimation sequence, see Chudoba [3]. It has been decided, that it is more time effective to perform specific code runs, than to modify the VORSTAB source code within the time frame given.

1. *Principal VORSTAB Run*: This block represents the IAO-sequence to execute VORSTAB with the principal input data set, which is the DCFC under investigation.
2. $\Delta\delta_I$ *Perturbation VORSTAB Runs*: Additional IAO VORSTAB runs are performed with CE perturbations, $\Delta\delta_i\Delta\delta_I$, for the estimation of control derivatives (α, β = constant). Each CE must be perturbed separately (separate computer run), to cleanly isolate the perturbation effects on the aerodynamic coefficients.
3. $\Delta\alpha_I$ *Perturbation VORSTAB Run*: An additional IAO VORSTAB run is required with $\Delta\alpha_i$ perturbations, for the estimation of the α-dependent derivatives (δ_i = constant).
4. Δu_I *Perturbation VORSTAB Run*: An additional IAO VORSTAB run is required to estimate the tuck derivative and others of interest (δ_i = constant).
5. Δxyz_I *Perturbaton VORSTAB Runs*: This IAO VORSTAB run symbolises further perturbation runs, to estimate derivatives of interest.
6. *Derivative Estimation*: Block 6 contains the estimation sequence of linearised derivatives for the above performed perturbation runs, by estimating the coefficient increments and division by the appropriate perturbation quantity (central difference derivatives).
7. *Aerodynamic Data Set*: During the first iteration loop, step 7 represents an untrimmed aerodynamic data set without thrust effects,[4] which will later converge towards a trimmed aerodynamic data set.

It must be noted, the above process estimates derivative information by perturbing the reference flight condition. As can be seen, no sideslip angle perturbation is required, since VORSTAB estimates the β-derivatives internally, see Lan [4].

[3]The selected aerodynamic estimation code VORSTAB predicts subsonic and supersonic aerodynamics. A different method needs to be employed if high supersonic and hypersonic aerodynamic data is required.

[4]Thrust effects are taken into account with the static (trim) 6-DOF EOM and the dynamic (small perturbation) 6-DOF EOM.

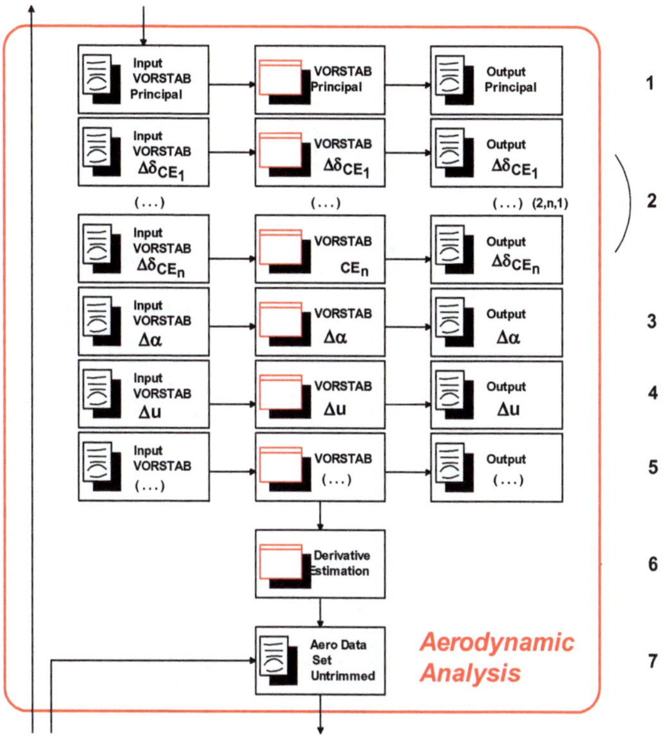

Fig. 5.2 *AeroMech* flowchart—aerodynamic analysis

5.2.1.3 Analysis—Stability and Control

The stability and control analysis sequence evaluates the generic set of DCFCs, by solving first the six degree-of-freedom (6-DOF) trim equations of motion (EOM), and in a second step the 6-DOF small perturbation EOM. This allows for the integration of static, dynamic, and manoeuvre stability requirements with control power, while taking design guidelines and certification requirements into account. Clearly, both flight condition complexity levels (1st-level and 2nd-level) are evaluated, using the same stability and control analysis sequence (an exception are the dynamic modes). The degree of inherent vehicle stability has a primary effect on hardware sizing, thus on performance and on certification. However, an integrated FCS design is clearly beyond the scope of the present investigation. Instead, the FCS is emulated using the equivalent derivative approach. This allows the investigation of aircraft configurations and concepts of any level of stability during conceptual design. The stability and control analysis sequence developed for *AeroMech* is shown in Fig. 5.3.

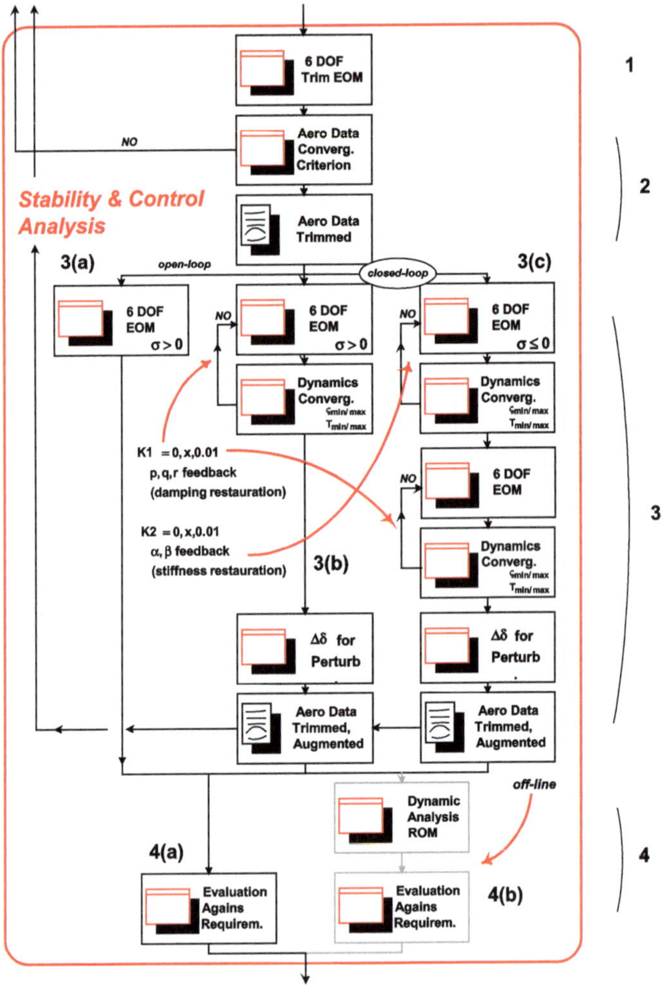

Fig. 5.3 *AeroMech* flowchart—stability and control analysis

1. *Static 6-DOF EOM*: The static (trim) 6-DOF EOMs are solved for individual DCFCs.[5] This process trims the aircraft for the particular DCFC as prepared in the input file. The *control power required* by each DCFC, taking design rules and certification constraints into account, and the *control power available*, are

[5]It is feasible to trim an indifferent or even unstable aircraft by solving the trim 6-DOF EOM, since no perturbation is disturbing the equilibrium. In contrast, it is not possible to solve the dynamic 6-DOF EOM for an unstable aircraft without a SAS in place, since the transient response to small perturbations about the reference flight condition is of divergent character.

evaluated here. Clearly, the activity of solving the trim EOMs must be seen as the primary definition phase for the flight vehicle's CEs.

2. *Trimmed Aerodynamic Data Set*: A convergency criterion checks, if the aerodynamic data set, used to solve the static 6-DOF EOM, is a trimmed or untrimmed data set. The criterion compares the CE deflection required to trim the aircraft for the particular DCFC with the CE deflection initially assumed to calculate the aerodynamic data set. In case of an untrimmed data set, the modified trim data (CE deflections to trim, trim angle-of-attack, etc.) calculated by the trim EOMs are fed back into the input file, and a new data set is estimated which is compatible with the overall aircraft trim state. It is expected that the trimmed aerodynamic data set is obtained after 1–2 iteration loops.
   Evaluating a DCFC with the trim EOMs produces some control power information for an initial sizing of the primary CEs. However, in case of a relaxed stable or unstable aircraft, the following steps are required to restore the inherent airframe stability characteristics (stiffness and damping) by using a FCS emulation, to enforce compliance with the constraints defined in the input file. Finally, the CE proposal is fine-tuned by evaluating the dynamic modes.
3. At first it is required to classify the type of FCS employed. Path 3(a) describes the open-loop aircraft ($\sigma_1 > 0$), whereas paths 3(b) and 3(c) simulate the closed-loop aircraft: 3(b) relaxed static stability ($\sigma_2 > 0$, $\sigma_2 < \sigma_1$), and 3(c) indifferent or unstable vehicle ($\sigma_3 = 0$, $\sigma_4 < 0$).
   3(a) Open-Loop Aircraft ($\sigma_1 > 0$): The 6-DOF small perturbation EOM are solved for the open-loop aircraft. Since no stability augmentation function logic is in place, the aerodynamic data set does not change.
   3(b) Closed-Loop Aircraft ($\sigma_2 > 0$): This path is valid for an aircraft with relaxed but still positive stability for either of the three axes. The stability augmentation system (SAS) for each axis has only damper function.[6] The coupled small perturbation 6-DOF EOMs are augmented using the basic control law, see Eq. (4.6), with classical feedback variables p, q, and r. The iteration starts with a gain $K_1 = 0$. The damping characteristics of either the Short Period Oscillation (SPO) or Dutch Roll (DR) mode are estimated and compared with the constraints defined in the input file. If not satisfactory, the gain is succeedingly increased until the dynamic characteristics comply with the design rules and certification requirements specified.
   The sequence estimates an appropriate gain, which restores the damping characteristics of the design in compliance with the constraints defined in the input file. It is clear that the pre-selection of a 'generic' control law, using classical feedback variables, has restrictions. However, this approach intends to augment inherent airframe stabilities for the purpose of estimating control power, rather

[6]In case of $\sigma > 0$, stiffness restoration is not obligatory compared to the case $\sigma \leq 0$, see 3(c). If required, paths 3(b) and 3(c) may be defined identical in structure. In the present context, however, path 3(b) has been chosen without stiffness augmentation for simplicity reasons, see Fig. 5.3.

than to assess the handling quality issue.[7] Therefore, the emulation of a SAS with only a simple gain should idealise even the most complex SAS to a satisfactory degree. As a consequence, the approach chosen delivers sufficient information for a follow-on selection of actuator bandwidth, -frequency, -weight, etc.

The next step estimates the control power requirement for stability augmentation (damping) in the three axes, since positive static stability is required for all flight phases. The following relations have been proposed by Lee et al. [5] as a first-order approximation only, to determine the control power required to satisfy the pitch-, yaw-, and roll accelerations imposed by the static stability and peak angle-of-attack and angle-of-sideslip deviations.

$$\dot{q} = f_1\{\delta_{CE}, Mach, Altitude, \alpha, \ldots\} = \frac{M_{\alpha_{SAS}}}{I_y}\Delta\alpha \tag{5.1}$$

$$\dot{r} = f_2\{\delta_{CE}, \dot{p}, Mach, Altitude, \alpha, \ldots\} = \frac{N_{\beta_{SAS}}}{I_x}\Delta\beta \tag{5.2}$$

$$\dot{p} = f\{\delta_{CE}, \dot{r}, Mach, Altitude, \alpha, \ldots\} = \frac{L_{\beta_{SAS}}}{I_x}\Delta\beta \tag{5.3}$$

For a given level of $\Delta\alpha$ and $\Delta\beta$ disturbance, the control power needed to restore those perturbations back to the original trim state can be obtained by solving Eqs. (5.1)–(5.3) for the CE deflection angles required. For more detail about the approximate formulation of the augmented derivatives $M_{\alpha\ SAS}$, $N_{\beta\ SAS}$, and $L_{\beta SAS}$ in (5.1)–(5.3), see Lee et al. [5]. Clearly, the formulation of the above equations for the asymmetric aircraft leads to the coupled 6-DOF dynamic EOM, where only a full dynamic simulation can comprehend the control power required for stability augmentation at step 4(a).

Having augmented the stability derivatives of interest, the aerodynamic data set requires modification. A feedback loop modifies the initial data set for the specific DCFC under investigation. Clearly, the trim state of the aircraft has changed with augmentation of the aerodynamic data set, resulting in modified dynamic stability characteristics. Thus, it is required to re-trim the aircraft by starting a follow-on iteration loop by solving the closed-loop 6-DOF trim EOM. The process converges to a trimmed and damping-augmented aerodynamic data set.

3(c) Closed-Loop Aircraft ($\sigma_3 = 0, \sigma_4 < 0$): This path is valid for indifferent or unstable aircraft. The stability augmentation system (SAS) has to restore stiffness *and* damping for the axis it applies to.

[7]The handling quality issue is a higher fidelity problem.

At first, the coupled small perturbation 6-DOF EOMs are augmented to restore stiffness using the basic control law with the classical feedback variables α and β, see Eq. (4.6). The iteration starts with a gain $K_2 = 0$. The stiffness characteristics are estimated and compared with the constraints defined in the input file. If not satisfactory, the gain is succeedingly increased until the stability characteristics comply with the design rules and certification requirements specified.

Next, the damping characteristics are restored in analogy to 3(b), using p, q, r feedback variables. The coupled small perturbation 6-DOF EOMs are augmented to restore damping using the basic control law with the classical feedback variables p, q, and r. The iteration starts with a gain $K_1 = 0$. The damping characteristics of either the SPO (Short Period Oscillation) or DR (Dutch Roll) mode are estimated and compared with the constraints defined in the input file. If not satisfactory, the gain is succeedingly increased until the dynamic characteristics comply with the design rules and certification requirements specified.

The next step estimates the control power requirement for stability augmentation (stiffness and damping) in analogy to 3(b).

Having augmented the stability derivatives of interest, the aerodynamic data set requires modification. A feedback modifies the initial data set for the DCFC under investigation. Clearly, the trim state of the aircraft has changed with augmentation of the aerodynamic data set. Thus, it is required to re-trim the aircraft by starting a follow-on iteration loop by solving the closed-loop 6-DOF trim EOM. The process converges to a trimmed stiffness- *and* damping augmented aerodynamic data set.

4. 4(a) 6-DOF EOM Dynamic Mode Evaluation (On-Line): Steps 3(a)–3(c) have solved the small perturbation 6-DOF EOM. The information generated is used to describe the transient response of the aircraft about the trimmed flight condition following a small input disturbance. Well-known techniques are used to solve and interpret the perturbed state 6-DOF EOM. No further detail is given about these techniques within the present research undertaking.

   4(b) Reduced-Order Model Dynamic Mode Evaluation (Off-Line): Reduced-order models (ROM) are solved in a parallel off-line mode. This information enables the designer to gain physical insights into the dynamic mode drivers. Clearly, this information does not contribute quantitatively, but it contributes qualitatively to increase physical visibility of the automated stability and control analysis sequence performed with step 4(a).

5.2.1.4 Output File

The output file is shown in Fig. 5.4. Any control power assessment study has to ensure harmonisation between control power and flight vehicle stability. The balance between those two stability and control key parameter-sets is considered the primary outcome provided by *AeroMech*.

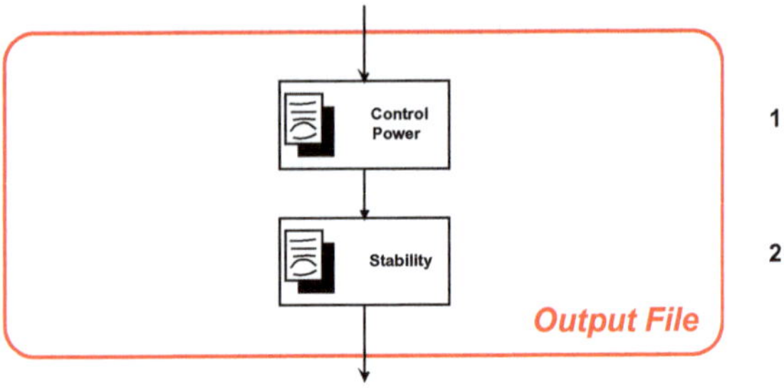

Fig. 5.4 *AeroMech* flowchart —output file

1. *Control Power*: At first, the primary control power assessment results are presented. Control power information is provided with (a) the volume coefficient (geometry), (b) stability derivative coefficients (aerodynamics), and (c) the CE deflection angle (operation) for a trimmed aircraft. This information is primarily used to define a layout for the CEs.
2. *Stability*: This file delivers the static-, dynamic-, and manoeuvre stability information.

The above information about control power and stability is available for each individual DCFC investigated. In a final step, the DCFC which requires the largest CE deflection for a given volume coefficient and aerodynamic efficiency, clearly defines the CE.

5.2.2 Synopsis of Process Logic, Information Flow, and Calculation Algorithms

Table 5.1 summarises *AeroMech*'s process logic and information flow along the complete flow chart as shown schematic in Fig. 5.5. Further information to process logic and information flow is given in Chap. 6 along selected case studies.

Table 5.1 Summary of *AeroMech* process logic and information flow

Step	Process
[1]	Interface to user or multi-disciplinary design environment (*AeroMech* input).
[2]	Input file complete; start calculation of aerodynamic data set.
[3]	Provide the initial untrimmed and unaugmented aerodynamic data set as input for stability and control calculations.
[4]	Iteration sequence to obtain a trimmed aerodynamic data set.
[5]	Restoration of airframe stability for closed-loop aircraft; final delivery of trimmed and augmented aerodynamic data set.
[6]	Complete analysis of control power and static-, dynamic-, and manoeuvre stability.
[7]	Interface to user or multi-disciplinary design environment (*AeroMech* output).

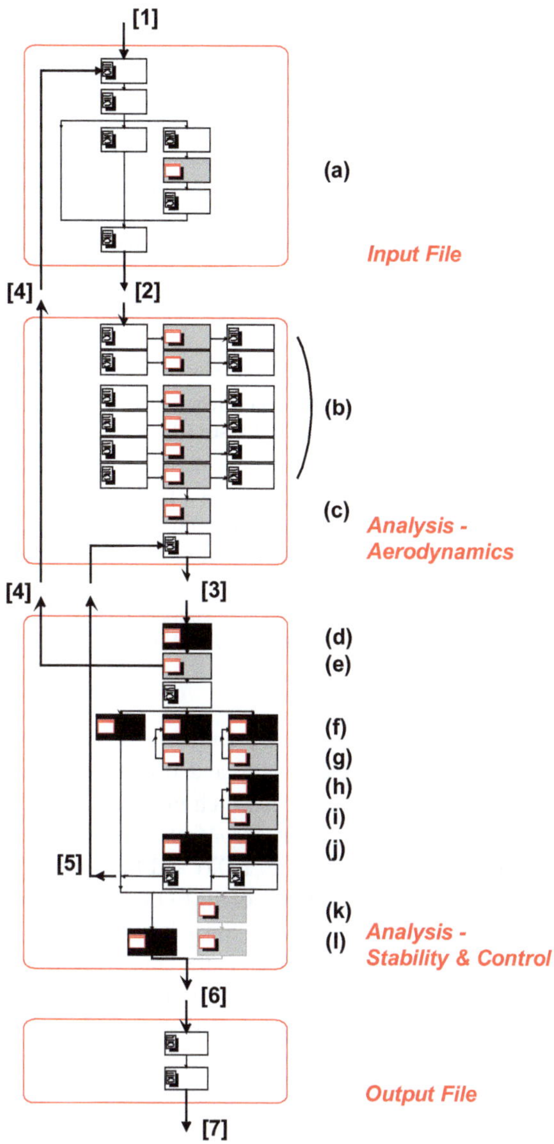

Fig. 5.5 *AeroMech* flowchart—illustration of information flow and emphasizing of calculation routines

Having presented the generic stability and control methodology concept *AeroMech*, the following reviews the key calculation algorithms pivotal in realising the idea of the generic method. As has been mentioned before, the limited time frame of the research undertaking has been utilised to concentrate predominantly on the development of the generic conception, rather than to compromise the initial research objective in favour of a non-generic executable software code.

Table 5.2 *AeroMech* calculation algorithms and development status

Step	Algorithm
(a)	Control allocation logic LOTS (Linear Optimum Trim Solution), see [2]; no further development.
(b)	Aerodynamic estimation utilising VORSTAB, see [4]; no further development.
(c)	Derivative estimation using central difference derivative approach; software development issue; no further development.
(d)	**Coupled 6-DOF trim EOM; derivation see this chapter and Appendix A.11.**
(e)	Convergency criterion; software development issue; no further development.
(f)	**Coupled 6-DOF small perturbation EOM; derivation see this chapter and Appendices A.9 and A.10.**
(g)	Stiffness and damping restoration; software development issue; no further development.
(h)	**Coupled 6-DOF small perturbation EOM; derivation see this chapter and Appendices A.9 and A.10 (see (f)).**
(i)	Damping restoration; software development issue; no further development.
(j)	**SAS control power allocation; 6-DOF dynamic EOM or Eqs.** (5.1) **to** (5.3)**; see this chapter and Appendices A.9 and A.10.**
(k)	Dynamic stability evaluation using reduced order models; see Table 4.9; no further development.
(l)	**Transient aircraft response; coupled 6-DOF small perturbation EOM; see this chapter and Appendices A.9 and A.10.**

Figure 5.5 highlights the calculation routines embedded in *AeroMech*. Table 5.2 summarises their functionality and whether those algorithms are developed in the framework of the present research undertaking.

Table 5.2 indicates that *AeroMech*'s key stability and control algorithms are the coupled 6-DOF trim (static) EOM and the coupled 6-DOF small perturbation (dynamic) EOM. The derivation of the EOM is summarised in Sect. 5.3 and presented in full in Appendices A.9–A.11. Since it has not been feasible to deliver an executable software version of *AeroMech* in the context of this research undertaking for reasons outlined before, Fig. 5.5 presents a functional concept ('skeleton') of the *AeroMech* methodology with the key calculation algorithms in place ('sparse matrix').

5.3 Algorithm—Stability and Control Mathematical Modelling

This chapter merges the understanding gained throughout Chaps. 1–4 into a feasible generic methodology concept (Sect. 5.2) and a mathematical algorithm (Sect. 5.3). Overall, the asymmetric-type aircraft has been selected as the development 'benchmark vehicle', because of its unequaled development potential, the inherent inclusion of the range of symmetric aircraft types, and the fact that the majority of critical flight conditions for the design of controls are asymmetric flight conditions. Clearly, the primary research aim is the development of an analytical framework with the capability to trim the aircraft in six degrees-of-freedom, to consider the problem of control allocation, and to assess the potential of relaxed static stability.

There are two general flight conditions for which solutions of the equations of motion (EOM) are of primary interest. The steady state EOM form the basis for studying vehicle *controllability* problems (control power), whereby the perturbed state EOM form the basis for studying aircraft *dynamic stability and response* problems and *automatic flight control theory and application*. The present context is only concerned with studying controllability, dynamic stability and response problems.

There exist three principal approaches in analytically modelling the asymmetric aircraft type:

(a) Decoupling of the longitudinal and lateral-directional motions and neglecting the cross-coupling terms finally leads to the classical 3-DOF approach.
(b) Separation of the analysis into the longitudinal and the lateral-directional motions without decoupling (inclusion of cross-coupling terms), see Thelander [6] and Maine [7].
(c) Formulation of the fully coupled 6-DOF EOM including primary cross-coupling effects.

The implications of the above three schemes are briefly discussed. All three approaches do not demand any particular computing power. The primary issue of interest is simplicity. It is a constant quest in aircraft conceptual design, that one strives for an analytical model complicated enough to adequately represent the system. Once a certain complexity level has been surpassed, extra complication in the model almost invariably degrades the result. The major complications one can foresee are twofold:

i. The aerodynamic estimation is not adequate for the complexity level selected.
ii. Excessive computation difficulties arise.

A combined longitudinal and lateral-directional model seems, at a first glance, far more complicated than two separate ones. However, it must be realised that both approaches might use the same aerodynamic data available, thus the only complication left is of computational character. If we assume that modern numerical methods are able to solve three equations simultaneously, then six equations do not pose a specific problem. Still, the 6-DOF approach provides more opportunity for things to go wrong. However, it should be recalled that modern CFD methods, finite element (FE) methods, or simulation software, are far more complex than the method developed here. The argumentation therefore has to concentrate on the issue, of how well the 3-DOF approach represents the physics of interest, or whether or not the 6-DOF approach provides more trustworthy information with more inherent potential in its approach for future applications (provision of a properly trimmed aircraft, consideration of all flight conditions of interest, consistent static and dynamic investigations, etc.).

It has been decided, to derive the underlying mathematical framework of the generic stability and control methodology *AeroMech* based on the 6-DOF static- and dynamic equations of motion (EOM). The trim EOM and the perturbed EOM are derived in Appendices A.9–A.11. With this analytical framework in place it is possible, to evaluate all design-constraining flight conditions (DCFCs) defined in Tables 4.12–4.14.

5.3.1 Steady State Equations of Motion

Steady state flight is characterised by having zero rates of change of the linear and angular velocity components with time relative to the body-fixed axis system in an atmosphere of constant density.

$$\dot{\vec{V}} = 0 \tag{5.4}$$

$$\dot{\vec{\omega}} = 0 \tag{5.5}$$

The following flight cases have been modelled for the asymmetric aircraft type:

1. *Steady State Straight Line Flight*;
2. *Steady State Turning Flight*;
3. *Steady State Pull-Up and Push-Over Flight*;
4. *Steady State Rolling Performance*;
5. *Quasi-Steady Take-Off Rotation Manoeuver.*

The quasi-steady take-off rotation flight case can not be considered a steady state flight case, since $\dot{q} \neq 0$. However, the instant at which the rotation is evaluated ($q = 0$) permits the grouping of this flight case with the steady state flight cases. The underlying equations for modelling the above steady state flight conditions are the <u>General Euler Equations Of Motion With Spinning Rotors</u> (derivation see Appendix A.9).

$$X_A + X_T - mg \sin\theta = m\left(\dot{u}^E + qw^E - rv^E\right) \tag{5.6a}$$

$$Y_A + Y_T + mg \cos\theta \sin\phi = m\left(\dot{v}^E + ru^E - pw^E\right) \tag{5.6b}$$

$$Z_A + Z_T + mg \cos\theta \cos\phi = m\left(\dot{w}^E + pv^E - qu^E\right) \tag{5.6c}$$

$$\begin{aligned} L_A + L_T = {} & I_x\dot{p} - I_{yz}\left(q^2 - r^2\right) - I_{zx}(\dot{r} + pq) - I_{xy}(\dot{q} - rp) \\ & - \left(I_y - I_z\right)qr + qh'_z - rh'_y \end{aligned} \tag{5.7a}$$

$$\begin{aligned} M_A + M_T = I_y\dot{q} - I_{zx}\left(r^2 - p^2\right) - I_{xy}(\dot{p} + qr) - I_{yz}(\dot{r} - pq) \\ - (I_z - I_x)rp + rh'_x - ph'_z \end{aligned} \tag{5.7b}$$

$$\begin{aligned} N_A + N_T = I_z\dot{r} - I_{xy}\left(p^2 - q^2\right) - I_{yz}(\dot{q} + rp) - I_{zx}(\dot{p} - qr) \\ - \left(I_x - I_y\right)pq + ph'_y - qh'_x \end{aligned} \tag{5.7c}$$

$$p = \dot{\phi} - \dot{\psi} \sin \theta \tag{5.8a}$$

$$q = \dot{\theta} \cos \phi + \dot{\psi} \sin \phi \cos \theta \tag{5.8b}$$

$$r = -\dot{\theta} \sin \phi + \dot{\psi} \cos \phi \cos \theta \tag{5.8c}$$

$$\dot{\phi} = p + q(\sin \phi + r \cos \phi) \tan \theta \tag{5.9a}$$

$$\dot{\theta} = q \cos \phi - r \sin \phi \tag{5.9b}$$

$$\dot{\psi} = (q \sin \phi + r \cos \phi) \sec \theta \tag{5.9c}$$

$$\begin{aligned} \dot{x}_E = u^E \cos \theta \cos \psi + v^E(\sin \phi \sin \theta \cos \psi - \cos \phi \sin \psi) \\ + w^E(\cos \phi \sin \theta \cos \psi + \sin \phi \sin \psi) \end{aligned} \tag{5.10a}$$

$$\begin{aligned} \dot{y}_E = u^E \cos \theta \sin \psi + v^E(\sin \phi \sin \theta \sin \psi + \cos \phi \cos \psi) \\ + w^E(\cos \phi \sin \theta \sin \psi - \sin \phi \cos \psi) \end{aligned} \tag{5.10b}$$

$$\dot{z}_E = -u^E \sin \theta + v^E \sin \phi \cos \theta + w^E \cos \phi \cos \theta \tag{5.10c}$$

The above equations contain the following assumptions:

1. The Earth is treated flat and stationary in inertial space, thus rotational velocity is neglected.
2. The equations are valid for any orthogonal axis system fixed at the c.g. of the aircraft (body axes).
3. The aircraft is a rigid body ($\dot{I}_B = 0$), having attached to it any number of rigid spinning rotors.
4. The spinning rotors have constant angular speed relative to the body axes $\left(\dot{\vec{h}}'_B = 0\right)$. The axis of any spinning rotor is fixed in direction relative to the body axes. This assumption is valid for thrust vectoring with a movable nozzle (usual), where the thrust vector alters direction but the axes of the spinning rotors stay constant.[8]

[8]The assumption of spinning rotors with fixed axes requires to be reviewed, when applied to the OFWC with engines pivoted dependent on wing sweep adjustment during flight.

5. The wind velocity is zero, so that $\vec{V}^E = \vec{V}$.

The usual assumptions like, (i) the existence of a plane of symmetry (C_{xz}), (ii) neglection of aerodynamic cross-coupling, (iii) the absence of rotor gyroscopic effects, have not been accepted in the present context.

5.3.1.1 Steady State Straight Line Flight

Steady state straight line flight is the simplest steady flight case, since all time derivatives are zero and there is no angular velocity of the body about its c.g. ($\vec{\omega} = 0$). The kinematic equations become trivial leading to the non-linear *6-DOF Trim EOM for Steady State Straight Line Flight* written in stability axes (derivation see Appendix A.11).

X-Force:

$$mg \sin\gamma = -\begin{bmatrix} C_{D_0} + C_{D_\alpha}\alpha + \sum_{j=1}^{n} C_{D_{i_{LoCE_j}}} i_{LoCE_j} + \sum_{k=1}^{m} C_{D_{\delta_{LoCE_k}}} \delta_{LoCE_k} \\ + \sum_{j=1}^{n} C_{D_{i_{DiCE_j}}} i_{DiCE_j} + \sum_{k=1}^{m} C_{D_{\delta_{DiCE_k}}} \delta_{DiCE_k} \\ + \sum_{j=1}^{n} C_{D_{i_{LaCE_j}}} i_{LaCE_j} + \sum_{k=1}^{m} C_{D_{\delta_{LaCE_k}}} \delta_{LaCE_k} \end{bmatrix} \bar{q}S + \sum_{i=1}^{n} T_i\left(\cos\phi_{T_i}\cos\psi_{T_i}\cos\alpha + \sin\phi_{T_i}\sin\alpha\right) - \sum_{i=1}^{n} \Delta X_{D_i} \tag{5.11a}$$

Y-Force:

$$-mg \sin\phi\cos\gamma = \begin{bmatrix} C_{y_0} + C_{y_\beta}\beta + \sum_{j=1}^{n} C_{y_{i_{LoCE_j}}} i_{LoCE_j} + \sum_{k=1}^{m} C_{y_{\delta_{LoCE_k}}} \delta_{LoCE_k} \\ + \sum_{j=1}^{n} C_{y_{i_{DiCE_j}}} i_{DiCE_j} + \sum_{k=1}^{m} C_{y_{\delta_{DiCE_k}}} \delta_{DiCE_k} \\ + \sum_{j=1}^{n} C_{y_{i_{LaCE_j}}} i_{LaCE_j} + \sum_{k=1}^{m} C_{y_{\delta_{LaCE_k}}} \delta_{LaCE_k} \end{bmatrix} \bar{q}S + \sum_{i=1}^{n} T_i \cos\phi_{T_i}\sin\psi_{T_i} \tag{5.11b}$$

Z-Force:

$$-mg\cos\gamma\cos\phi = -\begin{bmatrix} C_{L_0} + C_{L_\alpha}\alpha + \sum_{j=1}^{n} C_{L_{i_{LoCE_j}}} i_{LoCE_j} + \sum_{k=1}^{m} C_{L_{\delta_{LoCE_k}}} \delta_{LoCE_k} \\ + \sum_{j=1}^{n} C_{L_{i_{DiCE_j}}} i_{DiCE_j} + \sum_{k=1}^{m} C_{L_{\delta_{DiCE_k}}} \delta_{DiCE_k} \\ + \sum_{j=1}^{n} C_{L_{i_{LaCE_j}}} i_{LaCE_j} + \sum_{k=1}^{m} C_{L_{\delta_{LaCE_k}}} \delta_{LaCE_k} \end{bmatrix} \bar{q}S + \sum_{i=1}^{n} T_i\left(-\cos\phi_{T_i}\cos\psi_{T_i}\sin\alpha + \sin\phi_{T_i}\cos\alpha\right) \tag{5.11c}$$

L-Moment:

$$0 = \begin{bmatrix} C_{l_0} + C_{l_\beta}\beta + \sum_{j=1}^{n} C_{l_{i_{LoCE_j}}} i_{LoCE_j} + \sum_{k=1}^{m} C_{l_{\delta_{LoCE_k}}} \delta_{LoCE_k} \\ + \sum_{j=1}^{n} C_{l_{i_{DiCE_j}}} i_{DiCE_j} + \sum_{k=1}^{m} C_{l_{\delta_{DiCE_k}}} \delta_{DiCE_k} \\ + \sum_{j=1}^{n} C_{l_{i_{LaCE_j}}} i_{LaCE_j} + \sum_{k=1}^{m} C_{l_{\delta_{LaCE_k}}} \delta_{LaCE_k} \end{bmatrix} \bar{q}Sb + \sum_{i=1}^{n} T_i\cos\alpha\left(-\cos\phi_{T_i}\sin\psi_{T_i}z_T + \sin\phi_{T_i}y_T\right) + \sum_{i=1}^{n} T_i\sin\alpha\left(\cos\phi_{T_i}\sin\psi_{T_i}x_T - \cos\phi_{T_i}\cos\psi_{T_i}y_T\right) \tag{5.12a}$$

M-Moment:

$$0 = \begin{bmatrix} C_{m_0} + C_{m_\alpha}\alpha + \sum_{j=1}^{n} C_{m_{i_{LoCE_j}}} i_{LoCE_j} + \sum_{k=1}^{m} C_{m_{\delta_{LoCE_k}}} \delta_{LoCE_k} \\ + \sum_{j=1}^{n} C_{m_{i_{DiCE_j}}} i_{DiCE_j} + \sum_{k=1}^{m} C_{m_{\delta_{DiCE_k}}} \delta_{DiCE_k} \\ + \sum_{j=1}^{n} C_{m_{i_{LaCE_j}}} i_{LaCE_j} + \sum_{k=1}^{m} C_{m_{\delta_{LaCE_k}}} \delta_{LaCE_k} \end{bmatrix} \bar{q}Sb + \sum_{i=1}^{n} T_i\left(\cos\phi_{T_i}\cos\psi_{T_i}z_T - \sin\phi_{T_i}x_T\right) \pm \sum_{i=1}^{n} \Delta M_{D_i} \tag{5.12b}$$

N-Moment:

$$0 = \begin{bmatrix} C_{n_0} + C_{n_\beta}\beta + \sum_{j=1}^{n} C_{n_{i_{LoCE_j}}} i_{LoCE_j} + \sum_{k=1}^{m} C_{n_{\delta_{LoCE_k}}} \delta_{LoCE_k} \\ + \sum_{j=1}^{n} C_{n_{i_{DiCE_j}}} i_{DiCE_j} + \sum_{k=1}^{m} C_{n_{\delta_{DiCE_k}}} \delta_{DiCE_k} \\ + \sum_{j=1}^{n} C_{n_{i_{LaCE_j}}} i_{LaCE_j} + \sum_{k=1}^{m} C_{n_{\delta_{LaCE_k}}} \delta_{LaCE_k} \end{bmatrix} \bar{q}Sb$$
$$+ \sum_{i=1}^{n} T_i \sin\alpha\left(\cos\phi_{T_i}\sin\psi_{T_i} z_T - \sin\phi_{T_i} y_T\right)$$
$$+ \sum_{i=1}^{n} T_i \cos\alpha\left(\cos\phi_{T_i}\sin\psi_{T_i} x_T - \cos\phi_{T_i}\cos\psi_{T_i} y_T\right) \pm \sum_{i=1}^{n} \Delta N_{D_i} \tag{5.12c}$$

Equations (5.11a–c) and (5.12a–c) are non-linear equations with respect to the state variables α, ϕ, and γ. The above two sets of steady state equations form the basis for studying vehicle controllability design-aspects in rectilinear (straight-line) flight. The mathematical model is able to simulate the following steady state straight-line flight conditions for the full range of symmetric and anti-symmetric aircraft configurations and concepts:

(A) $\gamma \geq \leq 0$, $\beta = 0$, $\phi = 0$, *Thrust Symmetry* [horizontal flight, shallow climbs & dives, glides]

(B) $\gamma \geq \leq 0$, $\beta \geq \leq 0$, $\phi \geq \leq 0$, *Thrust Symmetry* [horizontal flight, shallow climbs & dives, glides, simulated crosswind condition due to $\beta >< 0$ during de-crab, crossed controls, certain systems failed]

(C) $\gamma \geq \leq 0$, $\beta \geq \leq 0$, $\phi \geq \leq 0$, *Thrust Asymmetry* [horizontal flight, shallow climbs & dives, glides, engine failure, simulated crosswind condition due to $\beta >< 0$ during de-crab, crossed controls, certain systems failed]

Equations (5.11a–c) and (5.12a–5.12c) can be solved for any combination of the following design- and state-variables for straight-line flight:

$$\alpha, \beta, \gamma, \phi, V, \bar{\rho}, \underbrace{\sum i_{LoCE,DiCE,LaCE}}_{\substack{\textit{variable incidence} \\ \textit{stabilizer}}}, \sum \delta_{LoCE,DiCE,LaCE}, T_i, x_{T_i}, y_{T_i}, z_{T_i}, \phi_{T_i}, \psi_{T_i}$$

Since there are six equations, several of the design- and state variables listed above have to be specified before the system can be solved using iterative matrix techniques. The following example flight cases indicate the potential of Eqs. (5.11a–c) and (5.12a–c).

Case (A) ($\gamma \geq \leq 0$, $\beta = 0$, $\phi = 0$, *Thrust Symmetry*)

(i) *Cruise Trim Drag—Utilising Aerodynamic CEs*

Pre-Selection: $\beta = 0, \gamma = 0, \phi = 0, \bar{\rho}, \sum_{j=1}^{n} i_j, \sum_{i=1}^{m-1} \delta_i, V, x_{T_i}, y_{T_i}, z_{T_i}, \phi_{T_i}, \psi_{T_i}$

Numerical Solution: $\alpha_{trim}, \delta_{LoCE_{i_{trim}}}, T_i$

Problem Description: This longitudinal case simulates the cruise condition with emphasis on LoCE margins, trim drag, mis-trim when pre-selecting LoCE stabiliser settings, etc.

(ii) *Cruise Trim Drag—Utilising Thrust Vectoring CEs*

Pre-Selection: $\beta = 0, \gamma = 0, \phi = 0, \bar{\rho}, \sum_{j=1}^{n} i_j, \sum_{i=1}^{m} \delta_i, \delta_{LoCE_{i_{trim}}} = 0, V, x_{T_i}, y_{T_i}, \psi_{T_i}$

Numerical Solution: $\alpha_{trim}, z_{T_i}, \phi_{T_i}, T_i$

Problem Description: Having trimmed the aircraft longitudinally the step before using aerodynamic CEs, the longitudinal trim drag is eliminated by trimming the aircraft altering the vertical position of the thrust line and the thrust-line inclination angle. Again, this case simulates the cruise condition with emphasis on LoCE margins, trim drag, mis-trim when pre-selecting a longitudinal CE setting.

(iii) *Minimum Control Speed*

Pre-Selection: $\alpha, \beta = 0, \gamma = 0, \phi = 0, \bar{\rho}, \sum_{j=1}^{n} i_j, \sum_{i=1}^{m-1} \delta_i, x_{T_i}, y_{T_i}, z_{T_i}, \phi_{T_i}, \psi_{T_i}$

Numerical Solution: $\delta_{LoCE_{i_{trim}}}, V, T_i$

Problem Description: This longitudinal case simulates minimum control speed, trim drag, etc.

Case (B) ($\gamma \geq \leq 0, \beta \geq \leq 0, \phi \geq \leq 0$, *Thrust Symmetry*)

(i) *Trim Drag During Straight Sideslip*

Pre-Selection: $\beta, \gamma = 0, \bar{\rho}, \sum_{j=1}^{n} i_j, \sum_{i=1}^{m-3} \delta_i, V, x_{T_i}, y_{T_i}, z_{T_i}, \phi_{T_i}, \psi_{T_i}$

Numerical Solution: $\alpha_{trim}, \phi_{trim}, \delta_{LoCE_{i_{trim}}}, \delta_{DiCE_{i_{trim}}}, \delta_{LaCE_{i_{trim}}}, T_i$

Problem Description: This asymmetric flight case evaluates the 6-DOF trim CE-settings required for straight sideslip to steepen, e.g., the glide slope. Depending on the constraints imposed, a characteristic diagram may be generated or the optimum solution for minimum trim drag is estimated iteratively.

(ii) *Crosswind Landing Using Aerodynamic CE*

Pre-Selection: $\beta, \gamma, \phi \geq \leq 0, V, \bar{\rho}, \sum_{j=1}^{n} i_j, \sum_{i=1}^{m-3} \delta_i, x_{T_i}, y_{T_i}, z_{T_i}, \phi_{T_i}, \psi_{T_i}$

Numerical Solution: $\alpha_{trim}, \delta_{LoCE_{i_{trim}}}, \delta_{DiCE_{i_{trim}}}, \delta_{LaCE_{i_{trim}}}, T_i$

Problem Description: This asymmetric flight case determines the CE deflections and thrust setting required for a pre-defined cross-wind component.

(iii) *Crosswind Landing Using Thrust Vectoring System*

Pre-Selection:

$$\beta, \gamma, \phi \geq \leq 0, V, \bar{\rho}, \sum_{j=1}^{n} i_j, \sum_{i=1}^{m} \delta_i, \delta_{LoCE_{i_{trim}}} = 0,$$

$$\delta_{DiCE_{i_{trim}}} = 0,$$

$$\delta_{LaCE_{i_{trim}}} = 0, x_{T_i}, y_{T_i}, z_{T_i}, \phi_{T_i}, \psi_{T_i}$$

Numerical Solution: $\alpha_{trim}, T_i, \phi_{T_i}, \psi_{T_i}$

Problem Description: This asymmetric flight case determines the thrust line angles required to trim the pre-defined cross-wind component.

Case (C) ($\gamma \geq \leq 0, \beta \geq \leq 0, \phi \geq \leq 0$, *Thrust Asymmetry*)

(i) *Trim Drag During One-Engine Inoperative (OEI) Cruise*

Pre-Selection: $\beta = 0, \gamma = 0, \bar{\rho}, \sum_{j=1}^{n} i_j, \sum_{i=1}^{m-3} \delta_i, T_i, x_{T_i}, y_{T_i}, z_{T_i}, \phi_{T_i}, \psi_{T_i}$

Numerical Solution: $\alpha_{trim}, \phi_{trim}, \delta_{LoCE_{i_{trim}}}, \delta_{DiCE_{i_{trim}}}, \delta_{LaCE_{i_{trim}}}, V$

Problem Description: This asymmetric flight case with thrust asymmetry evaluates the 6-DOF trim CE-settings required for flight with $\beta = 0$ (minimum drag configuration). The algorithm determines the maximum cruise speed possible for the remaining thrust at the altitude defined.

(ii) *Cross Wind Landing With One-Engine Inoperative (OEI)*

Pre-Selection: $\beta \geq \leq 0, \gamma \geq \leq 0, V, \bar{\rho}, \sum_{j=1}^{n} i_j,$

$\sum_{i=1}^{m-3} \delta_i, T_i, x_{T_i}, y_{T_i}, z_{T_i}, \phi_{T_i}, \psi_{T_i}$

Numerical Solution: $\alpha_{trim}, \phi_{trim}, \delta_{LoCE_{i_{trim}}}, \delta_{DiCE_{i_{trim}}}, \delta_{LaCE_{i_{trim}}}$

Problem Description: This asymmetric flight case (thrust asymmetry and cross wind component) evaluates the 6-DOF trim CE-settings required for this worst-case scenario. The algorithm determines the maximum cruise speed possible for the remaining thrust at the altitude defined.

(iii) *Rudder Hard-Over System Failure*

Pre-Selection: $\beta, \gamma = 0, \phi, V, \bar{\rho}, \sum_{j=1}^{n} i_j, \sum_{i=1}^{m-2} \delta_i, \delta_{DiCE_{i_{\max}}},$

$T_i, x_{T_i}, y_{T_i}, z_{T_i}, \phi_{T_i}$

Numerical Solution: $\alpha_{trim}, \delta_{LoCE_{i_{trim}}}, \delta_{LaCE_{i_{trim}}}, \psi_{T_i}$

Problem Description: This asymmetric flight case simulates a 'rudder hard-over' system failure. The thrust line toe angle is determined to check the survivability of this system failure case with the availability of a thrust vectoring system.

Any other than the above example combinations of the design- and state variables can be handled with the mathematical model. Trade studies can be performed, to investigate the effects of the design drivers[9] on meeting design guidelines and certification requirements.

In general, it is possible to model a non-zero flight-path angle, γ, which corresponds to an 'instantaneous' steady-state condition only, because of the density changes with varying altitude. As a consequence, only shallow climbs and dives are

[9] An example for a primary design driver is the positioning of the c.g. The trim solution is strongly influenced by altering this parameter, since all stability derivative coefficients change.

permissible. However, modelling the flight path angle is of particular relevance, since it is of interest to determine the rate of climb- and descent performance as a function of the design- and state variables. To remain within the band of validity, it is useful to define a rate-of-climb constrained simply by specifying $V sin\gamma$, which is the z-component of the velocity vector.

Clearly, Eqs. (5.11a–c) and (5.12a–c) can be solved, in theory, for six unknowns. However, the number of solutions obtained may be infinite. Finding practical solutions depends on placing adequate constraints on the design- and state variables, and/or by reducing the number of unknowns to a smaller number. Overall, the task of finding a practical solution is an iterative or optimisation task, requiring repeated solutions of the equations. The interpretation of the results may be performed by hand or may be supported (automated) using a mathematical optimiser environment to satisfy an objective function. Clearly, since the solution may not be unique, it is up to the design experience of the user to initially specify the steady state condition to the degree, so that the trim[10] algorithm converges to a practical, if not unique, solution.

For all example flight-cases presented above, a unique trim solution is attainable when defining/calculating realistic start values/solutions:

- The stabilizer incidence angles, i_k, and the flap deflection angles, δ_i, must be consistent with attached flow considerations (tail-stall, etc.).
- The aircraft angle of attack, α_{trim}, must be consistent with attached flow considerations over the wing.
- The flight path angle, γ, must be consistent with operational and model-validity constraints.
- The thrust force, T_i, must be consistent with power available and operational criteria.
- The airspeed, V, and density, $\bar{\rho}$, must be consistent with operational considerations.

The steady state asymmetric flight cases with and without thrust symmetry (Cases B and C), are of primary importance for the design of controls. When reviewing pertinent literature throughout aviation history related to asymmetric power conditions, cross wind effects, straight sideslipping flight, and adverse yaw compensation, the difficulty of evaluating the CE design problem at conceptual design level becomes obvious, see Table 5.3.

The design-oriented approach proposed by Roskam and Anemaat [15] is considered most promising for the conventional TAC. The approach developed in the present context is similar in philosophy but capable of considering the full 6-DOF problem, thus it avoids separating the longitudinal motion from the lateral-directional motion thereby including cross-coupling effects. Figure 5.6 illustrates possible CE design scenarios concerned with asymmetric flight with and without thrust symmetry.

---

[10]*AeroMech* considers only irreversible control systems, where the overall trim state does usually not imply hinge-moment trim (trim drag), and where force trim is not relevant.

Table 5.3 Design-oriented approaches to the analysis of asymmetric flight conditions

Implementation	Reference, year	Comments
Hartman	[8, 1938]	Wind tunnel investigation of One Engine Inoperative (OEI) flight conditions. Discussion of relevant design parameters, asymmetric flight with and without angle of sideslip, and the influence of power. The effects on stability, controllability, lift, and drag are estimated. Wind-tunnel investigation only without delivering of an analytical approach.
Archbold et al.	[9, 1945]	Development of an analytical method to estimate the size of the DiCE (fin and rudder) using a 1-DOF approach of the yawing moment equation. The analysis is restricted, for simplicity, to zero bank angle throughout the motion. A curve of maximum sideslip against DiCE size is the final result.
Yates	[10, 1947]	The aerodynamics of the problem of regaining and maintaining control after engine failure is discussed qualitatively. In a first part, the transient effects related to a sudden failure of an engine are discussed. The second part is concerned with steady flight under asymmetric power, and the final part evaluates the baulked landing with one or more engines dead. The complexity of the problem is presented without delivering the means to design for it. Some comments are made with respect to novel aircraft layouts.
Baker	[11, 1948]	A mathematical exposition of working rules for the choice of key lateral-directional stability derivatives is presented. The semi-empirical analytical framework presented is based on several assumptions valid for the TAC only. The 3-DOF approach presented evaluates the DiCE and the LaCE. Having presented the underlying assumptions, the balance equations are formulated, which are checked against quantified stability criteria.
Wright	[12, 1950]	The problems involved with flight on asymmetric engine power are evaluated from a pilot's perspective. This primarily qualitative description thoroughly defines 'Flight on Asymmetric Power'. A following section discusses the importance of the 'Safety Speed'. Stability issues are evaluated for flight under asymmetric conditions. Flight techniques are presented for engine failure on TO, in flight, approach and landing. The complexity of the problem is presented without delivering the means to design for it.
Pinsker	[13, 1967]	A criterion has been developed to define a minimum acceptable value for the directional stability derivative N_v. The theory has been in agreement with observations on the high-speed BAC 221. In particular, when the bank angle is constrained by aileron control, the lateral motion degenerates into a simple directional oscillation, dependent on the 'effective' directional stability parameter. The 3-DOF theory developed is based on the concept of a partially constrained motion.
Leyman et al.	[14, 1972]	The effects of engine failures in high-speed cruise are described. This highly configuration-specific discussion presents primarily wind tunnel and flight test results of Concorde. Any design-oriented interpretation depends

(continued)

Table 5.3 (continued)

Implementation	Reference, year	Comments
		primarily on the prediction quality of stability derivative information for asymmetric flight. Engine failures in supersonic flight of this FWC are described as innocuous. The multi-disciplinary complexity of asymmetric flight is vividly illustrated, but no generic design guidelines are presented.
Roskam et al.	[15, 1994]	A practical method is presented to analyse longitudinal and lateral-directional trim problems with All Engines Operating (AEO) and One Engine Inoperative (OEI). The analytical framework separates the longitudinal motion from the lateral-directional motion. The approach has been developed for conceptual design level, taking standard assumptions for the TAC into account. Clearly, this 3-DOF approach enables the evaluation of TAC-specific stability and control aspects.
Aly	[16, 1997]	This study evaluates the effects of side wind on the aerodynamic characteristics of an aircraft model in the wind tunnel. Focus has been on estimating the effects on aircraft performance in terms of lift, drag, sideforce, pitching-, rolling-, and yawing moment, which are reproduced quantitatively. The effects of thrust asymmetry have not been investigated. No design guideline is presented, having outlined the aerodynamic coupling effects of asymmetric flight.
Grasmeyer	[17, 1998]	This study describes the estimation of stability and control derivatives using primarily DATCOM, and the establishment of an engine-out constraint based on the required yawing moment coefficient. The use of thrust vectoring and circulation control to provide additional yawing moment is also described. The engine-out case is approached with a 2-DOF model. The aerodynamic data set is assembled with the classical component build-up technique. The method is non-generic in character, thus suitable for the TAC only.
Burcham et al.	[18, 1998]	A propulsion-controlled aircraft (PCA) system is presented in which computer-controlled engine thrust provides emergency flight control. Flight test results are presented of an F-15 and MD-11 landed without using any flight control surfaces. Studies have shown that engines on only one wing can provide some flight control capability if the lateral c.g. can be shifted towards the side of the aircraft that has the operating engines. The study illustrates the feasibility of PCA.

The following distinct flight cases are discussed in Fig. 5.6:

FLIGHT CASE (A): Lack of a sidewind component, $V_{wind} = 0$, and thrust symmetry result in performance-optimal flight from *cruise* to *flare*. The control authority required throughout is minimal.

FLIGHT CASE (B): The existence of a sidewind component, $V_{wind} \neq 0$, and thrust symmetry result in the 'crab-method' during cruise with zero

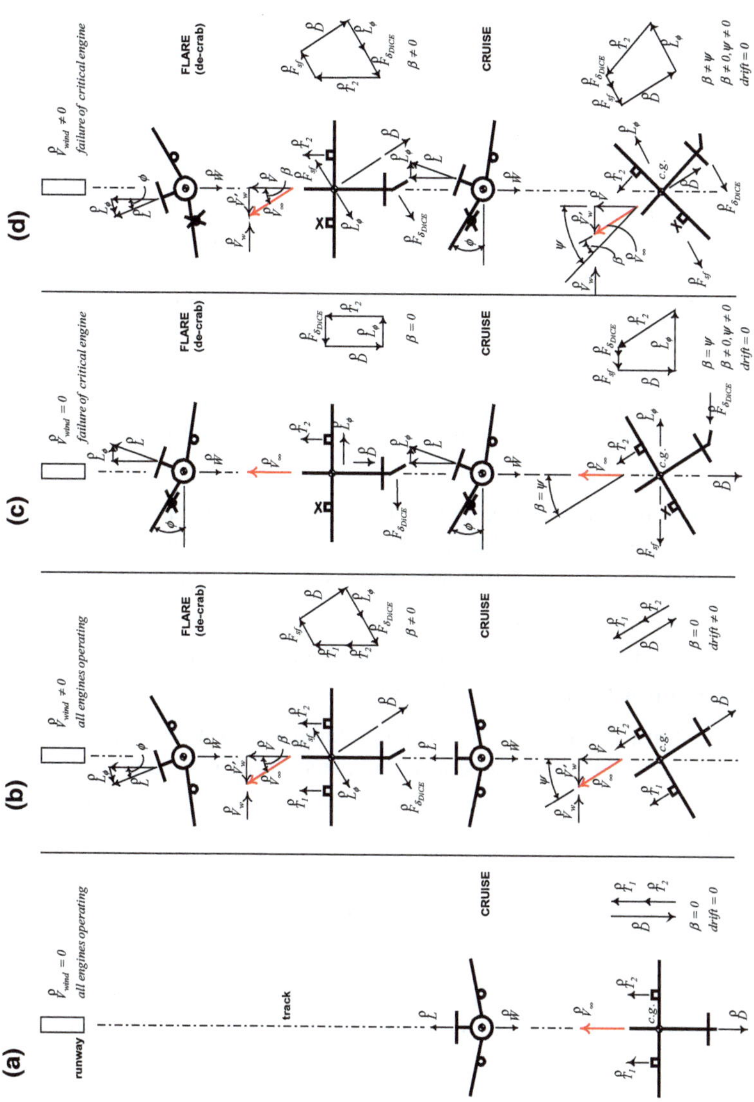

Fig. 5.6 Asymmetric-flight CE sizing scenarios qualitatively

sideslip, $\beta = 0$. Certification requirements require maintaining a straight course in crosswind conditions. The force-vector polygon indicates minimum CE deflection required. However, additional control power is required during the low dynamic pressure 'de-crab manoeuver' cross-wind landing, leading to $\beta \neq 0$.

FLIGHT CASE (C): Thrust asymmetry (critical engine failed) results in an aerodynamic sideslip angle $\beta \neq 0$ and a CE trim deflection during cruise (sideslip method). This flight case requires additional control power for trim at low dynamic pressure (low-speed) flight conditions and high thrust setting, since the sideslip angle becomes a maximum. Supplementary control power is required to de-crab the aircraft before touch-down (elimination of sideslip angle).

FLIGHT CASE (D): This flight case represents the worst case scenario with respect to control power required and control power available. The control power required to trim thrust asymmetry and to compensate for the cross wind component during the de-crab maneuver during the landing flare, usually surpasses CE-design limits.[11]

There is an infinite number of permutations of the sideslip parameters α, β, and ϕ, each individual combination influencing trim drag, ease of control, and comfort. The following flight cases with thrust asymmetry are of particular interest:

(a) $\beta = 0$, $\phi \neq 0$ [minimum drag solution but unfavorable comfort];
(b) $\beta \neq 0$, $\phi = 0$ [maximum drag solution with favorable comfort];
(c) $\beta \neq 0$, $\phi \neq 0$ [usually practiced solution].

Asymmetric thrust flight cases due to engine failure(s) on multi-engine aircraft result in drag increments on the inoperative engine(s), leading to additional forces and moments. The following force- and moment increments are taken into account in Eqs. (5.11a–c) and (5.12a–c):

$$\sum_{i=1}^{n} \Delta X_{D_i}, \sum_{i=1}^{n} \Delta M_{D_i}, \sum_{i=1}^{n} \Delta N_{D_i}$$

These force- and moment increments depend on the type of propulsive installation. It is usually acceptable to write for the total thrust-induced force and moments:

[11]The DiCE and LaCE of modern transonic aircraft are not designed to meet this flight case, since it would result in oversized CEs.

$$\underbrace{X_T}_{\substack{operative\\engine(s)}} - \underbrace{\sum_{i=1}^{n} \Delta X_{D_i}}_{in-operative\ engine(s)} \approx X_T - \underbrace{\sum_{i=1}^{n} X_T(\tau_X - 1)}_{thrust\ fraction\ of\ operative\ engine(s)} \tag{5.13a}$$

$$M_T \pm \sum_{i=1}^{n} \Delta M_{D_i} \approx M_T \pm \sum_{i=1}^{n} M_T(\tau_M - 1) \tag{5.13b}$$

$$N_T \pm \sum_{i=1}^{n} \Delta N_{D_i} \approx N_T \pm \sum_{i=1}^{n} N_T(\tau_N - 1) \tag{5.13c}$$

The terms τ_X, τ_M, and τ_N are larger than 1.0 and cover the effects of the propulsive installation. Roskam presents in [19] information for several propulsive installation schemes.

Sudden engine failure(s) lead to dynamic airframe motion-transients. These dynamic airframe responses are characterised by coupled longitudinal and lateral-directional motions. Before the steady state sideslip condition is attained, a dynamic overswing motion has to be arrested without risking CE stall. The extra control power demanded to cope with the dynamic overswing condition is accounted for by providing a sufficiently large control power margin.

It should be noted that the qualitative classification presented in Fig. 5.6 looses some of its meaning when considering asymmetric aircraft, in particular the OFWC (Oblique Flying-Wing Configuration). For the OFWC, steady state sideslip is the dominating operational flight condition. As a result, cross-wind landings may be performed without bank angle at all. Since the OFWC lacks the conventional fuselage and associated forces and moments (see force polygons in Fig. 5.6), the control power demand expected for cross wind landings is expected to be low.

5.3.1.2 Steady State Turning Flight

The following considers horizontal steady state turning flight, see Fig. 5.7.

Horizontal steady state turning flight is characterised by a constant rate of turn, leading to an angular velocity vector

$$\vec{\omega} = \vec{k}_1 \dot{\psi} \tag{5.14}$$

Since the turning manoeuver is assumed to take place in a horizontal plane, the stability x-axis lies in the same horizontal plane. The rate of turn vector $\dot{\psi}$ is perpendicular to the horizontal plane implying $p = 0$, leading to the non-linear <u>6-DOF Trim EOM for Steady State Turning Flight</u> written in stability axes (derivation see Appendix A.11).

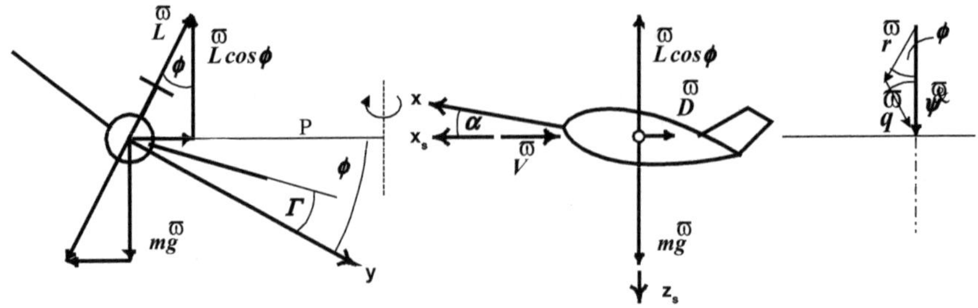

Fig. 5.7 Horizontal steady turning flight

X-Force:

$$-\frac{mg}{n}\tan\phi\,\sin\beta = -\begin{bmatrix} C_{D_0} + C_{D_\alpha}\alpha + C_{D_q}\frac{gn\bar{c}}{2V^2}\sin^2\phi + C_{D_r}\frac{g\bar{c}}{2V^2 n}\tan\phi \\ + C_{D_{i_{LoCE}}} i_{LoCE} + C_{D_{\delta_{LoCE}}}\delta_{LoCE} \\ + C_{D_{i_{DiCE}}} i_{DiCE} + C_{D_{\delta_{DiCE}}}\delta_{DiCE} \\ + C_{D_{i_{LaCE}}} i_{LaCE} + C_{D_{\delta_{LaCE}}}\delta_{LaCE} \end{bmatrix}\bar{q}S$$
$$+\sum_{i=1}^{n} T_i\left(\cos\phi_{T_i}\cos\psi_{T_i}\cos\alpha + \sin\phi_{T_i}\sin\alpha\right) - \sum_{i=1}^{n}\Delta X_{D_i} \tag{5.15a}$$

Y-Force:

$$mg\left(\frac{\tan\phi}{n} - \sin\phi\right) = \begin{bmatrix} C_{Y_0} + C_{Y_\beta}\beta + C_{Y_q}\frac{gnb}{2V^2}\sin^2\phi + C_{Y_r}\frac{gb}{2V^2 n}\tan\phi \\ + C_{Y_{i_{LoCE}}} i_{LoCE} + C_{Y_{\delta_{LoCE}}}\delta_{LoCE} \\ + C_{Y_{i_{DiCE}}} i_{DiCE} + C_{Y_{\delta_{DiCE}}}\delta_{DiCE} \\ + C_{Y_{i_{LaCE}}} i_{LaCE} + C_{Y_{\delta_{LaCE}}}\delta_{LaCE} \end{bmatrix}\bar{q}S \tag{5.15b}$$
$$+\sum_{i=1}^{n} T_i\left(\cos\phi_{T_i}\sin\psi_{T_i}\right)$$

Z-Force:

$$-mg\left(n\sin^2\phi - \cos\phi\right) = \begin{bmatrix} C_{L_0} + C_{L_\alpha}\alpha + C_{L_q}\frac{gn\bar{c}}{2V^2}\sin^2\phi + C_{L_r}\frac{g\bar{c}}{2V^2 n}\tan\phi \\ + C_{L_{i_{LoCE}}} i_{LoCE} + C_{L_{\delta_{LoCE}}}\delta_{LoCE} \\ + C_{L_{i_{DiCE}}} i_{DiCE} + C_{L_{\delta_{DiCE}}}\delta_{DiCE} \\ + C_{L_{i_{LaCE}}} i_{LaCE} + C_{L_{\delta_{LaCE}}}\delta_{LaCE} \end{bmatrix}\bar{q}S \tag{5.15c}$$
$$+\sum_{i=1}^{n} T_i\left(-\cos\phi_{T_i}\cos\psi_{T_i}\sin\alpha + \sin\phi_{T_i}\cos\alpha\right)$$

L-Moment:

$$-\frac{I_{yz}g^2(n^4\sin^4\phi-\tan^2\phi)}{V^2n^2}-\frac{(I_y-I_z)g^2\sin^3\phi}{V^2\cos\phi}+\frac{gnh'_z\sin^2\phi}{V}-\frac{gh'_y\tan\phi}{nV}$$
$$=\begin{bmatrix} C_{l_0}+C_{l_\beta}\beta+C_{l_q}\frac{gnb}{2V^2}\sin^2\phi+C_{l_r}\frac{gb}{2V^2n}\tan\phi \\ +C_{l_{i_{LoCE}}}i_{LoCE}+C_{l_{\delta_{LoCE}}}\delta_{LoCE} \\ +C_{l_{i_{DiCE}}}i_{DiCE}+C_{l_{\delta_{DiCE}}}\delta_{DiCE} \\ +C_{l_{i_{LaCE}}}i_{LaCE}+C_{l_{\delta_{LaCE}}}\delta_{LaCE}\end{bmatrix}\bar{q}Sb$$
$$+\sum_{i=1}^{n}T_i\cos\alpha\left(-\cos\phi_{T_i}\sin\psi_{T_i}z_T+\sin\phi_{T_i}y_T\right)$$
$$+\sum_{i=1}^{n}T_i\sin\alpha\left(\cos\phi_{T_i}\sin\psi_{T_i}x_T-\cos\phi_{T_i}\cos\psi_{T_i}y_T\right) \tag{5.16a}$$

M-Moment:

$$-I_{zx}\frac{g^2\tan^2\phi}{n^2u^2}-I_{xy}\frac{g^2\sin^2\phi\tan\phi}{u^2}+\frac{g\tan\phi}{nu}h'_x$$
$$=\begin{bmatrix} C_{m_0}+C_{m_\alpha}\alpha+C_{m_q}\frac{gn\bar{c}}{2V^2}\sin^2\phi+C_{m_r}\frac{g\bar{c}}{2V^2n}\tan\phi \\ +C_{m_{i_{LoCE}}}i_{LoCE}+C_{m_{\delta_{LoCE}}}\delta_{LoCE} \\ +C_{m_{i_{DiCE}}}i_{DiCE}+C_{m_{\delta_{DiCE}}}\delta_{DiCE} \\ +C_{m_{i_{LaCE}}}i_{LaCE}+C_{m_{\delta_{LaCE}}}\delta_{LaCE}\end{bmatrix}\bar{q}S\bar{c} \tag{5.16b}$$
$$+\sum_{i=1}^{n}T_i\left(\cos\phi_{T_i}\cos\psi_{T_i}z_T-\sin\phi_{T_i}x_T\right)$$
$$\pm\sum_{i=1}^{n}\Delta M_{D_i}$$

N-Moment:

$$
\begin{aligned}
& I_{xy}\frac{g^2n^2}{u^2}\sin^4\phi + I_{zx}\frac{g^2\sin^2\phi\tan\phi}{u^2} - \frac{gn\sin^2\phi}{u}h'_x \\
& = \begin{bmatrix} C_{n_0} + C_{n_\beta}\beta + C_{n_q}\frac{gnb}{2V^2}\sin^2\phi + C_{n_r}\frac{gb}{2V^2n}\tan\phi \\ + C_{n_{i_{LoCE}}} i_{LoCE} + C_{n_{\delta_{LoCE}}}\delta_{LoCE} \\ + C_{n_{i_{DiCE}}} i_{DiCE} + C_{n_{\delta_{DiCE}}}\delta_{DiCE} \\ + C_{n_{i_{LaCE}}} i_{LaCE} + C_{n_{\delta_{LaCE}}}\delta_{LaCE} \end{bmatrix}\bar{q}Sb \\
& + \sum_{i=1}^{n} T_i\cos\alpha(-\cos\phi_{T_i}\sin\psi_{T_i}z_T + \sin\phi_{T_i}y_T) \\
& + \sum_{i=1}^{n} T_i\sin\alpha(\cos\phi_{T_i}\sin\psi_{T_i}x_T - \cos\phi_{T_i}\cos\psi_{T_i}y_T) \\
& - \sum_{i=1}^{n}\Delta N_{D_i}
\end{aligned}
\tag{5.16c}
$$

Equations (5.15a–c) and (5.16a–c) are non-linear equations with respect to the state variables α, β, ϕ, and ψ. The above two sets of steady state equations form the basis for studying vehicle controllability design-aspects during horizontal turns, containing primary turn-performance parameters C_L, n, and T_i explicitly. The following horizontal turning flight conditions can be modelled for the full range of symmetric and asymmetric aircraft configurations and concepts:

(A) $\beta \geq \leq 0$, *Thrust Symmetry* [horizontal steady turns with all engines operative, crossed controls, certain systems failed]

(B) $\beta \geq \leq 0$, *Thrust Asymmetry* [horizontal steady turns with engine failure(s), crossed controls, certain systems failed]

Equations (5.15a–c) and (5.16a–c) can be solved for any combination of the following design- and state-variables for steady-state turning flight:

$$\alpha, \beta, \phi, \mathrm{n}, V, \bar{\rho}, \sum i_{LoCE,DiCE,LaCE}, \sum \delta_{LoCE,DiCE,LaCE}, T_i, x_{T_i}, y_{T_i}, z_{T_i}, \phi_{T_i}, \psi_{T_i}$$

Since there are six equations, several of the design- and state variables listed above have to be specified before the system can be solved using iterative matrix techniques. The following example flight cases indicate the potential of Eqs. (5.15a–c) and (5.16a–c).

Case (A) ($\beta \geq \leq 0$, *Thrust Symmetry*)

(i) *Load Factor Capability*

Pre-Selection: $\beta = 0, \phi, V, \bar{\rho}, \sum_{j=1}^{n} i_j, \sum_{i=1}^{m-3} \delta_i, x_{T_i}, y_{T_i}, z_{T_i}, \phi_{T_i}, \psi_{T_i}$

Numerical Solution: $\alpha_{trim}, n, \delta_{LoCE_{i_{trim}}}, \delta_{DiCE_{i_{trim}}}, \delta_{LaCE_{i_{trim}}}, T_i$

Problem Description: This longitudinal/lateral-directional flight case estimates the load factor capability with a given thrust setting.

(ii) *Horizontal Turn*

Pre-Selection: $\beta \neq 0, n, \phi, V, \bar{\rho}, \sum_{j=1}^{n} i_j, \sum_{i=1}^{m-2} \delta_i, x_{T_i}, y_{T_i}, z_{T_i}, \phi_{T_i}, \psi_{T_i}$

Numerical Solution: $\alpha_{trim}, \delta_{LoCE_{i_{trim}}}, \delta_{DiCE_{i_{trim}}}, \delta_{LaCE_{i_{trim}}}, T_i$

Problem Description: This is the flight case specific for an asymmetric aircraft configuration. The 6-DOF trim condition of the aircraft is determined.

Case (B) ($\beta \geq \leq 0$, *Thrust Asymmetry*)

(i) *Turn Into In-Operative Engine*

Pre-Selection: $\beta = 0, n, \phi, V, \bar{\rho}, \sum_{j=1}^{n} i_j, \sum_{i=1}^{m-3} \delta_i, x_{T_i}, y_{T_i}, z_{T_i}, \phi_{T_i}, \psi_{T_i}$

Numerical Solution: $\alpha_{trim}, \delta_{LoCE_{i_{trim}}}, \delta_{DiCE_{i_{trim}}}, \delta_{LaCE_{i_{trim}}}, T_i$

Problem Description: This worst-case flight condition is particularly demanding for the DiCE. It should be noted that the asymmetric aircraft exhibits different characteristics when turning to the left or to the right due to its geometric asymmetry. Then, either the left- or the right turn is the critical direction.

(ii) *Turn Performance at Altitude*

Pre-Selection: $\beta \neq 0, n, \phi, V, \bar{\rho}, \sum_{j=1}^{n} i_j, \sum_{i=1}^{m-2} \delta_i, x_{T_i}, y_{T_i}, z_{T_i}, \phi_{T_i}, \psi_{T_i}$

Numerical Solution: $\alpha_{trim}, \delta_{LoCE_{i_{trim}}}, \delta_{DiCE_{i_{trim}}}, T_i$

Problem Description: This flight case evaluates turn performance at high altitude with the remaining thrust available.

The steady pitch and yaw rates in this type of turning manoeuver are functions of load factor and bank angle. It must be noted that the climbing/descending steady turning flight demands additional longitudinal control power compared to the horizontal turning flight, see Etkin et al. [20] and Brüning et al. [21]. These cases are, however, not considered in the present context due to additional complexity involved.

If the turn is coordinated, then no net lateral acceleration acts on the aircraft. This condition implies, that in a steady level turn the aerodynamic sideforce in Eq. (5.15b) is equal to zero. However, the coordinated turn is considered a special application of Eqs. (5.15a–c) and (5.16a–c), thus, the algorithm presented allows steady 'skidding' sideslipping flight cases including thrust asymmetry.

5.3.1.3 Steady State Pull-up and Push-Over Flight

The steady state pull-up/push-over flight case is of primary interest at the bottom/top of the curved flight path with a horizontal flight path tangent in the *xz*-plane, see Fig. 5.8.

For symmetric flight we have $q \neq 0$. For pull-up and push-over flight with $\beta \neq 0$, we can write $q \neq 0$ and $r \neq 0$. It has been desirable to express the steady state angular rates as a function of the load factor *n*, leading to the non-linear *6-DOF Trim EOM for Quasi-Steady Pull-Up and Push-Over Flight* written in stability axes (derivation see Appendix A.11).

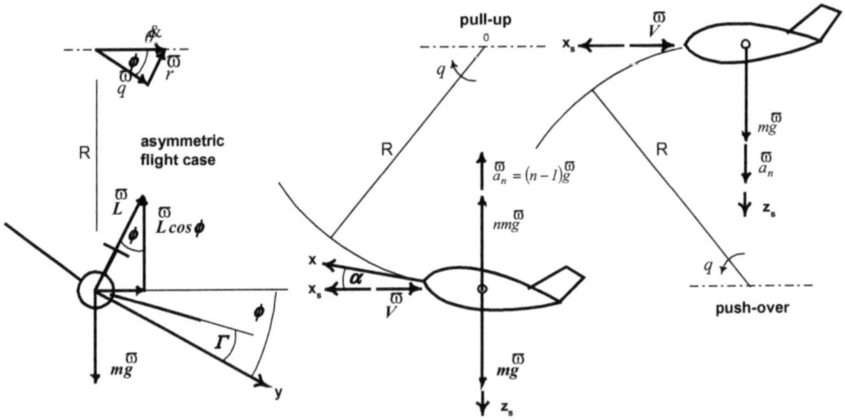

Fig. 5.8 Steady state pull-up and push-over flight

X-Force:

$$mg(n-1)\sin\phi\sin\beta = -\begin{bmatrix} C_{D_0} + C_{D_\alpha}\alpha + C_{D_q}\frac{g\bar{c}(n-1)\cos\phi}{2V^2} - C_{D_r}\frac{g\bar{c}(n-1)\sin\phi}{2V^2} \\ + C_{D_{i_{LoCE}}} i_{LoCE} + C_{D_{\delta_{LoCE}}}\delta_{LoCE} \\ + C_{D_{i_{DiCE}}} i_{DiCE} + C_{D_{\delta_{DiCE}}}\delta_{DiCE} \\ + C_{D_{i_{LaCE}}} i_{LaCE} + C_{D_{\delta_{LaCE}}}\delta_{LaCE} \end{bmatrix}\bar{q}S$$
$$+ \sum_{i=1}^{n} T_i\left(\cos\phi_{T_i}\cos\psi_{T_i}\cos\alpha + \sin\phi_{T_i}\sin\alpha\right) - \sum_{i=1}^{n}\Delta X_{D_i} \tag{5.17a}$$

Y-Force:

$$-mgn\sin\phi = \begin{bmatrix} C_{Y_0} + C_{Y_\beta}\beta + C_{Y_q}\frac{gb(n-1)\cos\phi}{2V^2} - C_{Y_r}\frac{gb(n-1)\sin\phi}{2V^2} \\ + C_{Y_{i_{LoCE}}} i_{LoCE} + C_{Y_{\delta_{LoCE}}}\delta_{LoCE} \\ + C_{Y_{i_{DiCE}}} i_{DiCE} + C_{Y_{\delta_{DiCE}}}\delta_{DiCE} \\ + C_{Y_{i_{LaCE}}} i_{LaCE} + C_{Y_{\delta_{LaCE}}}\delta_{LaCE} \end{bmatrix}\bar{q}S$$
$$+ \sum_{i=1}^{n} T_i\left(\cos\varphi_{T_i}\sin\psi_{T_i}\right) \tag{5.17b}$$

Z-Force:

$$-mgn\cos\phi = -\begin{bmatrix} C_{L_0} + C_{L_\alpha}\alpha + C_{L_q}\frac{g\bar{c}(n-1)\cos\phi}{2V^2} - C_{L_r}\frac{g\bar{c}(n-1)\sin\phi}{2V^2} \\ + C_{L_{i_{LoCE}}} i_{LoCE} + C_{L_{\delta_{LoCE}}}\delta_{LoCE} \\ + C_{L_{i_{DiCE}}} i_{DiCE} + C_{L_{\delta_{DiCE}}}\delta_{DiCE} \\ + C_{L_{i_{LaCE}}} i_{LaCE} + C_{L_{\delta_{LaCE}}}\delta_{LaCE} \end{bmatrix}\bar{q}S$$
$$+ \sum_{i=1}^{n} T_i\left(-\cos\phi_{T_i}\cos\psi_{T_i}\sin\alpha + \sin\phi_{T_i}\cos\alpha\right) \tag{5.17c}$$

L-Moment:

$$\begin{aligned}
& - I_{yz}\frac{g^2(n-1)^2\left(\cos^2\phi - \sin^2\phi\right)}{V^2} + \left(I_y - I_z\right)\frac{g^2(n-1)^2\cos\phi\,\sin\phi}{V^2} \\
& + \left(h'_z\cos\phi + h'_y\sin\phi\right)\frac{g(n-1)}{V} \\
& = \begin{bmatrix} C_{l_0} + C_{l_\beta}\beta + C_{l_q}\frac{gb(n-1)\cos\phi}{2V^2} - C_{l_r}\frac{gb(n-1)\sin\phi}{2V^2} \\ + C_{l_{i_{LoCE}}} i_{LoCE} + C_{l_{\delta_{LoCE}}}\delta_{LoCE} \\ + C_{l_{i_{DiCE}}} i_{DiCE} + C_{l_{\delta_{DiCE}}}\delta_{DiCE} \\ + C_{l_{i_{LaCE}}} i_{LaCE} + C_{l_{\delta_{LaCE}}}\delta_{LaCE} \end{bmatrix}\bar{q}S \\
& + \sum_{i=1}^{n} T_i\cos\alpha\left(-\cos\phi_{T_i}\sin\psi_{T_i}z_T + \sin\phi_{T_i}y_T\right) \\
& + \sum_{i=1}^{n} T_i\sin\alpha\left(\cos\phi_{T_i}\sin\psi_{T_i}x_T - \cos\phi_{T_i}\cos\psi_{T_i}y_T\right)
\end{aligned} \tag{5.18a}$$

M-Moment:

$$\begin{aligned}
& -\frac{g^2(n-1)^2\sin^2\phi}{V^2}I_{zx} + \frac{g^2(n-1)^2\cos\phi\,\sin\phi}{V^2}I_{xy} - \frac{g(n-1)\sin\phi}{V}h'_x \\
& = \begin{bmatrix} C_{m_0} + C_{m_\alpha}\alpha + C_{m_q}\frac{g\bar{c}(n-1)\cos\phi}{2V^2} - C_{m_r}\frac{g\bar{c}(n-1)\sin\phi}{2V^2} \\ + C_{m_{i_{LoCE}}} i_{LoCE} + C_{m_{\delta_{LoCE}}}\delta_{LoCE} \\ + C_{m_{i_{DiCE}}} i_{DiCE} + C_{m_{\delta_{DiCE}}}\delta_{DiCE} \\ + C_{m_{i_{LaCE}}} i_{LaCE} + C_{m_{\delta_{LaCE}}}\delta_{LaCE} \end{bmatrix}\bar{q}S\bar{c} \\
& + \sum_{i=1}^{n} T_i\left(\cos\phi_{T_i}\cos\psi_{T_i}z_T - \sin\phi_{T_i}x_T\right) \pm \sum_{i=1}^{n}\Delta M_{D_i}
\end{aligned} \tag{5.18b}$$

N-Moment:

$$\begin{aligned}
& I_{xy}\frac{g^2(n-1)^2\cos^2\phi}{V^2} - I_{zx}\frac{g^2(n-1)^2\cos\phi\,\sin\phi}{V^2} - h'_x\frac{g(n-1)\cos\phi}{V} \\
& = \begin{bmatrix} C_{n_0} + C_{n_\beta}\beta + C_{n_q}\frac{gb(n-1)\cos\phi}{2V^2} - C_{n_r}\frac{gb(n-1)\sin\phi}{2V^2} \\ + C_{n_{i_{LoCE}}} i_{LoCE} + C_{n_{\delta_{LoCE}}}\delta_{LoCE} \\ + C_{n_{i_{DiCE}}} i_{DiCE} + C_{n_{\delta_{DiCE}}}\delta_{DiCE} \\ + C_{n_{i_{LaCE}}} i_{LaCE} + C_{n_{\delta_{LaCE}}}\delta_{LaCE} \end{bmatrix}\bar{q}Sb \\
& + \sum_{i=1}^{n} T_i\cos\alpha\left(-\cos\phi_{T_i}\sin\psi_{T_i}z_T + \sin\phi_{T_i}y_T\right) \\
& + \sum_{i=1}^{n} T_i\sin\alpha\left(\cos\phi_{T_i}\sin\psi_{T_i}x_T - \cos\phi_{T_i}\cos\psi_{T_i}y_T\right) - \sum_{i=1}^{n}\Delta N_{D_i}
\end{aligned} \tag{5.18c}$$

Equations (5.17a–c) and (5.18a–c) are non-linear equations with respect to the state variables α, β, and ϕ. The above two sets of steady state equations form the basis for studying vehicle controllability design-aspects during pull-up and push-over flight in the *xz*-plane, containing load-factor specific parameters like C_L and n explicitly. The following pull-up/push-over flight conditions can be modelled for the full range of symmetric and asymmetric aircraft configurations and concepts:

(A) $\beta \geq \leq 0$, *Thrust Symmetry* [symmetric and asymmetric pull-up/push-over with all engines operative, certain systems failed]

(B) $\beta \geq \leq 0$, *Thrust Asymmetry* [asymmetric pull-up/push-over with engine failure(s), crossed controls, certain systems failed]

Equations (5.17a–c) and (5.18a–c) can be solved for any combination of the following design- and state-variables for pull-up or push-over flight:

$$\alpha, \beta, \phi, n, V, \bar{\rho}, \sum i_{LoCE,DiCE,LaCE}, \sum \delta_{LoCE,DiCE,LaCE}, T_i, x_{T_i}, y_{T_i}, z_{T_i}, \phi_{T_i}, \psi_{T_i}$$

Since there are six equations, several of the design- and state variables listed above have to be specified before the system can be solved using iterative matrix techniques. The following example flight cases indicate the potential of Eqs. (5.17a–c) and (5.18a–c).

Case (A) ($\beta \geq \leq 0$, *Thrust Symmetry*)

(i) *Load Factor Capability—CEV Manoeuver*

Pre-Selection: $\beta = 0, \phi, n, V, \bar{\rho}, \sum_{j=1}^{n} i_j, \sum_{i=1}^{m-3} \delta_i, x_{T_i}, y_{T_i}, z_{T_i}, \phi_{T_i}, \psi_{T_i}$

Numerical Solution: $\alpha_{trim}, \delta_{LoCE_{i_{trim}}}, \delta_{DiCE_{i_{trim}}}, \delta_{LaCE_{i_{trim}}}, T_i$

Problem Description: This longitudinal flight case estimates the load factor capability with a given thrust setting during the pull-up/push-over flight case. The control power available/required of the LoCE gets evaluated for the symmetric class of aircraft. The aircraft is trimmed at low speed, the variable incidence stabiliser commands pull, whereby the elevator has to trim the aircraft without risking CE stall. Clearly, it is not possible to simulate the true non-steady character of the 'CEV manoeuver' with the steady state formalism presented.

(ii) *Load Factor Capability—Asymmetric Flight Case*

Pre-Selection: $\beta \neq 0, \phi, V, \bar{\rho}, \sum_{j=1}^{n} i_j, \sum_{i=1}^{m-3} \delta_i, x_{T_i}, y_{T_i}, z_{T_i}, \phi_{T_i}, \psi_{T_i}$

Numerical Solution: $\alpha_{trim}, n, \delta_{LoCE_{i_{trim}}}, \delta_{DiCE_{i_{trim}}}, \delta_{LaCE_{i_{trim}}}, T_i$

Problem Description: This longitudinal/lateral-directional flight case estimates the load factor capability of the aircraft with a given thrust setting during the pull-up/push-over flight case. The control power available/required by the LoCE, DiCE, and LaCE gets evaluated for the symmetric aircraft in sideslip or the asymmetric aircraft in 'straight' or sideslipping flight.

Case (B) ($\beta \geq \leq 0$, *Thrust Asymmetry*)

(i) *Load Factor Capability With Engine(s) Inoperative*

Pre-Selection: $\beta \neq 0, \phi, V, \bar{\rho}, \sum_{j=1}^{n} i_j, \sum_{i=1}^{m-3} \delta_i, x_{T_i}, y_{T_i}, z_{T_i}, \phi_{T_i}, \psi_{T_i}$

Numerical Solution: $\alpha_{trim}, n, \delta_{LoCE_{i_{trim}}}, \delta_{DiCE_{i_{trim}}}, \delta_{LaCE_{i_{trim}}}, T_i$

Problem Description: This longitudinal/lateral-directional flight case estimates the load factor capability of the aircraft with a given thrust setting during pull-up or push-over. The trim settings of the primary CEs are determined with the appropriate load factor.

The load factor n in the above equations covers the following flight conditions at the vertical position (top or bottom) of the loop:

$n = 1$	non-manoeuvring flight
$n > 1$	pull-up
$n < 1$	push-over
$n = 0$	ballistic flight

Overall, the pull-up/push-over flight case evaluates control power required/available of primarily the LoCE, and determines the manoeuver point (m.p.) position of the aircraft. *AeroMech* uses this DCFC to evaluate control authority with given stability levels. Thus, the CE deflection and load factor are used as measures of merit for the evaluation of control authority for given stability levels.

5.3.1.4 Steady State Rolling Performance

Any discussion of this flight case requires consideration of the kinematics of the motion. Since any aircraft generally tends to follow the path of least resistance, there are two basic possibilities for roll:

(A) Roll about the wind axis (flight direction);
(B) Roll about the forward minimum inertia axis (principal axis).

Conventional aircraft configurations (e.g., TAC), having modest inertias and strong aerodynamic stiffness, tend to roll about the wind axis (flight path). Throughout this flight case the aircraft remains in trim, since the angle-of-attack remains at the trimmed value and the sideslip angle stays constant. In contrast, highly loaded slender aircraft with weaker stability characteristics tend to roll about the minimum inertia axis, resulting in an interchange of angle-of-attack and sideslip angles every 90° of roll. As a result, the motion is of oscillatory character. Clearly, the analysis of such motion has more physical significance, when the EOM are referred to principal axes. However, since the roll manoeuver primarily initiates course changes, the roll performance should be determined in stability axes, which is consistent with the previous choice of axis system (see Appendices A.9, A.10 and A.11).

The 'quasi-non-oscillatory' condition at the instant $\phi = 0$ applies to both aircraft types (slender and non-slender). Clearly, the roll subsidence mode is not a substitute for a real time simulation, which has to take the different kinematics of slender and non-slender aircraft into account. The complexity of the roll case becomes obvious, when observing the various roll-loading conditions (roll kinematics) involved: (a) steady level flight; (b) roll initiation; (c) steady roll rate; (d) roll arresting; or (e) reverse roll. It has been decided that steady state roll[12] is the convenient condition to be evaluated during conceptual design.

The inherent complexity of the aircraft configuration under investigation defines the analytical modelling framework, from 1-DOF sufficient for the conventional TAC to possibly 6-DOF for the asymmetric aircraft type, see Table 5.4.

Table 5.4 indicates that the non-slender symmetric aircraft (e.g., TAC) is the simplest aircraft configuration considered, whereby the slender asymmetric aircraft (e.g., OWC) is the most complicated one. Simplified conceptual design analysis is usually formulated for the non-slender symmetric aircraft, modelling roll performance as a single-degree-of-freedom problem (the rolling convergence is a motion of almost a single degree-of-freedom rotation about the stability x-axis). An exception to this are aircraft with highly swept low aspect ratio wings, where the roll-yaw-pitch coupling requires a complete 6-DOF simulation. Clearly, the 6-DOF analysis becomes obligatory when discussing asymmetric aircraft types.

It is possible to specify roll performance via the following figures of merits: (a) roll helix angle, $pb/2V$ (rad); (b) roll rate, p (deg/s); (c) roll acceleration, $\dot{p}$ (deg/s^2); (d) wing tip velocity, V_{tip} (m/s); (e) ratio C_l/C_L (–); or (f) roll mode time constant; time to bank, t_ϕ (deg/s). The criteria relevant in the present context are the roll helix angle, roll rate, and time to bank, which provide sufficient information for sizing the LaCEs.

---

[12]The steady state roll is achieved when the roll damping moment generated by the airframe is equal to the applied increment in rolling moment.

Table 5.4 Aircraft configuration complexity for roll analysis

	Non-slender symmetric aircraft	Slender symmetric aircraft	Non-slender asymmetric aircraft	Slender asymmetric aircraft
Roll about principal axis	No	Yes	No	Yes
Roll around stability axis	Yes	No	Yes	No
Oscillatory motion relative to wind axis	No	Yes	No	Yes
Aerodynamic CE coupling	No	No	Yes	Yes
Inertia coupling	No	No	Yes	Yes
Aircraft in trim during roll	Yes	No	No	No

The roll performance case is a special DCFC, because it usually requires a full simulation by solving the 6-DOF dynamic EOM. In the present context, the 6-DOF trim EOM are solved for a prescribed roll helix angle or roll rate, and a 1-DOF model estimates the time to bank. Thus, to ensure consistency with the approach taken so far, the following process is suggested in context with *AeroMech*:

(a) Solve 6-DOF trim EOM for a prescribed roll helix angle *pb/2V* (trim, physical visibility);
(b) Solve 1-DOF dynamic EOM for time to bank (step control input);
(c) Solve 6-DOF dynamic EOM (dynamic response analysis).

Steps (a) and (b) are presented below. Step (c) is enabled with the 6-DOF small perturbation EOM described in Sect. 5.3.2. The following presents the 6-DOF trim EOM, utilised to estimate the CE deflection required for a prescribed roll helix angle *pb/2V* with a value of, e.g., 0.07.[13] The EOM have to be solved for a trimmed condition, and in particular for the aileron deflection required to enforce the prescribed roll rate, p, see Fig. 5.9.

The flight condition of interest is horizontal flight ($\theta = \gamma = 0$). The aircraft performs a steady roll manoeuver and the situation of particular interest is, when the aircraft rolls through $\phi = 0$. However, $\phi \neq 0$ is permitted to trim the aircraft in y-direction (consider only small angles of ϕ). This steady state flight condition leads to the non-linear *6-DOF Trim EOM for Steady State Rolling Flight* written in stability axes (derivation see Appendix A.11).

[13]For a discussion of the Gilruth-Criterion, see Gilruth and Turner [22] and Abzug and Larrabee [23].

Fig. 5.9 Roll performance at $\phi = 0$

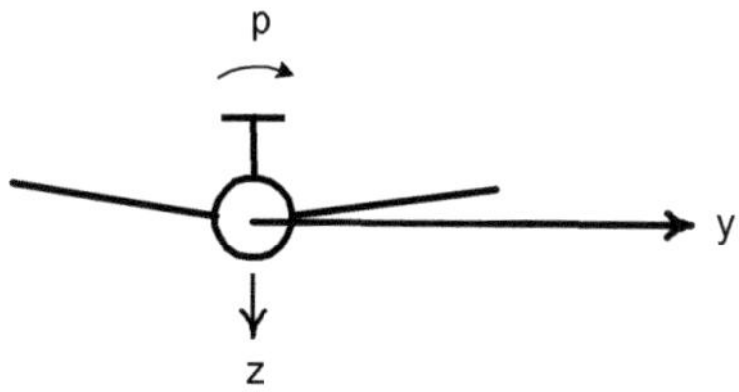

X-Force:

$$0 = -\begin{bmatrix} C_{D_0} + C_{D_\alpha}\alpha + C_{D_p}\frac{p\bar{c}}{2V} \\ + C_{D_{i_{LoCE}}} i_{LoCE} + C_{D_{\delta_{LoCE}}}\delta_{LoCE} \\ + C_{D_{i_{DiCE}}} i_{DiCE} + C_{D_{\delta_{DiCE}}}\delta_{DiCE} \\ + C_{D_{i_{LaCE}}} i_{LaCE} + C_{D_{\delta_{LaCE}}}\delta_{LaCE} \end{bmatrix}\bar{q}S + \sum_{i=1}^{n} T_i(\cos\phi_{T_i}\cos\psi_{T_i}\cos\alpha + \sin\phi_{T_i}\sin\alpha) \quad (5.19a)$$

Y-Force:

$$mg\sin\phi = \begin{bmatrix} C_{Y_0} + C_{Y_p}\frac{pb}{2V} \\ + C_{Y_{i_{LoCE}}} i_{LoCE} + C_{Y_{\delta_{LoCE}}}\delta_{LoCE} \\ + C_{Y_{i_{DiCE}}} i_{DiCE} + C_{Y_{\delta_{DiCE}}}\delta_{DiCE} \\ + C_{Y_{i_{LaCE}}} i_{LaCE} + C_{Y_{\delta_{LaCE}}}\delta_{LaCE} \end{bmatrix}\bar{q}S \quad (5.19b)$$

Z-Force:

$$-mg\cos\phi + mpV\sin\beta = -\begin{bmatrix} C_{L_0} + C_{L_\alpha}\alpha + C_{L_p}\frac{p\bar{c}}{2V} \\ + C_{L_{i_{LoCE}}} i_{LoCE} + C_{L_{\delta_{LoCE}}}\delta_{LoCE} \\ + C_{L_{i_{DiCE}}} i_{DiCE} + C_{L_{\delta_{DiCE}}}\delta_{DiCE} \\ + C_{L_{i_{LaCE}}} i_{LaCE} + C_{L_{\delta_{LaCE}}}\delta_{LaCE} \end{bmatrix}\bar{q}S + \sum_{i=1}^{n} T_i(-\cos\phi_{T_i}\cos\psi_{T_i}\sin\alpha + \sin\phi_{T_i}\cos\alpha) \quad (5.19c)$$

L-Moment:

$$0 = \begin{bmatrix} C_{l_0} + C_{l_p}\frac{pb}{2V} \\ + C_{l_{i_{LoCE}}} i_{LoCE} + C_{l_{\delta_{LoCE}}}\delta_{LoCE} \\ + C_{l_{i_{DiCE}}} i_{DiCE} + C_{l_{\delta_{DiCE}}}\delta_{DiCE} \\ + C_{l_{i_{LaCE}}} i_{LaCE} + C_{l_{\delta_{LaCE}}}\delta_{LaCE} \end{bmatrix}\bar{q}Sb \quad (5.20a)$$

M-Moment:

$$I_{zx}p^2 - ph'_z = \begin{bmatrix} C_{m_0} + C_{m_\alpha}\alpha + C_{m_p}\frac{p\bar{c}}{2V} \\ + C_{m_{i_{LoCE}}} i_{LoCE} + C_{m_{\delta_{LoCE}}} \delta_{LoCE} \\ + C_{m_{i_{DiCE}}} i_{DiCE} + C_{m_{\delta_{DiCE}}} \delta_{DiCE} \\ + C_{m_{i_{LaCE}}} i_{LaCE} + C_{m_{\delta_{LaCE}}} \delta_{LaCE} \end{bmatrix} \bar{q}S\bar{c} + \sum_{i=1}^{n} T_i\left(\cos\phi_{T_i}\cos\psi_{T_i} z_T - \sin\phi_{T_i} x_T\right) \tag{5.20b}$$

N-Moment:

$$-I_{xy}p^2 + ph'_y = \begin{bmatrix} C_{n_0} + C_{n_p}\frac{pb}{2V} \\ + C_{n_{i_{LoCE}}} i_{LoCE} + C_{n_{\delta_{LoCE}}} \delta_{LoCE} \\ + C_{n_{i_{DiCE}}} i_{DiCE} + C_{n_{\delta_{DiCE}}} \delta_{DiCE} \\ + C_{n_{i_{LaCE}}} i_{LaCE} + C_{n_{\delta_{LaCE}}} \delta_{LaCE} \end{bmatrix} \bar{q}Sb \tag{5.20c}$$

Equations (5.19a–c) and (5.20a–c) are non-linear equations with respect to the state variables α, β, and ϕ. The above two sets of steady state equations form the basis for studying vehicle controllability design-aspects during rolling flight. The following rolling flight conditions can be modelled for the full range of symmetric and asymmetric aircraft configurations and concepts:

(A) $\beta \geq \leq 0$, *Thrust Symmetry* [symmetric and asymmetric rolling with all engines operative, certain systems failed]

(B) $\beta \geq \leq 0$, *Thrust Asymmetry* [asymmetric rolling with engine failure(s), certain systems failed]

Equations (5.19a–c) and (5.20a–c) can be solved for any combination of the following design- and state-variables for steady-state rolling performance:

$$\alpha, \beta, \phi, V, \bar{\rho}, p, \sum i_{LoCE,DiCE,LaCE}, \sum \delta_{LoCE,DiCE,LaCE}, T_i, x_{T_i}, y_{T_i}, z_{T_i}, \phi_{T_i}, \psi_{T_i}$$

Since there are six equations, several of the design- and state variables listed above have to be specified before the system can be solved using iterative matrix techniques. The following example flight cases indicate the potential of Eqs. (5.19a–c) and (5.20a–c).

Case (A) ($\beta \geq \leq 0$, *Thrust Symmetry*)

(i) *Control Power Required to Attain Prescribed Roll Rate*

Pre-Selection: $\beta = 0, \phi, V, \bar{\rho}, \sum_{j=1}^{n} i_j, \sum_{i=1}^{m-3} \delta_i, x_{T_i}, y_{T_i}, z_{T_i}, \phi_{T_i}, \psi_{T_i}$

Numerical Solution: $\alpha_{trim}, \delta_{LoCE_{i_{trim}}}, \delta_{DiCE_{i_{trim}}}, \delta_{LaCE_{i_{trim}}}, T_i$

Problem Description: Having specified a roll rate, p, or helix angle, $pb/2V$, for the symmetric aircraft, the 6-DOF CE deflection angles required are estimated.

(ii) *Achievable Roll Rate With Control Power Available*

Pre-Selection: $\beta \neq 0, \phi, V, \bar{p}, \sum_{j=1}^{n} i_j, \sum_{i=1}^{m-2} \delta_i, x_{T_i}, y_{T_i}, z_{T_i}, \phi_{T_i}, \psi_{T_i}$

Numerical Solution: $\alpha_{trim}, \delta_{LoCE_{i_{trim}}}, \delta_{LoCE_{i_{trim}}}, p, T_i$

Problem Description: Having commanded maximum deflection of the LaCEs, the maximum attainable roll rate gets estimated. This case is valid for the symmetric aircraft with $\beta \neq 0$ and the asymmetric aircraft with symmetric thrust setting.

Case (**B**) ($\beta \geq \leq 0$, *Thrust Asymmetry*)

(i) *Control Power Required to Attain a Prescribed Roll Rate*

Pre-Selection: $\beta = 0, \phi, V, \bar{p}, p, \sum_{j=1}^{n} i_j, \sum_{i=1}^{m-3} \delta_i, x_{T_i}, y_{T_i}, z_{T_i}, \phi_{T_i}, \psi_{T_i}$

Numerical Solution: $\alpha_{trim}, \delta_{LoCE_{i_{trim}}}, \delta_{DiCE_{i_{trim}}}, \delta_{LaCE_{i_{trim}}}, T_i$

Problem Description: Having specified a roll rate, p, or helix angle, $pb/2V$, for the symmetric/asymmetric aircraft with thrust asymmetry, the 6-DOF CE deflection angles required are estimated.

(ii) *Achievable Roll Rate With Control Power Available*

Pre-Selection: $\beta \neq 0, \phi, V, \bar{p}, \sum_{j=1}^{n} i_j, \sum_{i=1}^{m-2} \delta_i, x_{T_i}, x_{T_i}, z_{T_i}, \phi_{T_i}, \psi_{T_i}$

Numerical Solution: $\alpha_{trim}, \delta_{LoCE_{i_{trim}}}, \delta_{DiCE_{i_{trim}}}, p, T_i$

Problem Description: Having commanded maximum deflection of the LaCEs, the maximum attainable roll rate gets estimated. This case with thrust asymmetry is valid for the symmetric aircraft with $\beta \neq 0$ and the asymmetric aircraft.

The estimation of time to bank is a complicated dynamic problem. The result is affected by FCS dynamics, control rate limiting, aeroelasticity, coupling effects typical for asymmetric aircraft types, etc. For simplicity reasons it has been decided, to consider the single-degree-of-freedom roll response for a specified control input, although rather optimistic results have to be expected compared to results provided by complicated transfer functions. The derivation in Appendix A.11 yields the following known single-degree-of-freedom model. The maximum steady state roll rate for the particular magnitude of LaCE step input is given with

$$p = -\frac{2V}{b}\frac{C_{l_{\delta_{LaCE}}}\delta_{LaCE}}{C_{l_p}} \tag{5.21a}$$

and for the helix angle follows

$$\frac{pb}{2V} = -\frac{C_{l_{\delta_{LaCE}}}\delta_{LaCE}}{C_{l_p}} \tag{5.21b}$$

The bank angle is obtained by integrating the roll rate given with (5.21a):

$$\int pdt = \phi(t) = -\frac{2V}{b}\frac{C_{l_{\delta_{LaCE}}}\delta_{LaCE}}{C_{l_p}}\left[t + \frac{1}{L_p}\left(1 - e^{L_p t}\right)\right] \tag{5.22}$$

The bank angle response given with Eq. (5.22) consists of a first term varying linearly with time, and a second term varying exponentially with time. The second term vanishes for infinite time, resulting in an overall linear bank angle response relationship (constant roll rate manoeuver) with time.[14]

In summary, the 6-DOF analysis estimates for a pre-defined roll helix angle relevant roll performance parameters $(V, C_{l_p}, C_{l_{\delta_{LaCE}}}, \delta_{LaCE})$, which are not dependent on roll dynamics. By feeding this information into Eq. (5.22), an estimate of the time-to-bank capability can be obtained. Clearly, the procedure couples the 6-DOF model with the 1-DOF analysis. The advantage is that the data provided to the 1-DOF analysis represents a trimmed aircraft, taking aerodynamic and inertia coupling effects into account. Thus, reasonably accurate results may be obtained without solving the dynamic EOM at this stage. As a result, physical visibility is maximum.

5.3.1.5 Quasi-Steady, Straight Take-off Rotation Manoeuvre

The take-off rotation manoeuver is not a steady state condition since $\dot{q} \neq 0$. The instant of interest during the manoeuver is, when nose-wheel lift-off is commanded, see Fig. 5.10.

This special DCFC models the instant, when the vehicle has reached lift-off speed, $V = V_{lift\text{-}off}$. It is then that the LoCE commands nose wheel lift-off. The pitching moment generated by the LoCE just balances the vehicle with the nose gear fully extended (no weight on the nose gear), thus no contact of the nose gear

[14]Note that the 1-DOF approximation alone does not consider a trimmed aircraft, since no rudder deflection is commanded to maintain a coordinated rolling motion, etc.

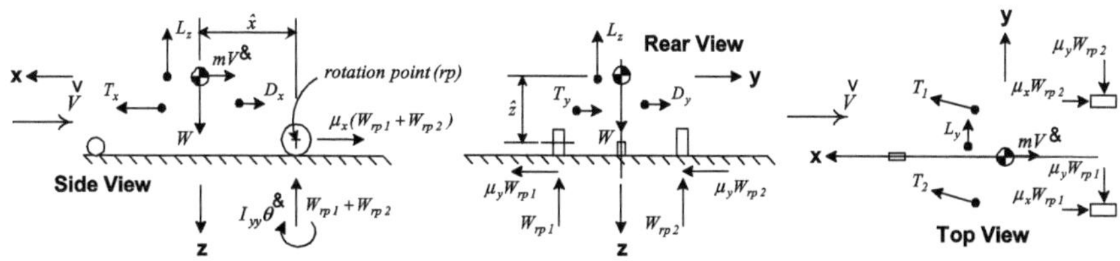

Fig. 5.10 Take-off rotation 'snap-shot'

with the runway. The rotational speed is assumed to be still zero, $q = 0$, but the angular acceleration, $\ddot{\theta}$, is maximum. The aircraft is in a forward acceleration process, $\dot{V} \neq 0$, with the load factor still one, $n = 1$.

Aircraft with a tricycle landing gear and taildraggers can be considered. A take-off run along a horizontal runway is modelled, $\gamma = 0$, since the runway slope has an insignificant effect on LoCE control power required. The c.g. is assumed to lie laterally between the main gear contact points; thus the aircraft weight is equally distributed. For the symmetric aircraft it is assumed that $\phi = 0$ and $\beta = 0$, whereby the asymmetric aircraft assumes $\phi = 0$ and $\beta \neq 0$.[15]

It is of primary interest to estimate the control power required to generate a predefined angular acceleration, $\ddot{\theta}$, about the main gear axel during initiation of the take-off rotation manoeuver with $\dot{\theta} = 0$. The take-off rotation problem is modelled without sideward drift,[16] thereby reducing the 6-DOF problem to the non-linear <u>5. DOF Trim EOM for the Quasi-Steady Take-Off Rotation Manoeuver</u> written in stability axis (derivation see Appendix A.11):

<u>X-Force</u>:

$$m\dot{V} + \mu_x W_{rp} = -\begin{bmatrix} C_{D_0} + C_{D_\alpha}\alpha \\ + C_{D_{i_{LoCE}}} i_{LoCE} + C_{D_{\delta_{LoCE}}} \delta_{LoCE} \\ + C_{D_{i_{DiCE}}} i_{DiCE} + C_{D_{\delta_{DiCE}}} \delta_{DiCE} \\ + C_{D_{i_{LaCE}}} i_{LaCE} + C_{D_{\delta_{LaCE}}} \delta_{LaCE} \end{bmatrix} \bar{q}S + \sum_{i=1}^{n} T_i \left(\cos\phi_{T_i} \cos\psi_{T_i} \cos\alpha + \sin\phi_{T_i} \sin\alpha \right) \quad (5.23a)$$

[15]The asymmetric aircraft is modelled with a perfectly aligned landing gear relative to the runway, whereby the airframe asymmetry is aerodynamically characterised with $\beta \neq 0$.

[16]The aircraft is arrested in y-direction.

Z-Force:

$$mg - W_{rp} = \begin{bmatrix} C_{L_0} + C_{L_\alpha}\alpha \\ + C_{L_{i_{LoCE}}} i_{LoCE} + C_{L_{\delta_{LoCE}}} \delta_{LoCE} \\ + C_{L_{i_{DiCE}}} i_{DiCE} + C_{L_{\delta_{DiCE}}} \delta_{DiCE} \\ + C_{L_{i_{LaCE}}} i_{LaCE} + C_{L_{\delta_{LaCE}}} \delta_{LaCE} \end{bmatrix} \bar{q}S - \sum_{i=1}^{n} T_i \left(-\cos\phi_{T_i} \cos\psi_{T_i} \sin\alpha + \sin\phi_{T_i} \cos\alpha \right) \tag{5.23b}$$

L-Moment:

$$I_{xy}\ddot{\theta} = -\begin{bmatrix} C_{l_0} + C_{l_\beta}\beta \\ + C_{l_{i_{LoCE}}} i_{LoCE} + C_{l_{\delta_{LoCE}}} \delta_{LoCE} \\ + C_{l_{i_{DiCE}}} i_{DiCE} + C_{l_{\delta_{DiCE}}} \delta_{DiCE} \\ + C_{l_{i_{LaCE}}} i_{LaCE} + C_{l_{\delta_{LaCE}}} \delta_{LaCE} \end{bmatrix} \bar{q}Sb \tag{5.24a}$$

M-Moment:

$$I_y\ddot{\theta} + W_{rp}(\mu_x\hat{z} + \hat{x}) = \begin{bmatrix} C_{m_0} + C_{m_\alpha}\alpha + C_{m_q}\frac{q\bar{c}}{2V} \\ + C_{m_{i_{LoCE}}} i_{LoCE} + C_{m_{\delta_{LoCE}}} \delta_{LoCE} \\ + C_{m_{i_{DiCE}}} i_{DiCE} + C_{m_{\delta_{DiCE}}} \delta_{DiCE} \\ + C_{m_{i_{LaCE}}} i_{LaCE} + C_{m_{\delta_{LaCE}}} \delta_{LaCE} \end{bmatrix} \bar{q}S\bar{c} + \sum_{i=1}^{n} T_i \left(\cos\phi_{T_i} \cos\psi_{T_i} z_T - \sin\phi_{T_i} x_T \right) \tag{5.24b}$$

N-Moment:

$$I_{xy}\dot{q} = -\begin{bmatrix} C_{n_0} + C_{n_\beta}\beta \\ + C_{n_{i_{LoCE}}} i_{LoCE} + C_{n_{\delta_{LoCE}}} \delta_{LoCE} \\ + C_{n_{i_{DiCE}}} i_{DiCE} + C_{n_{\delta_{DiCE}}} \delta_{DiCE} \\ + C_{n_{i_{LaCE}}} i_{LaCE} + C_{n_{\delta_{LaCE}}} \delta_{LaCE} \end{bmatrix} \bar{q}Sb \tag{5.24c}$$

Equations (5.23a, b) and (5.24a–c) are non-linear equations with respect to the state variable α. The above two sets of quasi-steady equations form the basis for studying vehicle controllability design-aspects during initiation of the take-off rotation manoeuver. The following flight conditions can be modelled for the full range of symmetric and asymmetric aircraft configurations and concepts:

(A) $\beta = 0/\beta \neq 0$, *Thrust Symmetry* [take-off rotation with all engines operative; valid for the symmetric/asymmetric types of aircraft]

(B) $\beta = 0/\beta \neq 0$, *Thrust Asymmetry* [take-off rotation with engine(s) inoperative; valid for the symmetric/asymmetric types of aircraft]

Equations (5.23a, b) and (5.24a–c) can be solved for any combination of the following design- and state-variables for the quasi-steady, straight take-off rotation manoeuver:

$$\alpha, \beta, V, \dot{V}, \bar{p}, \ddot{\theta}, \sum i_{LoCE,DiCE,LaCE}, \sum \delta_{LoCE,DiCE,LaCE}, T_i, x_{T_i}, y_{T_i}, z_{T_i}, \phi_{T_i}, \psi_{T_i}$$

Since there are five equations, several of the design- and state variables listed above have to be specified before the system can be solved using iterative matrix techniques. The following example flight cases indicate the potential of Eqs. (5.23a, b) and (5.24a–c).

Case (A) ($\beta = 0/\beta \neq 0$, *Thrust Symmetry*)

(i) *Control Power Required to Attain Prescribed Pitch Acceleration (Symmetric Aircraft)*

Pre-Selection: $\alpha, \beta = 0, V, \dot{V}, \bar{p}, \ddot{\theta}, \sum_{j=1}^{n} i_j, \sum_{i=1}^{m-1} \delta_i, x_{T_i}, y_{T_i}, z_{T_i}, \phi_{T_i}, \psi_{T_i}$

Numerical Solution: $\delta_{LoCE_{i_{trim}}}, T_i$

Problem Description: Having pre-selected an angular acceleration, $\ddot{\theta}$, for the symmetric aircraft with thrust symmetry, the LoCE deflection angles required are estimated.

(ii) *Control Power Required to Attain Prescribed Pitch Acceleration (Asymmetric Aircraft)*

Pre-Selection: $\alpha, \beta \neq 0, V, \dot{V}, \bar{p}, \ddot{\theta}, \sum_{j=1}^{n} i_j, \sum_{i=1}^{m-3} \delta_i, x_{T_i}, y_{T_i}, z_{T_i}, \phi_{T_i}, \psi_{T_i}$

Numerical Solution: $\delta_{LoCE_{i_{trim}}}, \delta_{DiCE_{i_{trim}}}, \delta_{LaCE_{i_{trim}}}, T_i$

Problem Description: Having pre-selected an angular acceleration, $\ddot{\theta}$, for the asymmetric aircraft with thrust symmetry, the CE deflection angles required are estimated. Clearly, the landing gear is symmetrically aligned with the runway to avoid lateral drift, whereby the airframe is of asymmetric geometry layout.

Case (B) ($\beta = 0/\beta \neq 0$, *Thrust Asymmetry*)

(i) *Control Power Required to Attain Prescribed Pitch Acceleration (Symmetric Aircraft)*

Pre-Selection: $\alpha,\ \beta = 0,\ V,\ \dot{V},\ \bar{\rho},\ \ddot{\theta},\ \sum_{j=1}^{n} i_j,\ \sum_{i=1}^{m-3} \delta_i,\ x_{T_i},\ y_{T_i},\ z_{T_i},\ \phi_{T_i},\ \psi_{T_i}$

Numerical Solution: $\delta_{LoCE_{i_{trim}}},\ \delta_{DiCE_{i_{trim}}},\ \delta_{LaCE_{i_{trim}}},\ T_i$

Problem Description: Having pre-selected an angular acceleration, $\ddot{\theta}$, for the symmetric aircraft, the CE deflection angles required are estimated. The thrust asymmetry condition needs to be trimmed to maintain $\beta = 0$.

(ii) *Control Power Required to Attain Prescribed Pitch Acceleration (Asymmetric Aircraft)*

Pre-Selection: $\alpha,\ \beta \neq 0,\ V,\ \dot{V},\ \bar{\rho},\ \ddot{\theta},\ \sum_{j=1}^{n} i_j,\ \sum_{i=1}^{m-3} \delta_i,\ x_{T_i},\ y_{T_i},\ z_{T_i},\ \phi_{T_i},\ \psi_{T_i}$

Numerical Solution: $\delta_{LoCE_{i_{trim}}},\ \delta_{DiCE_{i_{trim}}},\ \delta_{LaCE_{i_{trim}}},\ T_i$

Problem Description: Having pre-selected an angular acceleration, $\ddot{\theta}$, for the asymmetric aircraft, the CE deflection angles required are estimated ($\beta \neq 0$ indicates the geometric asymmetry of the airframe under investigation; it is, however, not meant that the asymmetric aircraft skids with a sideslip angle of the landing gear relative to the runway, thereby violating the assumption of no lateral skid).

5.3.2 Small Perturbation Equations of Motion

The derivation of the coupled 6-DOF small perturbation EOM has been presented in Appendices A.9 and A.10. The simplifying assumptions classically made in its derivation for conceptual design application usually lead to the decoupled set of longitudinal- and lateral-directional equations. Since this approach has shown not to be feasible in the present context, the following assumptions have not been accepted: (i) the existence of a plane of symmetry (C_{xz}), (ii) the neglect of aerodynamic cross-coupling, (iii) the absence of rotor gyroscopic effects. For the most general case of the asymmetric aircraft, there exists no pure longitudinal motion, since (i) no plane of symmetry is assumed, and (ii) rotor gyroscopic effects are included. The absence of pure lateral-directional motions is a direct result of (i) taking rotor gyroscopic effects into account, and (ii) the inclusion of aerodynamic cross-coupling effects. In addition, having linearised the 6-DOF EOM does not imply decoupled EOM.

The *State Vector Form of the Open-Loop Linear 6-DOF Small Perturbation EOM*, written in vector/matrix notation, is presented in Eqs. (5.25a), (5.25b), and (5.25c). When comparing the open-loop coupled EOM with the decoupled set derived by Etkin and Reid in [20], the influence of inertia- and aerodynamic cross-coupling becomes obvious. Equations (5.25a), (5.25b), and (5.25c) form the basis for aircraft stability and response discussions, and for automatic flight control theory and applications.

The closed-loop EOM are obtained by incorporating the control law given with Eq. (4.6) into the open-loop EOM.

$$\underbrace{C_{x_{y_{SAS}}}}_{Equivalent\ Stability\ Derivative} = \underbrace{C_{x_{y_{Airframe}}}}_{Inherent\ Stability\ Derivative} + \underbrace{C_{x_\delta}}_{Control\ Derivative} \underbrace{K}_{Gain} \tag{4.6}$$

Classical feedback variables have been selected for stiffness restoration (attitude feedback) and damping restoration (rate feedback), see Table 5.5. The reader is referred back to Sect. 4.4.2.5 for further comments related to the selection of feedback variables.

The *State Vector Form of the Closed-Loop Linear 6-DOF Small Perturbation EOM*, written in vector/matrix notation, is presented in Eq. (5.26). Note that the control vector $\vec{c}$ in Eq. (5.25c) contains the individual CE deflection angles δ_i, whereas the control vector $\vec{c}$ in Eq. (5.26) contains pilot manouever commands $\hat{\delta}_i$ instead.

It should be noted, the closed-loop dynamic EOM, see Eq. (5.26), need to be solved at first with the control gains, K_i, all set to zero. The response properties of

Table 5.5 Inherent airframe stability augmentation

Feedback variables	Feedback gain	Command variables
Stiffness restauration		
w	K_w	δd_{LoCE}
v	K_v	δd_{DiCE}
Damping restauration		
p	K_p	δd_{LaCE}
q	K_q	δd_{LoCE}
r	K_r	δd_{DiCE}

the 'open-loop' aircraft need to be checked and compared against the quantified requirements, leading to the iterative sequence to determine appropriate gains, K_i. If the airframe is unstable, then it is required to define start-values for the control gains, K_i.

There exists no immediate short-cut for specifying gross FCS properties without solving the closed-loop dynamic EOM for the asymmetric aircraft type.[17] However, the gain constants K_i do relate to FCS characteristics like actuator frequency, -bandwidth, -size, -weight, -cost, and others involved. Clearly, it is recommended that a follow-on study entitled *'Integrated FCS Design for Conceptual Design'* be undertaken, to investigate the relationships between the gain constants K_i and design-relevant information, leading to an integrated FCS-design. Overall, the FCS design-issue is of particular importance for the conceptual design of unstable vehicles.

The techniques for solving the dynamic EOM are described elsewhere, see for example Cook [24] and Stevens and Lewis [25].

Open-Loop, Coupled 6-Degree Of Freedom Small Perturbation Equations of Motion, Eq. (5.25a).

---

[17]Some simplification can be expected by solving reduced order transfer functions (reduced order models) for symmetric type of aircraft, see Sect. 4.4.2.7.

$$\underbrace{\begin{bmatrix} \Delta\dot{u} \\ \dot{v} \\ \dot{w} \\ \dot{p} \\ \dot{q} \\ \dot{r} \\ \dot{\varphi} \\ \Delta\dot{\theta} \\ \dot{\psi} \end{bmatrix}}_{\dot{\vec{x}}} = \underbrace{\begin{bmatrix} \quad \end{bmatrix}}_{A} \bullet \underbrace{\begin{bmatrix} \Delta u \\ v \\ w \\ p \\ q \\ r \\ \varphi \\ \Delta\theta \\ \psi \end{bmatrix}}_{\vec{x}} + \underbrace{\begin{bmatrix} \dfrac{\Delta X_{CE}}{m} \\ \dfrac{\Delta Y_{CE}}{m} \\ \dfrac{\Delta Z_{CE}}{(m - Z_{\dot{w}})} \\ \Delta L_{CE} I_L^* + \Delta M_{CE} I_M^* + \Delta N_{CE} I_N^* + \dfrac{\Delta Z_{CE} M_{\dot{w}} I_M^*}{(m - Z_{\dot{w}})} \\ \Delta L_{CE} I_L^{**} + \Delta M_{CE} I_M^{**} + \Delta N_{CE} I_N^{**} + \dfrac{\Delta Z_{CE} M_{\dot{w}} I_M^{**}}{(m - Z_{\dot{w}})} \\ \Delta L_{CE} I_L^{***} + \Delta M_{CE} I_M^{***} + \Delta N_{CE} I_N^{***} + \dfrac{\Delta Z_{CE} M_{\dot{w}} I_M^{***}}{(m - Z_{\dot{w}})} \\ 0 \\ 0 \\ 0 \end{bmatrix}}_{B\vec{c}} \tag{5.25a}$$

Open-Loop, Coupled 6-Degree Of Freedom Small Perturbation Equations of Motion, Eq. (5.25a) cont.

$$
\underbrace{\begin{bmatrix}
\frac{X_u}{m} & \frac{X_v}{m} & \frac{X_w}{m} & \frac{X_p}{m} & \frac{X_q}{m} & \frac{X_r}{m} & 0 & -g\cos\theta_0 & 0 \\
\frac{Y_u}{m} & \frac{Y_v}{m} & \frac{Y_w}{m} & \frac{Y_p}{m} & \frac{Y_q}{m} & \frac{Y_r}{m}-u_0 & g\cos\theta_0 & 0 & 0 \\
\frac{Z_u}{(m-Z_{\dot w})} & \frac{Z_v}{(m-Z_{\dot w})} & \frac{Z_w}{(m-Z_{\dot w})} & \frac{Z_p}{(m-Z_{\dot w})} & \frac{Z_q+mu_0}{(m-Z_{\dot w})} & \frac{Z_r}{(m-Z_{\dot w})} & 0 & \frac{-mg\sin\theta_0}{(m-Z_{\dot w})} & 0 \\
\begin{aligned} &L_u I_L^{*} \\ &+\left(M_u+\frac{M_{\dot w}Z_u}{(m-Z_{\dot w})}\right)I_M^{*} \\ &+N_u I_N^{*}\end{aligned} &
\begin{aligned} &L_v I_L^{*} \\ &+\left(M_v+\frac{M_{\dot w}Z_v}{(m-Z_{\dot w})}\right)I_M^{*} \\ &+N_v I_N^{*}\end{aligned} &
\begin{aligned} &L_w I_L^{*} \\ &+\left(M_w+\frac{M_{\dot w}Z_w}{(m-Z_{\dot w})}\right)I_M^{*} \\ &+N_w I_N^{*}\end{aligned} &
\begin{aligned} &L_p I_L^{*} \\ &+\left(M_p+h'_z+\frac{M_{\dot w}Z_p}{(m-Z_{\dot w})}\right)I_M^{*} \\ &+\left(N_p-h'_y\right)I_N^{*}\end{aligned} &
\begin{aligned} &\left(L_q-h'_z\right)I_L^{*} \\ &+\left(M_q+\frac{M_{\dot w}Z_q+M_{\dot w}mu_0}{(m-Z_{\dot w})}\right)I_M^{*} \\ &+\left(N_q+h'_x\right)I_N^{*}\end{aligned} &
\begin{aligned} &\left(L_r+h'_y\right)I_L^{*} \\ &+\left(M_r-h'_x+\frac{M_{\dot w}Z_r}{(m-Z_{\dot w})}\right)I_M^{*} \\ &+N_r I_N^{*}\end{aligned} &
0 & \frac{-mgM_{\dot w}I_M^{*}\sin\theta_0}{(m-Z_{\dot w})} & 0 \\
\begin{aligned} &L_u I_L^{**} \\ &+\left(M_u+\frac{M_{\dot w}Z_u}{(m-Z_{\dot w})}\right)I_M^{**} \\ &+N_u I_N^{**}\end{aligned} &
\begin{aligned} &L_v I_L^{**} \\ &+\left(M_v+\frac{M_{\dot w}Z_v}{(m-Z_{\dot w})}\right)I_M^{**} \\ &+N_v I_N^{**}\end{aligned} &
\begin{aligned} &L_w I_L^{**} \\ &+\left(M_w+\frac{M_{\dot w}Z_w}{(m-Z_{\dot w})}\right)I_M^{**} \\ &+N_w I_N^{**}\end{aligned} &
\begin{aligned} &L_p I_L^{**} \\ &+\left(M_p+h'_z+\frac{M_{\dot w}Z_p}{(m-Z_{\dot w})}\right)I_M^{**} \\ &+\left(N_p-h'_y\right)I_N^{**}\end{aligned} &
\begin{aligned} &\left(L_q-h'_z\right)I_L^{**} \\ &+\left(M_q+\frac{M_{\dot w}Z_q+M_{\dot w}mu_0}{(m-Z_{\dot w})}\right)I_M^{**} \\ &+\left(N_q+h'_x\right)I_N^{**}\end{aligned} &
\begin{aligned} &\left(L_r+h'_y\right)I_L^{**} \\ &+\left(M_r-h'_x+\frac{M_{\dot w}Z_r}{(m-Z_{\dot w})}\right)I_M^{**} \\ &+N_r I_N^{**}\end{aligned} &
0 & \frac{-mgM_{\dot w}I_M^{**}\sin\theta_0}{(m-Z_{\dot w})} & 0 \\
\begin{aligned} &L_u I_L^{***} \\ &+\left(M_u+\frac{M_{\dot w}Z_u}{(m-Z_{\dot w})}\right)I_M^{***} \\ &+N_u I_N^{***}\end{aligned} &
\begin{aligned} &L_v I_L^{***} \\ &+\left(M_v+\frac{M_{\dot w}Z_v}{(m-Z_{\dot w})}\right)I_M^{***} \\ &+N_v I_N^{***}\end{aligned} &
\begin{aligned} &L_w I_L^{***} \\ &+\left(M_w+\frac{M_{\dot w}Z_w}{(m-Z_{\dot w})}\right)I_M^{***} \\ &+N_w I_N^{***}\end{aligned} &
\begin{aligned} &L_p I_L^{***} \\ &+\left(M_p+h'_z+\frac{M_{\dot w}Z_p}{(m-Z_{\dot w})}\right)I_M^{***} \\ &+\left(N_p-h'_y\right)I_N^{***}\end{aligned} &
\begin{aligned} &\left(L_q-h'_z\right)I_L^{***} \\ &+\left(M_q+\frac{M_{\dot w}Z_q+M_{\dot w}mu_0}{(m-Z_{\dot w})}\right)I_M^{***} \\ &+\left(N_q+h'_x\right)I_N^{***}\end{aligned} &
\begin{aligned} &\left(L_r+h'_y\right)I_L^{***} \\ &+\left(M_r-h'_x+\frac{M_{\dot w}Z_r}{(m-Z_{\dot w})}\right)I_M^{***} \\ &+N_r I_N^{***}\end{aligned} &
0 & \frac{-mgM_{\dot w}I_M^{***}\sin\theta_0}{(m-Z_{\dot w})} & 0 \\
0 & 0 & 0 & 1 & 0 & \tan\theta_0 & 0 & 0 & 0 \\
0 & 0 & 0 & 0 & 1 & 0 & 0 & 0 & 0 \\
0 & 0 & 0 & 0 & 0 & \sec\theta_0 & 0 & 0 & 0
\end{bmatrix}}_{A}
$$

Expanding the system matrix $\boldsymbol{B}$ and the control vector $\vec{c}$ in (5.25a) yields (5.25b). Note, that the Δ in the derivative of the state vector, $\dot{\vec{x}}$, and the state vector itself, $\vec{x}$, indicates, that the reference value is not zero.

Open-Loop, Coupled 6-Degree of Freedom Small Perturbation Equations of Motion, Eq. (5.25b).

$$\underbrace{\begin{bmatrix} \Delta\dot{u} \\ \dot{v} \\ \dot{w} \\ \dot{p} \\ \dot{q} \\ \dot{r} \\ \dot{\phi} \\ \Delta\dot{\theta} \\ \dot{\psi} \end{bmatrix}}_{\dot{\vec{x}}} = \underbrace{\begin{bmatrix} \quad \end{bmatrix}}_{A} \bullet \underbrace{\begin{bmatrix} \Delta u \\ v \\ w \\ p \\ q \\ r \\ \phi \\ \Delta\theta \\ \psi \end{bmatrix}}_{\vec{x}} + \underbrace{\begin{bmatrix} \frac{X_{\delta_{LoCE}}}{m} & \frac{X_{\delta_{DiCE}}}{m} & \frac{X_{\delta_{LaCE}}}{m} & \frac{X_{\tau}}{m} \\ \frac{Y_{\delta_{LoCE}}}{m} & \frac{Y_{\delta_{DiCE}}}{m} & \frac{Y_{\delta_{LaCE}}}{m} & \frac{Y_{\tau}}{m} \\ \frac{Z_{\delta_{LoCE}}}{(m-Z_{\dot{w}})} & \frac{Z_{\delta_{DiCE}}}{(m-Z_{\dot{w}})} & \frac{Z_{\delta_{LaCE}}}{(m-Z_{\dot{w}})} & \frac{Z_{\tau}}{(m-Z_{\dot{w}})} \\ L_{\delta_{LoCE}} I_L^{*} + M_{\delta_{LoCE}} I_M^{*} + N_{\delta_{LoCE}} I_N^{*} + \frac{Z_{\delta_{LoCE}} M_{\dot{w}} I_M^{*}}{(m-Z_{\dot{w}})} & L_{\delta_{DiCE}} I_L^{*} + M_{\delta_{DiCE}} I_M^{*} + N_{\delta_{DiCE}} I_N^{*} + \frac{Z_{\delta_{DiCE}} M_{\dot{w}} I_M^{*}}{(m-Z_{\dot{w}})} & L_{\delta_{LaCE}} I_L^{*} + M_{\delta_{LaCE}} I_M^{*} + N_{\delta_{LaCE}} I_N^{*} + \frac{Z_{\delta_{LaCE}} M_{\dot{w}} I_M^{*}}{(m-Z_{\dot{w}})} & L_{\tau} I_L^{*} + M_{\tau} I_M^{*} + N_{\tau} I_N^{*} \\ L_{\delta_{LoCE}} I_L^{**} + M_{\delta_{LoCE}} I_M^{**} + N_{\delta_{LoCE}} I_N^{**} + \frac{Z_{\delta_{LoCE}} M_{\dot{w}} I_M^{**}}{(m-Z_{\dot{w}})} & L_{\delta_{DiCE}} I_L^{**} + M_{\delta_{DiCE}} I_M^{**} + N_{\delta_{DiCE}} I_N^{**} + \frac{Z_{\delta_{DiCE}} M_{\dot{w}} I_M^{**}}{(m-Z_{\dot{w}})} & L_{\delta_{LaCE}} I_L^{**} + M_{\delta_{LaCE}} I_M^{**} + N_{\delta_{LaCE}} I_N^{**} + \frac{Z_{\delta_{LaCE}} M_{\dot{w}} I_M^{**}}{(m-Z_{\dot{w}})} & L_{\tau} I_L^{**} + M_{\tau} I_M^{**} + N_{\tau} I_N^{**} \\ L_{\delta_{LoCE}} I_L^{***} + M_{\delta_{LoCE}} I_M^{***} + N_{\delta_{LoCE}} I_N^{***} + \frac{Z_{\delta_{LoCE}} M_{\dot{w}} I_M^{***}}{(m-Z_{\dot{w}})} & L_{\delta_{DiCE}} I_L^{***} + M_{\delta_{DiCE}} I_M^{***} + N_{\delta_{DiCE}} I_N^{***} + \frac{Z_{\delta_{DiCE}} M_{\dot{w}} I_M^{***}}{(m-Z_{\dot{w}})} & L_{\delta_{LaCE}} I_L^{***} + M_{\delta_{LaCE}} I_M^{***} + N_{\delta_{LaCE}} I_N^{***} + \frac{Z_{\delta_{LaCE}} M_{\dot{w}} I_M^{***}}{(m-Z_{\dot{w}})} & L_{\tau} I_L^{***} + M_{\tau} I_M^{***} + N_{\tau} I_N^{***} \\ 0 & 0 & 0 & 0 \\ 0 & 0 & 0 & 0 \\ 0 & 0 & 0 & 0 \end{bmatrix}}_{B} \bullet \underbrace{\begin{bmatrix} \delta_{LoCE_i} \\ \delta_{DiCE_i} \\ \delta_{LaCE_i} \\ \tau_i \end{bmatrix}}_{\vec{c}} \tag{5.25b}$$

Writing Eq. (5.25b) in concise form yields Eq. (5.25c). The full-state feedback form of Eq. (5.25c) is given with Eq. (5.26).

Open-Loop, Coupled 6-Degree of Freedom Small Perturbation Equations of Motion in concise form, Eq. (5.25c).

$$\underbrace{\begin{bmatrix} \Delta\dot{u} \\ \dot{v} \\ \dot{w} \\ \dot{p} \\ \dot{q} \\ \dot{r} \\ \dot{\phi} \\ \Delta\dot{\theta} \\ \dot{\psi} \end{bmatrix}}_{\vec{\dot{x}}} = \underbrace{\begin{bmatrix} x_u & x_v & x_w & x_p & x_q & x_r & 0 & x_\theta & 0 \\ y_u & y_v & y_w & y_p & y_q & y_r & y_\varphi & 0 & 0 \\ z_u & z_v & z_w & z_p & z_q & z_r & 0 & z_\theta & 0 \\ l_u & l_v & l_w & l_p & l_q & l_r & 0 & l_\theta & 0 \\ m_u & m_v & m_w & m_p & m_q & m_r & 0 & m_\theta & 0 \\ n_u & n_v & n_w & n_p & n_q & n_r & 0 & n_\theta & 0 \\ 0 & 0 & 0 & 1 & 0 & \tan\theta_0 & 0 & 0 & 0 \\ 0 & 0 & 0 & 0 & 1 & 0 & 0 & 0 & 0 \\ 0 & 0 & 0 & 0 & 0 & \sec\theta_0 & 0 & 0 & 0 \end{bmatrix}}_{A} \underbrace{\begin{bmatrix} \Delta u \\ v \\ w \\ p \\ q \\ r \\ \phi \\ \theta \\ \psi \end{bmatrix}}_{\vec{x}} + \underbrace{\begin{bmatrix} x_{\delta_{LoCE}} & x_{\delta_{DiCE}} & x_{\delta_{LaCE}} & x_\tau \\ y_{\delta_{LoCE}} & y_{\delta_{DiCE}} & y_{\delta_{LaCE}} & y_\tau \\ z_{\delta_{LoCE}} & z_{\delta_{DiCE}} & z_{\delta_{LaCE}} & z_\tau \\ l_{\delta_{LoCE}} & l_{\delta_{DiCE}} & l_{\delta_{LaCE}} & l_\tau \\ m_{\delta_{LoCE}} & m_{\delta_{DiCE}} & m_{\delta_{LaCE}} & m_\tau \\ n_{\delta_{LoCE}} & n_{\delta_{DiCE}} & n_{\delta_{LaCE}} & n_\tau \\ 0 & 0 & 0 & 0 \\ 0 & 0 & 0 & 0 \\ 0 & 0 & 0 & 0 \end{bmatrix}}_{B} \underbrace{\begin{bmatrix} \delta_{LoCE} \\ \delta_{DiCE} \\ \delta_{LaCE} \\ \tau \end{bmatrix}}_{\vec{c}} \tag{5.25c}$$

Closed-Loop, Coupled 6-Degree of Freedom Small Perturbation Equations of Motion in concise form, Eq. (5.26).

$$
\underbrace{\begin{bmatrix} \Delta\dot{u} \\ \dot{v} \\ \dot{w} \\ \dot{p} \\ \dot{q} \\ \dot{r} \\ \dot{\varphi} \\ \Delta\dot{\theta} \\ \dot{\psi} \end{bmatrix}}_{\dot{\vec{x}}} = \underbrace{\begin{bmatrix}
x_u & x_v - K_v x_{\delta_{DiCE}} & x_w - K_w x_{\delta_{LoCE}} & x_p - K_p x_{\delta_{LaCE}} & x_q - K_q x_{\delta_{LoCE}} & x_r - K_r x_{\delta_{DiCE}} & 0 & x_\theta & 0 \\
y_u & y_v - K_v y_{\delta_{DiCE}} & y_w - K_w y_{\delta_{LoCE}} & y_p - K_p y_{\delta_{LaCE}} & y_q - K_q y_{\delta_{LoCE}} & y_r - K_r y_{\delta_{DiCE}} & y_\varphi & 0 & 0 \\
z_u & z_v - K_v z_{\delta_{DiCE}} & z_w - K_w z_{\delta_{LoCE}} & z_p - K_p z_{\delta_{LaCE}} & z_q - K_q z_{\delta_{LoCE}} & z_r - K_r z_{\delta_{DiCE}} & 0 & z_\theta & 0 \\
l_u & l_v - K_v l_{\delta_{DiCE}} & l_w - K_w l_{\delta_{LoCE}} & l_p - K_p l_{\delta_{LaCE}} & l_q - K_q l_{\delta_{LoCE}} & l_r - K_r l_{\delta_{DiCE}} & 0 & l_\theta & 0 \\
m_u & m_v - K_v m_{\delta_{DiCE}} & m_w - K_w m_{\delta_{LoCE}} & m_p - K_p m_{\delta_{LaCE}} & m_q - K_q m_{\delta_{LoCE}} & m_r - K_r m_{\delta_{DiCE}} & 0 & m_\theta & 0 \\
n_u & n_v - K_v n_{\delta_{DiCE}} & n_w - K_w n_{\delta_{LoCE}} & n_p - K_p n_{\delta_{LaCE}} & n_q - K_q n_{\delta_{LoCE}} & n_r - K_r n_{\delta_{DiCE}} & 0 & n_\theta & 0 \\
0 & 0 & 0 & 1 & 0 & \tan\theta_0 & 0 & 0 & 0 \\
0 & 0 & 0 & 0 & 1 & 0 & 0 & 0 & 0 \\
0 & 0 & 0 & 0 & 0 & \sec\theta_0 & 0 & 0 & 0
\end{bmatrix}}_{A}
$$

$$
\times \underbrace{\begin{bmatrix} \Delta u \\ v \\ w \\ p \\ q \\ r \\ \varphi \\ \theta \\ \psi \end{bmatrix}}_{\vec{x}} + \underbrace{\begin{bmatrix}
x_{\delta_{LoCE}} & x_{\delta_{DiCE}} & x_{\delta_{LaCE}} & x_\tau \\
y_{\delta_{LoCE}} & y_{\delta_{DiCE}} & y_{\delta_{LaCE}} & y_\tau \\
z_{\delta_{LoCE}} & z_{\delta_{DiCE}} & z_{\delta_{LaCE}} & z_\tau \\
l_{\delta_{LoCE}} & l_{\delta_{DiCE}} & l_{\delta_{LaCE}} & l_\tau \\
m_{\delta_{LoCE}} & m_{\delta_{DiCE}} & m_{\delta_{LaCE}} & m_\tau \\
n_{\delta_{LoCE}} & n_{\delta_{DiCE}} & n_{\delta_{LaCE}} & n_\tau \\
0 & 0 & 0 & 0 \\
0 & 0 & 0 & 0 \\
0 & 0 & 0 & 0
\end{bmatrix}}_{B} \underbrace{\begin{bmatrix} \hat{\delta}_{LoCE} \\ \hat{\delta}_{DiCE} \\ \hat{\delta}_{LaCE} \\ \hat{\tau} \end{bmatrix}}_{\vec{c}} \tag{5.26}
$$

5.4 Summary of Results

The current chapter has presented the process logic of the methodology concept *AeroMech*. It has introduced the key calculation algorithms instrumental in performing design-oriented stability and control analysis for the range of symmetric and asymmetric aircraft configurations and concepts. Overall, it has been a stringent development requirement to derive a generic stability and control methodology with focus on full functionality and simplicity.

The input file devised defines the range of DCFCs (Design-Constraining Flight Conditions) of interest and company-specific design guidelines and certification requirements. The problem of control allocation for indeterminate aircraft is considered, and solutions suitable for conceptual design application are selected. It is here that the aerodynamic model of the aircraft geometry is set up. Overall, the input file is arranged to enforce user interaction with the program in such a way as to keep the user from overlooking critical conditions.

It has been a primary aim to establish a set of generic aerodynamics- and stability and control analysis algorithms. Clearly, the same methods are utilised for the range of aircraft configurations and concepts under investigation (generic models). Furthermore, the same methods are able to evaluate stability and control aspects through two successive complexity levels, both residing at conceptual design. It has been vital to ensure a consistent analytical approach for both steps, to avoid method switching and implications like correlating results.

All design decision making is based on the evaluation of the trimmed aircraft in six degrees-of-freedom (6-DOF), an exception being the take-off rotation manoeuver. The trim EOM are solved to determine control power available/required for a range of pre-defined critical flight conditions (DCFC). The dynamic EOM are solved to estimate the gain constants required to restore stiffness and damping for reduced stability to unstable airframes. The dynamic stability characteristics are analysed using the results generated by the dynamic EOM. An off-line analysis sequence evaluates the dynamic stability characteristics of the vehicle with reduced order models. This trend-information enables the designer to gain physical insights into the mode drivers.

The output file finally contains design-relevant information, which ensures the balance between control power and static-, manoeuver-, and dynamic stability. Control power available/required is defined with (a) the volume coefficient (geometry), (b) stability derivative coefficients (aerodynamics), and (c) the CE deflection angle required (operation).

References

1. Chudoba, B., "Stability & Control Aerospace Vehicle Design and Test Condition Matrix," Technical Report EF-039/96, Daimler-Benz Aerospace Airbus, September 1996.
2. Goodrich, K.H., Sliwa, S.M., and Lallman, F.J., "A Closed-Form Trim Solution Yielding Minimum Trim Drag for Airplanes With Multiple Longitudinal-Control Effectors," NASA TP 2907, NASA Langley Research Center, May 1989.
3. Chudoba, B., "Software Defining Specification for the VORSTAB Aerodynamic Prediction Program," CoA Report NFP0105, Department of Aerospace Technology, College of Aeronautics, Cranfield University, June 1999.
4. Lan, C.E., "Methods of Analysis in The VORSTAB Code (Version 3.1)," The University of Kansas, May 1993.
5. Lee, H.P., Chang, M., and Kaiser, M.K., "Flight Dynamics and Stability and Control Characteristics of the X-33 Vehicle," AIAA Paper 98-4410, *Guidance, Navigation, and Control Conference and Exhibit*, Boston, MA, 10-12 August 1998.
6. Thelander, J.A., "Aircraft Motion Analysis," Technical Documentary Report FDL-TDR-64-70, March 1965.
7. Maine, R.E., "Aerodynamic Derivatives for an Oblique Wing Aircraft Estimated From Flight Data by Using a Maximum Likelihood Technique," NASA TP 1336, NASA Dryden Flight Research Center, October 1978.
8. Hartman, E.P., "Wind-Tunnel Tests of a 2-Engine Airplane Model as a Preliminary Study of Flight Conditions Arising on the Failure of One Engine," NACA TN No. 646, NACA, April 1938.
9. Archbold, E.J.N. and McKenzie, K.T., "Response in Yaw," Paper, Vol. 50, Issue 424, The Royal Aeronautical Society, The Aeronautical Journal, April 1946, pp. 275–286.
10. Yates, A.H., "Control in Flight Under Asymmetric Power," Aircraft Engineering, September 1947, pp. 287–290.
11. Baker, F.B., "Choice of Fin Area and Dihedral," Aircraft Engineering, March 1948, pp. 87–88.
12. Wright, I.E., "Flight on Asymmetric Engine Power," Aircraft Engineering, December 1950, pp. 350–374.

13. Pinsker, W.J.G., "Directional Stability in Flight With Bank Angle Constraint as a Condition Defining a Minimum Acceptable Value for n_v," RAE TR 67127, Ministry of Aviation, Royal Aircraft Establishment, RAE Farnborough, 30 June 1967.
14. Leyman, C.S. and Scotland, R.L., "The Effect of Engine Failure at Supersonic Speeds on a Slender Aircraft – Predicted and Actual," AGARD-P-119, April 1972.
15. Roskam, J. and Anemaat, W., "An Easy Way to Analyze Longitudinal and Lateral-Directional Trim Problems with AEO or OEI," SAE Technical Paper 941143, Aerospace Atlantic Conference & Exposition, 01 April 1994.
16. Aly, M.S., "Contribution of Side Wind on Aircraft Aerodynamic Characteristics," AIAA Paper 97-3650, *22nd Atmospheric Flight Mechanics Conference*, New Orleans, LA, 11–13 August 1997.
17. Grasmeyer, J., "Stability and Control Derivative Estimation and Engine-Out Analysis," VPI-AOE-254, Department of Aerospace and Ocean Engineering, Virginia Polytechnic Institute and State University, January 1998.
18. Burcham, F.W., Maine, T.A., Burken, J.J., and Bull, J., "Using Engine Thrust for Emergency Flight Control: MD-11 and B-747 Results," NASA TM-1998-206552, NASA Dryden Flight Research Center, May 1998.
19. Roskam, J., "Airplane Flight Dynamics and Automatic Flight Controls – Part I," Third Edition, *DARcorporation*, 1995.
20. Etkin, B. and Reid, L.D., "Dynamics of Flight – Stability and Control," Third Edition, *John Wiley & Sons, Inc.*, 1996.
21. Brüning, G., Hafer, X., and Sachs, G., "Flugleistungen," Third Edition, *Springer*, 1993.
22. Gilruth, R.R. and Turner, W.N., "Lateral Control Required for Satisfactory Flying Qualities Based on Flight Tests of Numerous Airplanes," NACA Report 715, NACA, 1941.
23. Abzug, M.J. and Larrabee, E.E., "Airplane Stability and Control – A History of the Technologies That Made Aviation Possible," First Edition, Cambridge Aerospace Series, *Cambridge University Press*, 1997.
24. Cook, M.V., "Flight Dynamics Principles," First Edition, *Arnold*, 1997.
25. Stevens, B.L. and Lewis, F.L., "Aircraft Control and Simulation," First Edition, *John Wiley & Sons, Inc.*, 1992.

Chapter 6
AeroMech Feasibility

When using the expression 'generic',[1] the present development does not claim to contribute to the on-going search for the *Theory of Everything* in the micro-cosmos, see Weinberg [2], or the macro-cosmos, see Ferguson [3]. Instead, the generic character of *AeroMech* aims to provide the aircraft conceptual design environment with a method, capable on one hand to efficiently investigate current design trends, but flexible enough to unlock advanced vehicle design potential.

6.1 Introduction

The generic methodology concept *AeroMech* has been developed to assess stability and control characteristics of conventional and unconventional winged aircraft at conceptual design level, guided and constrained by design guidelines and quantified certification requirements. The transformation of *AeroMech* into an executable program has been beyond the scope of this time-restricted research period. Instead, emphasis has been placed on conceptually achieving the highest degree of generic functionality and simplicity possible in the time frame allocated, whilst avoiding to compromise the initial research objectives with time-consuming software development activities. Although individual *AeroMech* calculation subroutines have been prepared and executed in a stand-alone mode, it has been decided to stay with the functional description of *AeroMech* in the present report, circumventing stand-alone numerical results without direct context.

[1]**generic** *adj* shared by or including a whole group or class; not specific [1].

B. Chudoba, *Stability and Control of Conventional and Unconventional Aerospace Vehicle Configurations*, Springer Aerospace Technology,
https://doi.org/10.1007/978-3-030-16856-8_6

6.2 Demonstration of Process Logic

The following virtual case studies affirm the theoretical feasibility of *AeroMech*. The different paths possible through the methodology are demonstrated schematically, visualising process flow, data handling, design parameters, numerical processes, altogether using a consistent set of calculation routines. Figures 6.1, 6.2, 6.3 and 6.4 illustrate the general process logic applicable to the engineering extremes in aerospace vehicle design. The 'conventional' TAC (Tail-Aft Configuration) represents the class of highly evolved B707-type transonic transport aircraft, eventually with some degree of relaxed inherent static stability. This highly non-integrated aircraft configuration may be mutated into the TFC (Tail-First Configuration) and the indeterminate TSC (Three-Surface Configuration). At the other end of the design range is the 'unconventional' asymmetric aircraft type, the OFWC (Oblique Flying-Wing Configuration). This promising, highly integrated, inherently unstable, fully coupled aircraft inherits the full spectrum of aircraft conceptual design complication conceivable.

The presentation of the 'fully-fledged' *AeroMech* calculation capability illustrated in Figs. 6.1, 6.2, 6.3 and 6.4 applies to both ends of the flight vehicle complexity spectrum, represented by the TAC and the OFWC (the TAC and OFWC 'bookend' the FWC, TFC, TSC, JWC, etc.). Clearly, the algorithm can be consistently applied to both engineering extremes during the conceptual design phase, confirming, in part, the generic character of the method. Having available a tool capable of assessing the control power required/available for design- and certification-relevant flight cases, being able to trim the aircraft in 6-DOF of either stable or unstable layout, it is obviously possible to simplify the analysis if demanded. When enforcing the standard conceptual design assumptions for the TAC, the pertinent equations in Chap. 5 and Appendix 9 to 11 do collapse to the known, functionally restricted, format. However, there is no intention to demonstrate the standard approach in the present context. Instead, the presentation of the overall process logic illustrated in Figs. 6.1, 6.2, 6.3 and 6.4 indicates commonalties and peculiarities, when applied to the relaxed stability TAC and the unstable OFWC. In detail, the input-file definition for the symmetric TAC is analogous to the input file specified for the asymmetric OFWC. The aerodynamic estimation sequence of the TAC may not consider the full spectrum of coupling derivatives. However, it should be noted that cross-coupling derivatives are required for the symmetric aircraft when investigating asymmetric flight conditions. The TAC usually avoids the control allocation problem in all three axes. For an A380-type of aircraft having 3% manoeuver margin [7], only damping restoration in the longitudinal and directional axes may be required as a first measure.

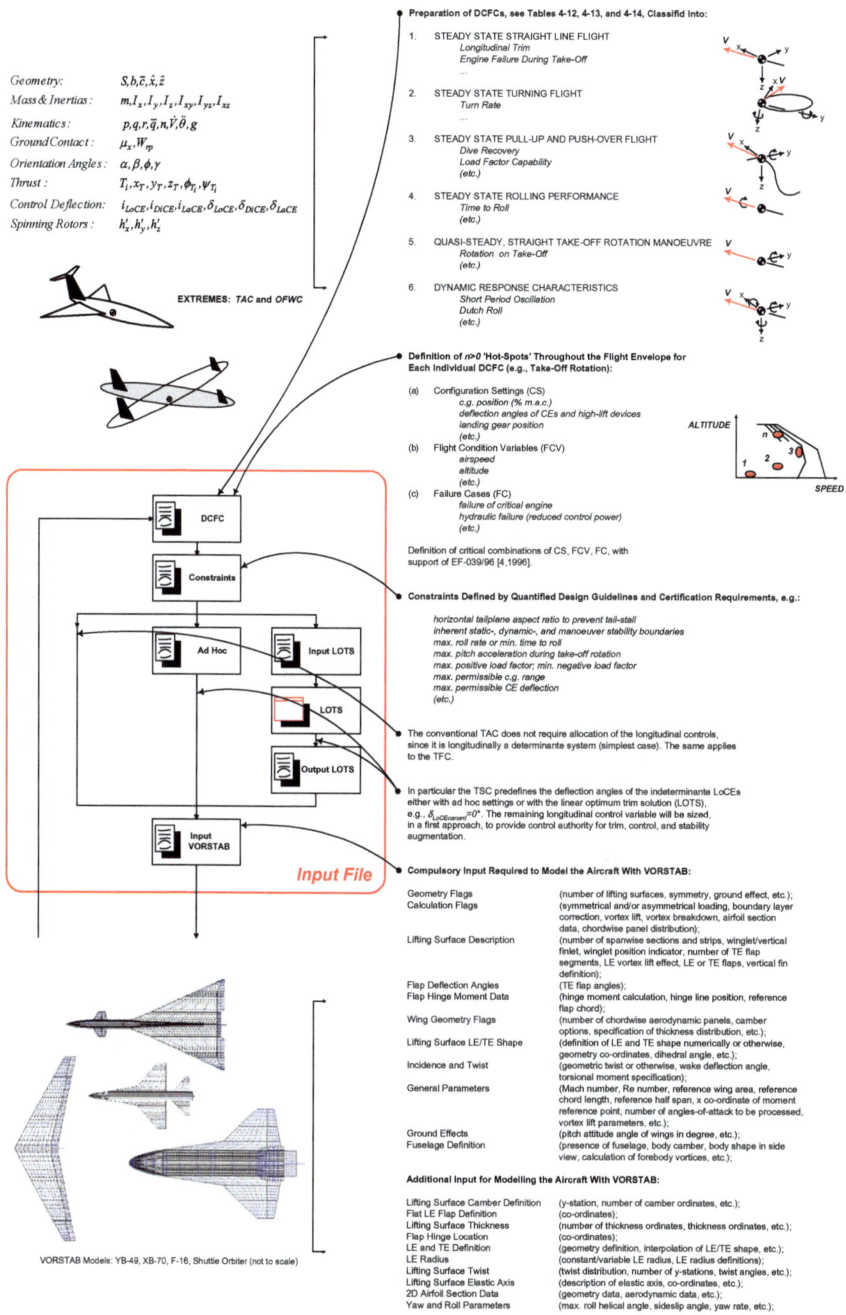

Fig. 6.1 *AeroMech* TAC to OFWC input file definition schematic

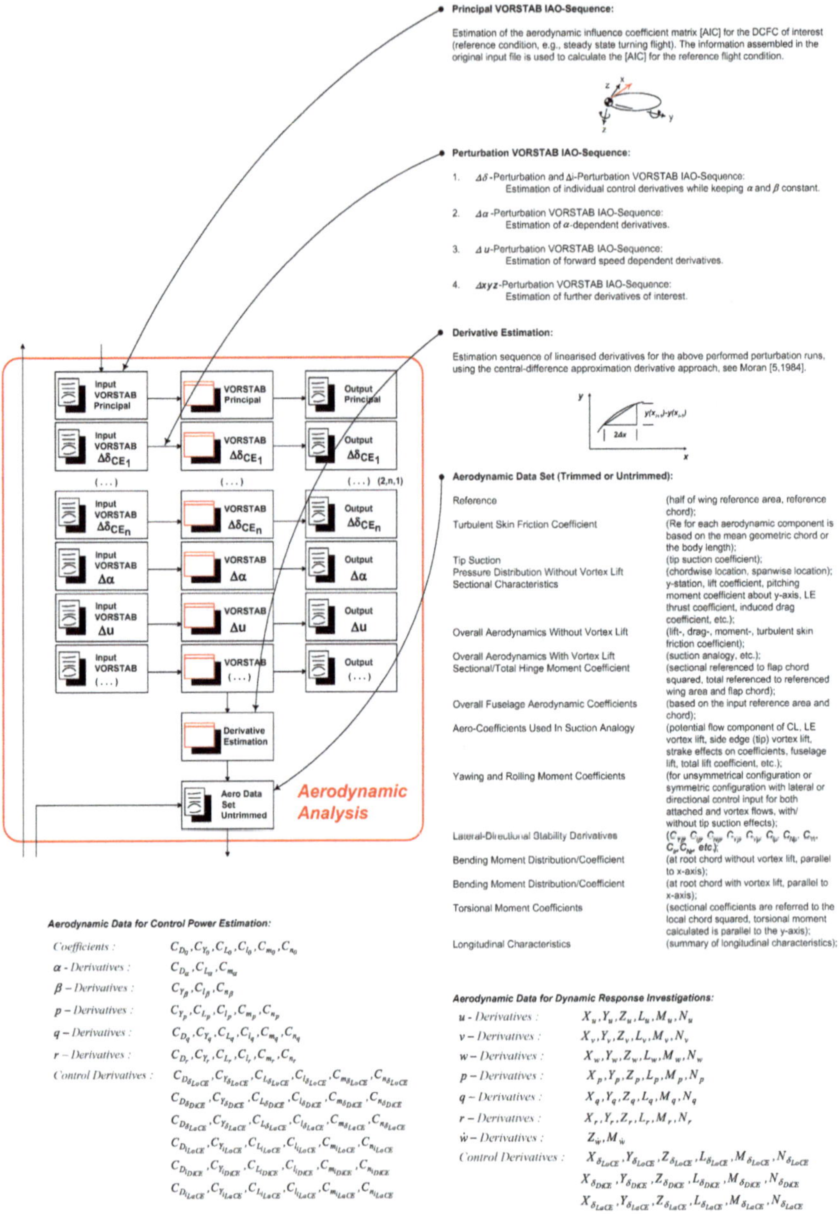

Fig. 6.2 *AeroMech* TAC to OFWC aerodynamic analysis schematic

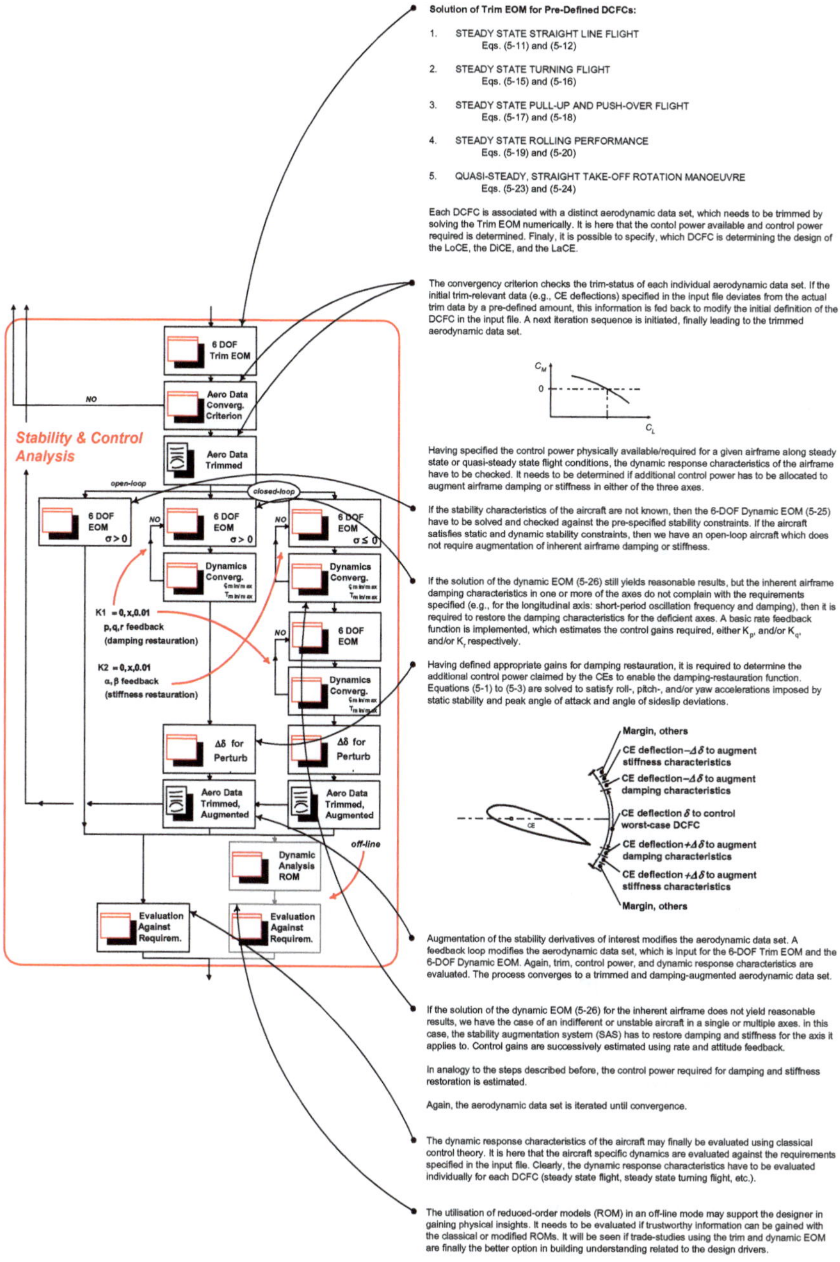

Fig. 6.3 *AeroMech* TAC to OFWC stability and control analysis schematic

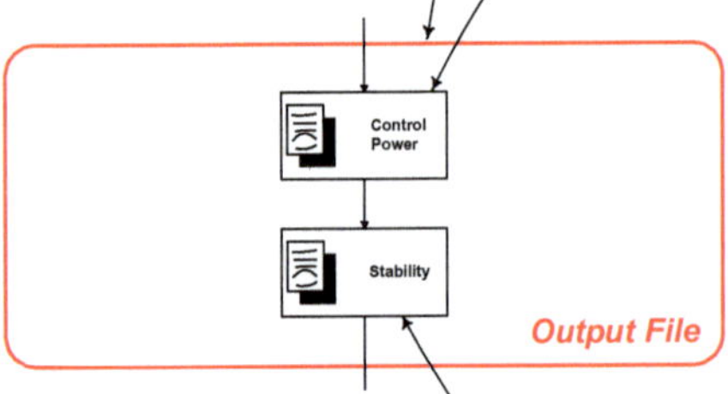

- The stability and control assessment at aircraft conceptual design level evaluates the balance between control power and the flight vehicle static-, dynamic-, and manoeuvre stabilities available and required in all three axes. Stability and control information is provided for each individual DCFC:

1. STEADY STATE STRAIGHT LINE FLIGHT
   Eqs. (5-11) and (5-12)
2. STEADY STATE TURNING FLIGHT
   Eqs. (5-15) and (5-16)
3. STEADY STATE PULL-UP AND PUSH-OVER FLIGHT
   Eqs. (5-17) and (5-18)
4. STEADY STATE ROLLING PERFORMANCE
   Eqs. (5-19) and (5-20)
5. QUASI-STEADY, STRAIGHT TAKE-OFF ROTATION MANOEUVRE
   Eqs. (5-23) and (5-24)
6. DYNAMIC RESPONSE CHARACTERISTICS
   Eqs. (5-25) and (5-26)

- The 6-DOF Trim EOM deliver the primary information specifying the control power demands and control power availability status:

(a) Volume Coefficient (Geometry)

(b) Stability Derivatives (Aerodynamics)

(c) Control Effector (CE) Deflection Angles (Operation)

The information delivered belongs to a trimmed vehicle in 6-DOF, taking weight, inertia, kinematic, and thrust effects into acount. The aircraft design environment has the choice, to optimise the control power availability either with the vehicle geometry choice (configuration, concept, etc.), aerodynamic shape characteristics, and operational features (trimmable stabilizer, all moving CE, control allocation, etc.).

Having evaluated all relevant DCFC, a ranking of the criticality of the individual DCFCs can be undertaken. Clearly, taken the same misson requirements, this ranking differs dependent on the choice of aircraft configuration and concept.

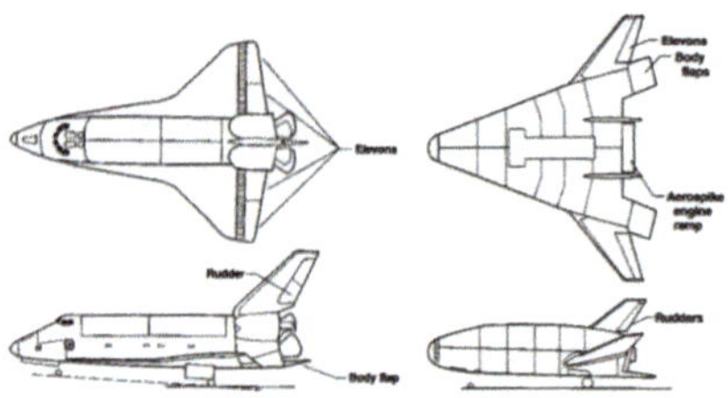

Delta-Winged Shuttle Orbiter *Lifting Body X-33 (SSTO)*

Obviously, a healthy balance between control power and inherent stabilities is sought. However, the availability of sufficient control power is of primary importance, since any stability level can be realised having sufficient control authority available.

- The vehicle stability characteristics, open- or closed-loop, are delivered for each individual DCFC investigated:

(a) **STATIC STABILITY**
   Longitudinal
   Directional
   'Lateral'
   etc.

(b) **DYNAMIC STABILITY**
   Short-Period Mode
   Phugoid Mode
   Dutch Roll Mode
   Spiral Mode
   Roll Mode
   etc.

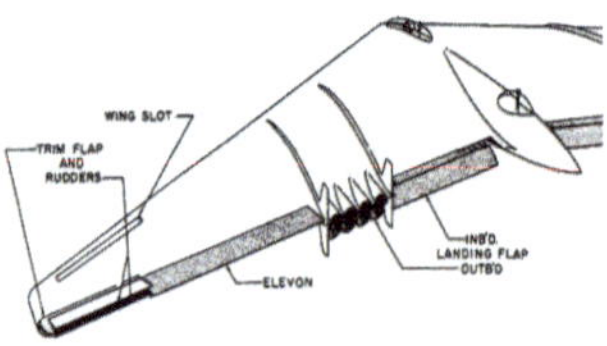

CHALLENGE: *YB-49 Integrated Controls Design, see Ref.* [6,1988]

Clearly, *AeroMech* does not intend to replace the stability and control expert in the disciplinary department. *AeroMech* gives the conceptual design engineer an instrument with the capability, to improve the stability and control quality of the initial design concepts at the same rank as the other primary design disciplines. In the multidisciplinary design environment, this provides a much better starting point, especially when observing the overall effects of stability and control on aircraft safety and in particular on the design of novel aircraft configurations.

Fig. 6.4 *AeroMech* TAC to OFWC output file definition schematic

6.3 Validation and Integration of *AeroMech*

6.3.1 Data Availability to Enable Validation and Calibration of AeroMech

It has often been argued, that it is not feasible to develop a design-methodology dealing with 'novel' vehicles. The main argument supporting this hypothesis is the opinion, that there is a permanent lack of data available enabling adequate validation and calibration. Clearly, novelty can not necessarily be originated by interpolating or extrapolating statistical databases. The outcomes of such an approach are the well-known vehicle design handbooks and guidelines, which are primarily based on semi-empirical and empirical relations.

The developments undertaken in the present context have followed two main principles. At first, strive for a physically correct and highly flexible modelling algorithm. Secondly, learn as much as possible from more than one-hundred years of flying vehicle design experience. The outcome of the second task has been a dedicated conceptual design data-base system (DBS), leading to the aircraft knowledge-based system (KBS), see Chapt. 2.5. The DBS and the KBS have enabled the author to invalidate the hypothesis expressed before. The author is in the position to validate and calibrate *AeroMech* along the range of 'conventional' and 'unconventional' aircraft configurations and concepts discussed throughout the report. To recall, Fig. 2–9 in Chap. 2 has assembled a selection of case studies, where the availability of data in the DBS and KBS comprising company-internal and public-domain information, has been judged adequate for a thorough testing phase of *AeroMech*. Table 6.1 presents one example case study for each aircraft configuration type with two selected public domain references.

6.3.2 Integration of AeroMech *into an Aerospace Vehicle Design Synthesis Environment*

As has been outlined before, the multi-disciplinary design-effects of the stability and control methodology on the overall flight vehicle are best demonstrated, when having integrated *AeroMech* into an advanced aircraft conceptual design synthesis environment. The following briefly sketches some of the integration implications having integrated *AeroMech* into, e.g., an industry environment.

The engineering design process, from conceptual to detail design and flight test, has to harmonise technical capability and design flexibility with efficient knowledge and people management. Figure 6.5 shows, how *AeroMech* will be integrated into such dynamic aircraft development engineering organisation.

The conceptual design team operates at the highest level of design-abstraction, supported by a computer-based aircraft conceptual design synthesis methodology.

Table 6.1 Public-domain *AeroMech* Validation sweep

Aircraft	Configuration	Concept Choice				References (Examples)	Validation	
		Lift source	Volume supply	Control effector	Propulsion		*Items*	*Remarks*
X-15	TAC	Low-aspect ratio wing	Distinct fuselage	All-moving LoCE, all-moving DiCE, differential LoCE, thrusters	Single rocket, mid-fuselage	[8, 9]	Full range	Fuselage dominated hypersonic design; representative SSTO technology-contributing research vehicle configuration; very good range of data availability in public domain: wind tunnel, flight test, analytical predictions
XB-70	TFC	Delta wing	Distinct fuselage and wing	Elevons (LoCE and LaCE), all-moving DiCE, trimmable canard	6 turbo jets, low-wing	[10, 11]	Full range	Wing dominated supersonic design using extensive aerodynamic shaping; highly interesting due to variable wing geometry, canard layout, high speed range; very good range of data availability in public domain: wind tunnel, flight test, analytical predictions
P-180	TSC	Unswept high-aspect ratio wing	Distinct fuselage and wing	LoCE, DiCE, LaCE, canard with landing flap	2 turbo props, mid-wing	[12, 13]	Aerodynamics, performance, stability and control	Multi-wing transonic design; representative for an advanced commercial transport configuration with outstanding performance potential; example for a well-performed integrated TSC-design; good range of data available to the author: wind tunnel and analytical predictions

(continued)

Table 6.1 (continued)

Aircraft	Configuration	Concept Choice				References (Examples)	Validation	
		Lift source	Volume supply	Control effector	Propulsion		*Items*	*Remarks*
XB-35	FWC	Swept high-aspect ratio wing	Wing only	Elevons (LoCE, LaCE), DiCE, drag rudders (DiCE)	4 turbo props, mid-wing	[14, 15]	Full range	Fully integrated/blended transonic design; direct relevance to current blended-wing body (BWB) research activities; primary case study for inherently unstable designs; very good range of data availability to the author: wind tunnel, flight test, analytical predictions
Space Shuttle	FWC	Delta wing	Distinct fuselage	Elevons (LoCE, LaCE), body flap (LoCE), DiCE, thrust vectoring, thrusters	3 rockets, aft body	[16, 17]	Full range	Fuselage dominated hypersonic design; first truly operational delta winged reentry vehicle of limited cross-range capability; aerodynamic, thrust, and thruster controls; very good (outstanding) range of data availability in public domain: wind tunnel, flight test, analytical predictions
AD-1	OWC	High-aspect ratio wing, variable sweep	Distinct fuselage	LoCE, DiCE, LaCE	2 turbo jets, aft fuselage	[18, 19]	Full range	Open-loop subsonic design; availability of inherent airframe stability and control and handling qualities assessment results; predestined case study for integrated FCS design; very good range of data availability in public domain: wind tunnel, flight test, analytical predictions

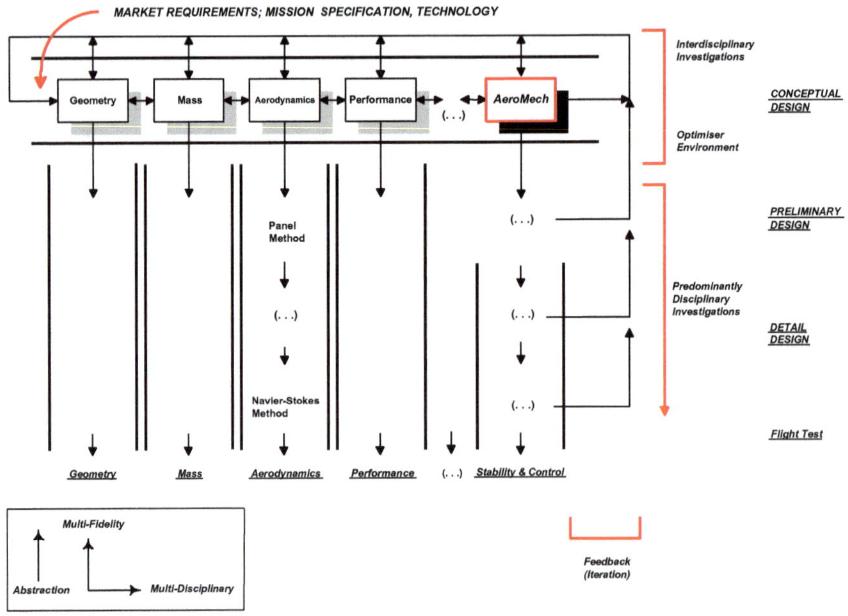

Fig. 6.5 Functional integration of *AeroMech* into an aircraft development engineering organisation

Overall, the multidisciplinary design space of an aircraft project is evaluated and finally defined at conceptual design level.

As Fig. 6.5 indicates, the individual analysis routines at conceptual design level are linked 'horizontally' (multi-disciplinary investigations), requiring consistent (single fidelity) calculation routines. In contrast, design contributions at preliminary and detail design level are of rather disciplinary character. To ensure an efficient overall design process, rapid feedback communications have to link the specialist departments with the conceptual design team. Such feedback ensures, that stability and control detail design solutions and others, proposed by the specialist departments, are checked and balanced in the multidisciplinary context.

It is to be expected, that the stability and control design-decision making process will be significantly supported and accelerated, having *AeroMech* integrated into the design synthesis environment. *AeroMech*'s role in either the stand-alone or automated operating mode will be, to harmonise the stability and control design parametrics with the gross aerospace vehicle design parametrics, leading to a well behaved and safe flight vehicle.

6.4 Summary of Results

In this chapter, the author's developments have been put together to form the basis for a generic conceptual design process, by which the aircraft configuration and concept is shaped to provide good stability and control characteristics, ultimately leading to a safe aircraft.

The theoretical feasibility of *AeroMech* has been affirmed by qualitatively discussing engineering extremes in aerospace vehicle design, being the 'conventional' symmetric tail-aft configuration (TAC), and the 'unconventional' asymmetric oblique flying-wing configuration (OFWC). The *AeroMech* algorithm has not been simplified for demonstrating compliance with the traditional approach. Instead, the 'fully-fledged' process logic has been visualised, indicating the design-benefit to be expected when applied to the range of aircraft configurations, from the TAC to the OFWC.

The input, aerodynamic analysis, stability and control analysis, and output sequences have been summarised using a clearly laid out format. The process logic of *AeroMech* has been demonstrated, assuming maximum modelling complexity throughout the discussion. Clearly, the same maximum complexity level can be consistently applied to both vehicle extremes, indicating the generic character of the method. It may be permissible to simplify the analysis for the range of symmetric aircraft configurations, in particular the TAC. However, this approach is not feasible for the spectrum of asymmetric aircraft and the symmetric aircraft types in asymmetric flight conditions.

The importance of having available suitable aircraft data, which enable a thorough validation and calibration process of *AeroMech*, has been stressed. Having constructed the dedicated aircraft conceptual design data-base system (DBS) and the knowledge-based system (KBS) for the range of aircraft configurations and concepts, both systems will show again their overall importance and potential when performing the final validation/calibration of *AeroMech*. Clearly, the availability of trustworthy data has invalidated the initial fear of not being able to test a generic method conception.

The final section has outlined some implications associated with integration of *AeroMech* as a stand-alone or automated module into an aerospace vehicle design synthesis environment. The transformation of *AeroMech* into an executable software package and the integration into an advanced aerospace vehicle synthesis system have been beyond this time-restricted research undertaking. Although beyond the coverage of the present book, the software *AeroMech* has been developed, tested and applied by Pippalapalli in 2004 [20], Coleman in 2007 [21] and Omoragbon in 2010 [22] in the AVD Laboratory at The University of Texas at Arlington.

References

1. Hornby, A.S., "Oxford Advanced Learner's Dictionary of Current English," Fourth Edition, *Oxford University Press*, 1990.
2. Weinberg, S., "Unified Theories of Elementary Particle-Interaction," Vol. 231, No. 1, Scientific American, July 1974, pp. 50–59.
3. Ferguson, K., "Stephen Hawking – Quest for a Theory of Everything," Reprinted, *Bantam Books*, 1992, 1996.
4. Chudoba, B., "Stability & Control Aerospace Vehicle Design and Test Condition Matrix," Technical Report EF-039/96, Daimler-Benz Aerospace Airbus, September 1996.
5. Moran, J., "An Introduction to Theoretical and Computational Aerodynamics," First Edition, *John Wiley & Sons*, 1984.
6. Kohn, L.J., "Pilot's Handbook – The Flying Wings of Northrop," *Aviation Publications*, Washington, 1988.
7. Wissel, W.D., "A380 Programmstatus," Presentation at the Technical University Munich, 22 February 2001.
8. Matranga, G.J., "Analysis of X-15 Landing Approach and Flare Characteristics Determined from the First 30 Flights," NASA TN D-1057, NASA Dryden Flight Research Center, April 1961.
9. Dana, W., "The X-15 Airplane - Lessons Learned," AIAA Paper 93–0309, *31st Aerospace Sciences Meeting*, Reno, NV, 11-14 January 1993.
10. Wolowicz, C.H. and Yancey, R.B., "Summary of Stability and Control Characteristics of the XB-70 Airplane," NASA TM X-2933, NASA Dryden Flight Research Center, October 1973.
11. Petersen, R.H., "The Effects of Wing-Tip Droop on the Aerodynamic Characteristics of a Delta-Wing Aircraft at Supersonic Speeds," NASA TM X-363, May 1960.
12. Sacco, G., "P-180: Reasons and Evolution of an Unconventional Aerospace Vehicle Design," I.A.M. Rinaldo Piaggio S.p.A., The Michigan State University, October 1989.
13. de'Pompeis, R., Cinquetti, P. and Martini P.I.S., "Development and Certification Flight Test on the Piaggio P.180 Avanti Aircraft: A General Overview," SAE Paper 91–1003, *General, Corporate & Regional Aviation Meeting & Exposition*, 01 April 1991.
14. Kamm, R.W. and Pepoon, P.W., "Spin-Tunnel Tests of a 1/57.33-Scale Model of the Northrop XB-35 Airplane," NACA L-739, NACA Wartime Report, April 1944.
15. Sivells, J.C. and Burgess, J., "Tests in the NACA 19-Foot Pressure Tunnel of a 1/10.75-Scale Model of the Northrop XB-35 Tailless Airplane," NACA MR ARC 7391, February 1943.
16. Anon., "Aerodynamic Design Data Book, Orbiter Vehicle 102, Vol. 1," SD72-SH-0060, Vol. 1 M, Space Systems Group, Rockwell International, April 1979.
17. Young, J.C. and Underwood, J.M., "The Development of Aerodynamic Uncertainties for the Space Shuttle Orbiter," AIAA Paper 82–0563, *12th Aerodynamic Testing Conference*, Williamsburg, VA, 22-24 March 1982.
18. Sim, A.G. and Curry, R.E., "Flight-Determined Aerodynamic Derivatives of the AD-1 Oblique-Wing Research Airplane," NASA TP-2222, NASA Ames Research Center, 01 October 1984.
19. White, W.L. and Bowman, J.S., "Spin-Tunnel Investigation of a 1/13-Scale Model of the NASA AD-1 Oblique-Wing Research Aircraft," NASA TM 83236, NASA Langley Research Center, 01 February 1982.
20. Pippalapalli, K.K., "AeroMech – A Conceptual Design Stability and Control Analysis Program," M.S. Thesis, AVD Laboratory, The University of Oklahoma, Oklahoma, 2004.
21. Coleman, G.J., "Aircraft Conceptual Design – A Generic Stability and Control Tool for Flight Vehicle Conceptual Design: AeroMech Software Development," M.S. Thesis, AVD Laboratory, The University of Texas at Arlington, Arlington, Texas, May 2007.
22. Omoragbon, A., "An Integration of a Modern Flight Control System Design Technique into a Conceptual Design Stability and Control Tool, AeroMech," M.S. Thesis, AVD Laboratory, The University of Texas at Arlington, Arlington, Texas, May August 2010.

Chapter 7
Conclusions

This book has been written about 128 years after Otto Lilienthal made his first controlled gliding flight, and some 116 years after the Wright brothers accomplished their first controlled powered flight. The period since those early pioneers has been filled with exceptionally interesting activities, having led to breathtaking breakthroughs in aeronautics and access to space. There is no doubt, the future promises challenges, which would seem revolutionary to the early pioneers and us alike. The quest for the new era of transonic, supersonic, and hypersonic air transportation, inexpensive access to space for commercial and scientific purpose, has not yet been answered.

When comparing the flight performance of the early flight vehicles with those of the most advanced aerospace vehicles in operation today, technological advances have multiplied the initial performance expectations. However, common to all flight vehicles is the stringent demand to be stable and controllable, both characteristics being the primary measure for flight safety. As an example, the highly blended or integrated breed of advanced hypersonic flight vehicles does promise exciting performance improvements, while resulting in true stability and control design challenges. Since the provision of satisfactory stability and control characteristics invariably compromises flight performance to some extent, it becomes essential in today's unforgiving environment, to implement performance-optimal stability and control design-solutions into the initial flight vehicle conception.

This text aims to contribute to the engineering toolbox required, to efficiently unlock hidden design potential in state-of-the-art and advanced flight vehicle proposals.

B. Chudoba, *Stability and Control of Conventional and Unconventional Aerospace Vehicle Configurations*, Springer Aerospace Technology,
https://doi.org/10.1007/978-3-030-16856-8_7

7.1 Contributions and Conclusion Summary

This work has demonstrated several new ideas and methods for the assessment of stability and control characteristics of conventional and unconventional aircraft configurations and concepts during the conceptual design phase. The following summarises these contributions along the individual book chapters.

Chapter 1: It has been the principal aim of the present research undertaking, to advance aircraft conceptual design tool maturity with respect to current and future aerospace vehicle demands. The motivation for doing so has been justified by identifying *Today's Aerospace vVehicle dDesign Problem* leading to the *New Aerospace vVehicle dDesign Challenge* . One major contributor to the above situation is the overall lack of adequately representing stability and control at the conceptual design level. Based on the understanding gained, the research project aims, scope, and objectives are defined, and are repeated here for convenience:

1. *Development of a generic aerospace conceptual design methodology with the primary objective to size the vehicle's stability and control surfaces, thereby reducing overall development risk and today's prolonged vehicle design cycle periods, whilst improving vehicle safety and performance* .
2. *Widening of the project engineer's design freedom by creation of a generic or configuration & concept independent methodology which enables control surface sizing of subsonic to hypersonic aerospace vehicle designs of conventional and unconventional configuration layout.*
3. *In conventional aircraft conceptual design procedures, design for performance is done before design for stability and control. This improved methodology shall enable evaluation of stability and control in parallel with performance during the conceptual and preliminary design phase of future efficient aerospace vehicles* .
4. *Transformation of flight mechanics as today's advanced analysis discipline (disconnected from design) to a generic design discipline by harmonisation of the complex balance between control power and inherent airframe stabilities (static-, dynamic-, and manoeuvre stability).*
5. *Integration of flight test and certification aspects relevant to the design of controls into conceptual aerospace vehicle design.*
6. *Assemblage, extraction, management and inclusion of appropriate aerospace vehicle design data, design information, and design knowledge to enable an informed approach with the consequent intent "... things should be as simple as possible, but no simpler...".*

Chapter 2: The above research objectives have triggered the definition of an ambitious research strategy, demanding extensive utilisation of design data, -information,—knowledge, and expertise, due to the wide scope of flight vehicles considered. This knowledge utilisation activity consequently initiated the development of a dedicated aerospace vehicle conceptual design literature *Data-Base System* (DBS) and of a *Knowledge-Based System* (KBS), both being unique in conception, contents, and scope. Clearly, the DBS and KBS have favourably served as the knowledge-foundation for all follow-on development steps.

Chapter 3: The ingredients and pecularities of the aircraft conceptual design process, relevant in the context of developing a generic stability and control methodology, have been identified and analysed. It has been here that the airworthiness problems of conventional and advanced aircraft have been discussed. The aerodynamic tools essential for configuration aerodynamic estimation at the conceptual design level are distinguished. Furthermore, the potential and limitations of past and present aircraft conceptual design synthesis methodologies have been identified, since these are the environments where the stability and control methodology *AeroMech* will be integrated. To recall, nearly all aerospace vehicle conceptual design environments do consider stability and control as a secondary rather than a primary design discipline. This fact is even more surprising when identifying its overall importance on flight safety, flight operation, flight performance and certification. Clearly, there exists a wide discrepancy between the sophisticated approach of the modern flight dynamicist during detail design compared to the traditional approach in use during conceptual design. Realising the multitude of performance-driven advancements in the field of modern aerospace vehicle design like relaxed static stability, fuel transfer, control allocation, and advanced configuration layouts, it becomes obvious that the traditional stability and control methods in use have stagnated in their evolution over more than half of a century. It is the clear aim of the present research undertaking to bridge the apparent gap between conceptual design work and detail design work.

Chapter 4: The ability to define the problem solving capabilities of *AeroMech* naturally depends on knowing the true impact of stability and control on aerospace vehicle design in the first place. With this intention in mind, the current chapter has presented an attempt to identify, isolate, and interpret relevant design parameters required for the development of the generic stability and control methodology *AeroMech*. Four primary impact-subdisciplines to the design of controls are discussed: (1) *Geometry and Mass Properties*: An attempt has been made to consistently define flight vehicle configurations and concepts. (2) *Aerodynamics*: After having defined a generic set of gross aerodynamic design parameters, a suitable configuration-aerodynamics estimation technique has been selected. Clearly, the capability and potential of the aerodynamic method selected is pivotal in realising the generic capability of *AeroMech*. (3) *Stability and Control*: The techniques used for the design of controls have been reviewed, realising that the methods in use have stagnated in development although the flight vehicle has continued to evolve. Several concepts and technologies related to stability and control have been evaluated for suitability within *AeroMech*. (4) *Flight Evaluation Expertise*: Considerable effort has been invested in closing the loop between aircraft conceptual design and flight test. JAR/FAR 25 certification-relevant formulations have been reviewed, which inevitably guide the design of control effectors. A generic set of design-critical flight conditions has been defined with the support of A340 and Concorde flight test schedules while taking the range of relevant certification requirements into account. This generic set of design-constraining flight conditions has been grouped into two successive calculation phases, both being relevant for conceptual design studies.

Chapter 5: Having assembled a generic set of design parameters and processes for the range of aircraft types under investigation, the generic stability and control methodology concept *AeroMech* is introduced. The input-, analysis-, and output process logic is illustrated. The key calculation algorithms for performing design-oriented stability and control analysis of the range of symmetric and asymmetric aircraft configurations and concepts are presented. In summary, the algorithm is capable of trimming the aircraft in six degrees-of-freedom, the control allocation problem is handled, airframes of stable, indifferent, or unstable layout can be evaluated, the aerodynamic control effectors are defined by specifying the volume coefficient, stability derivatives, and the deflection angles required to comply with pre-defined design- and certification requirements.

Chapter 6: The theoretical feasibility of *AeroMech* has been affirmed by qualitatively discussing engineering extremes in aerospace vehicle design, being the 'conventional' symmetric tail-aft configuration (TAC), and the 'unconventional' asymmetric oblique flying-wing configuration (OFWC). The *AeroMech* algorithm has not been simplified for demonstrating compliance with the traditional approach. Instead, the 'fully-fledged' process logic has been visualised, indicating the design-benefit to be expected when applied to the range of aircraft configurations, from the TAC to the OFWC. Although the software development phase of *AeroMech* had to be excluded from the current research undertaking, it is important to discuss the availability of suitable aircraft data to enable a thorough validation and calibration process. Having constructed the dedicated aircraft conceptual design DBS and KBS for the range of aircraft configurations and concepts, both systems will show again their overall importance and potential when performing the final validation/calibration of *AeroMech*. Clearly, the availability of trustworthy data has invalidated the initial fear of not being able to test a generic method. The final section of this book briefly outlines some implications associated with integration of *AeroMech* as a stand-alone or automated module into an aerospace vehicle design synthesis environment.

Summary: The initial research objectives have been satisfied. The research undertaking has demonstrated a feasible approach to develop generic calculation algorithms and it delivers a generic stability and control methodology concept named *AeroMech*. The capability inherent in *AeroMech* truly opens the design space to new solutions previously hidden by the biases of classic designs and approaches during conceptual design. The transformation of the *AeroMech* methodology into an executable software package has been beyond the scope of the original research period as documented in this book. (Note: the software *AeroMech* has been developed from 2004 through 2010 in the AVD Laboratory at The University of Texas at Arlington.) The first of its kind, the conceptual design aeronautical *Data-Base System* (DBS) and *Knowledge-Based System* (KBS), as devised for this research undertaking, both are well suited to enable the development of a range of conceptual design level disciplinary calculation routines of generic character.

"The beauty of a capability such as AeroMech is that it could mature with the design and be an ever-ready tool for use at all stages. The result would be a more mature design at "Configuration Freeze" and a well developed and correlated tool for use in squeezing out the last bit of stability and control capability from a "Frozen" design." Gerald C. Blausey, Lockheed Martin, 1998

7.2 Recommendations for Future Work

(1) Transformation of the *AeroMech* conception into an executable program including stand-alone validation and calibration.
(2) Integration of *AeroMech* into a computer-based advanced aircraft conceptual design synthesis system and extensive application to realised aircraft and new aircraft proposals.
(3) Improvements of *AeroMech*:

   (a) Include the effects of aeroelasticity.
   (b) Include the effects of icing conditions.
   (c) Develop/integrate a control allocation logic for lateral staggering of control effectors.
   (d) Advance the vortex lattice method (high-lift devices, derivative estimation).
   (e) Develop the logic for an integrated flight control system (FCS) design.

(4) Advance the computer-based aircraft conceptual design *Data-Base System* (DBS) and *Knowledge-Based System* (KBS).

Appendix

A.1 DBS—File Structure

Figure A.1 presents the overall file-structure of the computer-based literature *Data-Base System* (DBS). As an example, a screenshot of the file *Fwc.doc* is given at the bottom, containing several hundred references dealing with the flying-wing configuration (FWC).

B. Chudoba, *Stability and Control of Conventional and Unconventional Aerospace Vehicle Configurations*, Springer Aerospace Technology,
https://doi.org/10.1007/978-3-030-16856-8

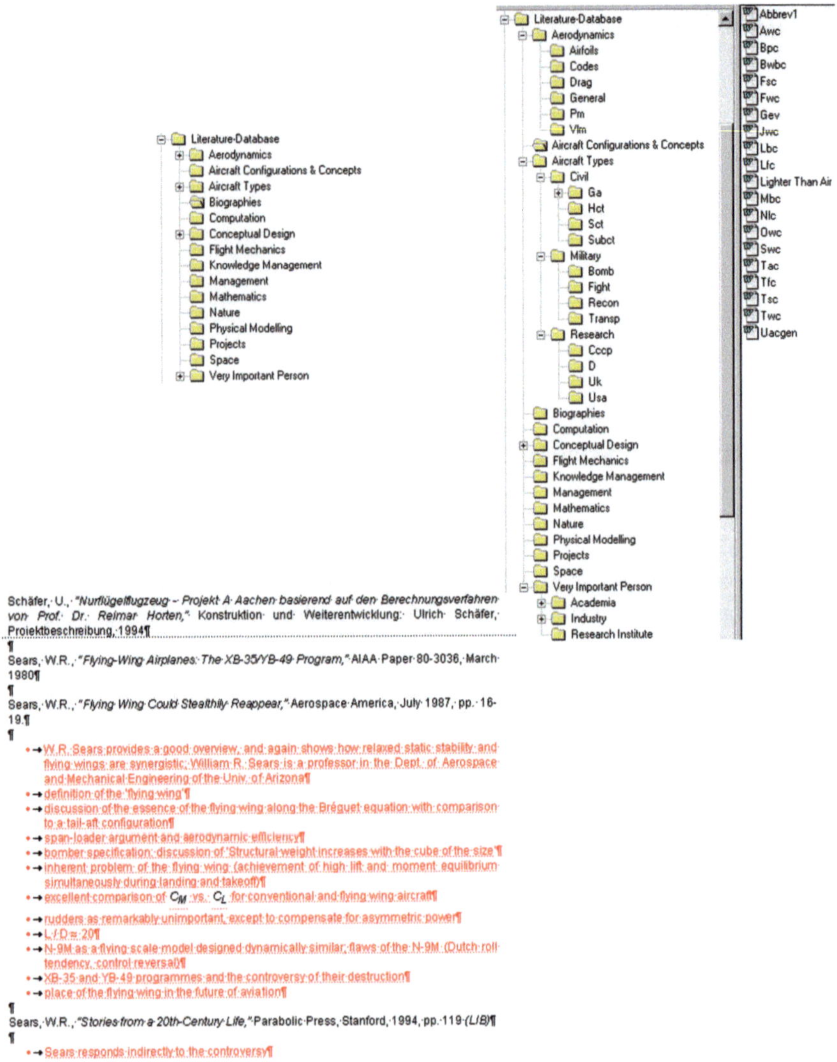

Fig. A.1 File structure of the literature Data-Base System (DBS) and a screenshot of the FWC.doc flying-wing file

A.2 DBS—Table of Contents of 'S&C Characteristics of Subsonic, Supersonic, and Hypersonic Aircraft Configurations'[1]

[1]Chudoba, B., "Stability and Control Characteristics of Subsonic, Supersonic, and Hypersonic *Aircraft* Configurations," CoA Report NFP0103, Department of Aerospace Technology, College of Aeronautics, Cranfield University, April 2001.

A.3 KBS—Table of Contents of 'Aircraft Configuration Characterisation for Project Flight Mechanics'[2]

[2]Chudoba, B., "Aircraft Configuration Characterisation For 'Project Flight Mechanics'," Issue 1, CoA Report NFP0106, DoAT, College of Aeronautics, Cranfield University, April 2001.

Longitudinal Motion

FLIGHT CHARACTER - Design Constraining Flight Conditions

TRIM & CONTROL

1 g Flight (Equilibrium / Performance Case) ....

Low Speed Performance

Minimum Control Speed, Take-Off Climb (V_{MCA}) (144) ....

Minimum Control Speed, Approach and Landing (V_{MCG}) (145) ....

Minimum Control Speed, Approach and Landing (V_{MCL}) (146) ....

Minimum Speed at High Incidence ($V_{MIN}.V_{\alpha_{max\,1g}}$) (149) ....

Go Around on 4 Engines Without Ground Effect (90) ....

Stall Demonstration in Free Flight ....

C.G. Range Performance

Longitudinal Trim (161) ....

Fuel Transfer Failures (Trim Tank Failure) (168) ....

Trim Curves (169) ....

Subsonic Cruise With Trim Jam (170) ....

Aft C.G. Clearance - C.G. Changes at Constant Speed and Height (182) ....

Double Hydraulic Failure (295-315) ....

Effect of C.G. Error on Normal Laws (318) ....

Effect of C.G. Error on Direct Laws (319) ....

C.G. Shifting Speed by Fuel Transfer System ....

Emergency Descent with Partial Loss of Forward C.G. Transfer Facility (127) ....

Nose Wheel Load at Break Release ....

Power Application Performance

Transonic and Supersonic Climb with Thrust Reversal in Flight (19) ....

Emergency Descent with Reverse Thrust Operating, Partial Loss of C.G. Transfer Facility (131)

Power Application (8) ....

Response to Rapid Changes in Thrust (13) ....

Supersonic Cruise Response to Sudden Throttle Changes (24) ....

Shut Over ....

Secondary Controls Performance

Slats / Flaps Failures (320-324) ....

Landing at 6° Slope (69) ....

Primary Control Failures

Foreplane Runaway ....

Hydraulic System Failure ....

Foreplane Failing to a Fixed Position ....

(>,<) 1 g Flight (Manoeuvring / Special Case) ....

Low Speed Performance

Rotation on Take-Off (Nosewheel Lift-Off) (11) ....

Rotation on Landing with Ground Effect ....

Nose Wheel Touch Down at Landing ....

Speed Recovery (5) ....

Mandatory Over-Shoot Operation (14) ....

High Speed Performance

Load Factor Capability in Straight Flight (2) ....

Longitudinal Handling at V_{REF} (12) ....

Transonic and Supersonic Climb Longitudinal Manoeuvrability (20) ....

Transonic and Supersonic Elevator Angle per g in turns (104) ....

Supersonic Rapid Power Reduction ....

Supersonic Elevator Angle per g in Turns (114) ....

Pull-Up at V_D ....

STABILITY

Static

Stability Performance

Longitudinal Stability ....

Dynamic

Mode Performance

Phugoid Mode (203) ....

Short Period Oscillation Mode (203) ....

Dynamic Longitudinal Response ....

AERODYNAMIC CHARACTER - Stability and Control Derivatives

u-Derivatives

C_{D_u}, C_{T_u} [Variation of Drag/Thrust Coefficient with Dimensionless Speed (Speed Damping)] ....

C_{L_u} [Variation of Lift Coefficient with Dimensionless Speed] ....

C_{m_u} [Variation of C_M with Dimensionless Speed (Speed Damping, Tuck Derivative)] ....

u/t-Derivatives

$C_{D_{\dot{u}}}, C_{T_{\dot{u}}}$ [Variation of Drag / Thrust Coefficient with Rate of Change of Dimensionless Speed] ....

$C_{L_{\dot{u}}}$ [Variation of Lift Coefficient with Rate of Change of Dimensionless Speed] ....

$C_{m_{\dot{u}}}$ [Variation of Pitching Moment Coefficient with Rate of Change of Dimensionless Speed] ....

w (α)-Derivatives

$C_{D_\alpha}, C_{T_\alpha}$ [Variation of Drag / Thrust Coefficient with Angle-of-Attack (Induced Drag Derivative)] ....

C_{L_α} [Variation of Lift Coefficient with Angle of Attack (Lift Curve Slope)] ....

C_{m_α} [Variation of C_M with Angle of Attack (Longitudinal Static Stability, Pitch Stiffness)] ....

w/t (α/t)-Derivatives

$C_{D_{\dot{\alpha}}}, C_{T_{\dot{\alpha}}}$ [Variation of Drag / Thrust Coefficient with Rate of Change of Dim. Angle-of-Attack] ....

$C_{L_{\dot{\alpha}}}$ [Variation of Lift Coefficient with Rate of Change of Dimensionless Angle-of-Attack] ....

$C_{m_{\dot{\alpha}}}$ [Variation of Pitching Moment Coefficient with Rate of Change of Dim. Angle-of-Attack] ....

q-Derivatives

C_{D_q}, C_{T_q} [Variation of Drag / Thrust Coefficient with Dimensionless Pitch Rate] ....

C_{L_q} [Variation of Lift Coefficient with Dimensionless Pitch Rate] ....

C_{m_q} [Variation of Pitching Moment Coefficient with Dimensionless Pitch Rate (Pitch Damping)] ....

q/t-Derivatives

$C_{D_{\dot{q}}}, C_{T_{\dot{q}}}$ [Variation of Drag / Thrust Coefficient with Rate of Change of Dimensionless Pitch Rate] .

$C_{L_{\dot{q}}}$ [Variation of Lift Coefficient with Rate of Change of Dimensionless Pitch Rate]

$C_{m_{\dot{q}}}$ [Variation of Pitching Moment Coefficient with Rate of Change of Dimensionless Pitch Rate]

δ_{PC}-Derivatives

$C_{D_{\delta PC}}$, $C_{T_{\delta PC}}$ [Variation of Drag / Thrust Coefficient with Primary Controls Deflection Angle]

$C_{Y_{\delta PC}}$ *(long)* [Variation of Sideforce Coefficient with Primary Controls Deflection Angle]

$C_{L_{\delta PC}}$ [Variation of Lift Coefficient with Primary Controls Deflection Angle (Control Effectiveness)]

$C_{l_{\delta PC}}$ *(long)* [Variation of Rolling Moment Coefficient with Primary Controls Deflection Angle]

$C_{m_{\delta PC}}$ [Variation of Pitching Moment Coefficient with Primary Controls Deflection Angle]

$C_{n_{\delta PC}}$ *(long)* [Variation of Yawing Moment Coefficient with Primary Controls Deflection Angle]

δ_{SC}-Derivatives

$C_{D_{\delta SC}}$, $C_{T_{\delta SC}}$ [Variation of Drag / Thrust Coefficient with Secondary CE Deflection]

$C_{Y_{\delta SC}}$ *(long)* [Variation of Sideforce Coefficient with Secondary CE Deflection]

$C_{L_{\delta SC}}$ [Variation of Lift Coefficient with Secondary CE Deflection]

$C_{l_{\delta SC}}$ *(long)* [Variation of Rolling Moment Coefficient with Secondary CE Deflection]

$C_{m_{\delta SC}}$ [Variation of Pitching Moment Coefficient with Secondary CE Deflection]

$C_{n_{\delta SC}}$ *(long)* [Variation of Yawing Moment Coefficient with Secondary CE Deflection]

δ_{C}-Derivatives

$C_{D_{\delta C}}$, $C_{T_{\delta C}}$ [Variation of Drag / Thrust Coefficient with Configuration Setting Angle]

$C_{Y_{\delta C}}$ *(long)* [Variation of Sideforce Coefficient with Configuration Setting Angle]

$C_{L_{\delta C}}$ [Variation of Lift Coefficient with Configuration Setting Angle]

$C_{l_{\delta C}}$ *(long)* [Variation of Rolling Moment Coefficient with Configuration Setting Angle]

$C_{m_{\delta C}}$ [Variation of Pitching Moment Coefficient with Configuration Setting Angle]

$C_{n_{\delta C}}$ *(long)* [Variation of Yawing Moment Coefficient with Configuration Setting Angle]

FLOW CHARACTER - Flow Phenomena

Non-Linear Aerodynamics

Main Lifting Surface Stall Characteristics

HCE Stall Characteristics

Aeroelasticity

Tuck

Pitch-Up

High Angle-of-Attack Departure

Flutter

Speed Brakes

Aerodynamic Interference

Apparent Mass Effects

ADDITIONAL GROUNDS

High-Risk Flights

Landing Gear Location

C.G. Range

Engine Nacelle Strike Constraint

Tail Strike Constraint

Thrust Vectoring

Control Allocation (Trim Drag)

Wing Positioning

Airframe-Propulsion Interaction

Ride Quality

High-Lift Devices

Performance Considerations

Lateral / Directional Motion

FLIGHT CHARACTER - Design Constraining Flight Conditions

TRIM & CONTROL

1 g Flight (Equilibrium / Performance Case)

Asymmetric Flight Performance

- Minimum Control Speed, Approach and Landing, 2 Engines Out (V_{MCL-2}) (148)
- Engine Failures During TO
- Directional Control, 2 Engines Inoperative (33)
- Landing from Approach Slope 4° with 2 Engines Failed on Same Side (81)
- Engine Failed on Ground
- Trim 2 Engines Inoperative (164)
- Demonstration of Maximum Cross Wind on Landing (3 Engines) (248)
- During Ground Run (Landing) No Reverse Thrust on 2 Adjacent Engines (257)
- Supersonic Cruise with Failure of Critical Engine and Malfunction of Adjacent Engine (56)
- Directional Control, 1 Engine Inoperative (32)
- Double Hydraulic Failure (295-315)
- Transonic and Supersonic Straight Sideslips (46)

Secondary Controls Performance

- Slats / Flaps Failures (320-324)
- Landing from Approach Slope 4° with 2 Engines Failed on Same Side (81)

(>,<) 1 g Flight (Manoeuvring / Special Case)

Roll Performance

- Lateral Control All Engines Operating (36)
- Lateral Control, 2 Engines Inoperative (39)
- High Speed Characteristics; Lateral Control Capacity (45)
- Transonic and Supersonic Aileron Rates of Roll (47)
- Supersonic Cruise Roll Manoeuvrability with All Engines Operating (54)
- Dynamic Engine Out
- Windup Turns
- Coordinated Velocity Axis Roll

Primary Control Failures

- Foreplane Failure to a Fixed Position

Power Application Performance

- Supersonic Cruise with Failure of Critical Engine and Malfunction of the Adjacent Engine (56)

STABILITY

Static

Stability Performance

- Directional Stability
- Lateral Stability

Dynamic

Mode Performance

- Dutch Roll Mode (203)
- Roll Mode (203)
- Spiral Mode (203)
- Dynamic Lateral Response

<u>Coupling Performance</u>

Inertia Coupling: Pitch Due to Velocity Axis Roll ....

Inertia Coupling: Yaw Due to Loaded Roll ....

AERODYNAMIC CHARACTER - Stability and Control Derivatives

u-Derivatives

C_{Y_u} [Variation of Sideforce Coefficient with Dimensionless Speed] ....

C_{l_u} [Variation of Rolling Moment Coefficient with Dimensionless Speed] ....

C_{n_u} [Variation of Yawing Moment Coefficient with Dimensionless Speed] ....

u/t-Derivatives

$C_{Y_{\dot{u}}}$ [Variation of Sideforce Coefficient with Rate of Change of Dimensionless Speed] ....

$C_{l_{\dot{u}}}$ [Variation of Rolling Moment Coefficient with Rate of Change of Dimensionless Speed] ....

$C_{n_{\dot{u}}}$ [Variation of Yawing Moment Coefficient with Rate of Change of Dimensionless Speed] ....

v (β)-Derivatives

C_{D_β}, C_{T_β} [Variation of Drag / Thrust Coefficient with Angle of Sideslip] ....

C_{Y_β} [Variation of C_Y with Angle of Sideslip (Static Lateral Stability)] ....

C_{L_β} [Variation of C_L with Angle of Sideslip] ....

C_{l_β} [Variation of C_l with Angle of Sideslip (Static Lateral Stability; Effective Dihedral)] ....

C_{m_β} [Variation of C_m with Angle of Sideslip] ....

C_{n_β} [Variation of C_n with Angle of Sideslip (Static Directional Stability)] ....

v/t (β/t)-Derivatives

$C_{D_{\dot{\beta}}}, C_{T_{\dot{\beta}}}$ [Variation of Drag / Thrust Coefficient with Rate of Change of Dimensionless β] ....

$C_{Y_{\dot{\beta}}}$ [Variation of Sideforce Coefficient with Rate of Change of Dimensionless β] ....

$C_{L_{\dot{\beta}}}$ [Variation of Lift Coefficient with Rate of Change of Dimensionless β] ....

$C_{l_{\dot{\beta}}}$ [Variation of Rolling Moment Coefficient with Rate of Change of Dimensionless β] ....

$C_{m_{\dot{\beta}}}$ [Variation of C_m with Rate of Change of Dimensionless β] ....

$C_{n_{\dot{\beta}}}$ [Variation of C_n with Rate of Change of Dimensionless Angle of Sideslip] ....

w (α)-Derivatives

C_{Y_α} [Variation of Sideforce Coefficient with Angle of Attack] ....

C_{l_α} [Variation of Rolling Moment Coefficient with Angle of Attack] ....

C_{n_α} [Variation of Yawing Moment Coefficient with Angle of Attack] ....

w/t (α/t)-Derivatives

$C_{Y_{\dot{\alpha}}}$ [Variation of Sideforce Coefficient with Rate of Change of Dimensionless Angle of Attack] ....

$C_{l_{\dot{\alpha}}}$ [Variation of Rolling Moment Coeff. with Rate of Change of Dimensionless Angle of Attack] ....

$C_{n_{\dot{\alpha}}}$ [Variation of Yawing Moment Coeff. with Rate of Change of Dimensionless Angle of Attack] ....

p-Derivatives

C_{D_p}, C_{T_p} [Variation of Drag / Thrust Coefficient with Dimensionless Roll Rate] ....

C_{Y_p} [Variation of Sideforce Coefficient with Dimensionless Roll Rate] ....

C_{L_p} [Variation of Lift Coefficient with Dimensionless Roll Rate] ....

C_{l_p} [Variation of Rolling Moment Coefficient with Dimensionless Roll Rate (Damping in Roll)] ....

C_{m_p} [Variation of Pitching Moment Coefficient with Dimensionless Roll Rate] ....

C_{n_p} [Variation of Yawing Moment Coefficient with Dimensionless Roll Rate] ....

p/t-Derivatives

$C_{D_{\dot{p}}}, C_{T_{\dot{p}}}$ [Variation of Drag / Thrust Coefficient with Rate of Change of Dimensionless Roll Rate] ....

$C_{Y_{\dot{p}}}$ [Variation of Sideforce Coefficient with Rate of Change of Dimensionless Roll Rate] ....

$C_{L_{\dot{p}}}$ [Variation of Lift Coefficient with Rate of Change of Dimensionless Roll Rate] ....

$C_{l_{\dot{p}}}$ [Variation of Rolling Moment Coefficient with Rate of Change of Dimensionless Roll Rate] ....

$C_{m_{\dot{p}}}$ [Variation of Pitching Moment Coefficient with Rate of Change of Dimensionless Roll Rate] ....

$C_{n_{\dot{p}}}$ [Variation of Yawing Moment Coefficient with Rate of Change of Dimensionless Roll Rate] ....

q-Derivatives

C_{Y_q} [Variation of Sideforce Coefficient with Dimensionless Pitch Rate]

C_{l_q} [Variation of Rolling Moment Coefficient with Dimensionless Pitch Rate]

C_{n_q} [Variation of Yawing Moment Coefficient with Dimensionless Pitch Rate]

q/t-Derivatives

$C_{Y_{\dot{q}}}$ [Variation of C_Y with Rate of Change of Dimensionless Pitch Rate]

$C_{l_{\dot{q}}}$ [Variation of C_l with Rate of Change of Dimensionless Pitch Rate]

$C_{n_{\dot{q}}}$ [Variation of C_n with Rate of Change of Dimensionless Pitch Rate]

r-Derivatives

C_{D_r}, C_{T_r} [Variation of Drag / Thrust Coefficient with Dimensionless Yaw Rate]

C_{Y_r} [Variation of Sideforce Coefficient with Dimensionless Yaw Rate]

C_{L_r} [Variation of Lift Coefficient with Dimensionless Yaw Rate]

C_{l_r} [Variation of Rolling Moment Coefficient with Dimensionless Yaw Rate]

C_{m_r} [Variation of Pitching Moment Coefficient with Dimensionless Yaw Rate]

C_{n_r} [Variation of C_n with Dimensionless Yaw Rate (Damping in Yaw)]

r/t-Derivatives

$C_{D_{\dot{r}}}, C_{T_{\dot{r}}}$ [Variation of Drag / Thrust Coefficient with Rate of Change of Dimensionless Yaw Rate]

$C_{Y_{\dot{r}}}$ [Variation of Sideforce Coefficient with Rate of Change of Dimensionless Yaw Rate]

$C_{L_{\dot{r}}}$ [Variation of Lift Coefficient with Rate of Change of Dimensionless Yaw Rate]

$C_{l_{\dot{r}}}$ [Variation of Rolling Moment Coefficient with Rate of Change of Dimensionless Yaw Rate]

$C_{m_{\dot{r}}}$ [Variation of C_m with Rate of Change of Dimensionlesss Yaw Rate]

$C_{n_{\dot{r}}}$ [Variation of C_n with Rate of Change of Dimensionless Yaw Rate]

δ_{PC}-Derivatives

$C_{D_{\delta PC}\,(lat,\,dir)}, C_{T_{\delta PC}\,(lat,\,dir)}$ [Variation of Drag / Thrust Coefficient with Primary CE Deflection]

$C_{Y_{\delta PC}}$ [Variation of Sideforce Coefficient with Primary Controls Deflection Angle]

$C_{L_{\delta PC}\,(lat,\,dir)}$ [Variation of Lift Coefficient with Primary CE Deflection (Control Effectiveness)]

$C_{l_{\delta PC}}$ [Variation of Rolling Moment Coefficient with Primary Controls Deflection Angle]

$C_{m_{\delta PC}\,(lat,\,dir)}$ [Variation of Pitching Moment Coefficient with Primary CE Deflection]

$C_{n_{\delta PC}}$ [Variation of Yawing Moment Coefficient with Primary Controls Deflection Angle]

δ_{SC}-Derivatives

$C_{D_{\delta SC}\,(lat,\,dir)}, C_{T_{\delta SC}\,(lat,\,dir)}$ [Variation of Drag / Thrust Coefficient with Secondary CE Deflection]

$C_{Y_{\delta SC}}$ [Variation of Sideforce Coefficient with Secondary Controls Deflection Angle]

$C_{L_{\delta SC}\,(lat,\,dir)}$ [Variation of Lift Coefficient with Secondary Controls Deflection Angle]

$C_{l_{\delta SC}}$ [Variation of Rolling Moment Coefficient with Secondary Controls Deflection Angle]

$C_{m_{\delta SC}\,(lat,\,dir)}$ [Variation of Pitching Moment Coefficient with Secondary CE Deflection Angle]

$C_{n_{\delta SC}}$ [Variation of Yawing Moment Coefficient with Secondary Controls Deflection Angle]

δ_C-Derivatives

$C_{D_{\delta C}\,(lat,\,dir)}, C_{T_{\delta C}\,(lat,\,dir)}$ [Variation of Drag / Thrust Coefficient with Configuration Setting Angle]

$C_{Y_{\delta C}}$ [Variation of Sideforce Coefficient with Configuration Setting Angle]

$C_{L_{\delta C}\,(lat,\,dir)}$ [Variation of Lift Coefficient with Configuration Setting Angle]

$C_{l_{\delta C}}$ [Variation of Rolling Moment Coefficient with Configuration Setting Angle]

$C_{m_{\delta C}\,(lat,\,dir)}$ [Variation of Pitching Moment Coefficient with Configuration Setting Angle]

$C_{n_{\delta C}}$ [Variation of Yawing Moment Coefficient with Configuration Setting Angle]

FLOW CHARACTER - Flow Phenomena

Gust

Non-Linear Aerodynamics

Speed Brake

Aeroelasticity

ADDITIONAL GROUNDS

Manoeuvre Load Alleviation (MLA) ....

C.G. Range ....

Engine Nacelle Strike Constraint ....

Airframe-Propulsion Intercations ....

Inherent Flight Characteristics ....

A.4 Classification Scheme for Fixed Wing Aircraft Configurations and Concepts

Configurations			Symmetric					Asymmetric		
			TAC	TFC	TSC	FWC	…	OWC	OFWC	…
Concepts			(Tail Aft)	(Tail first)	(Three Surface)	(Flying Wing)	(…)	(Oblique Wing)	(Oblique Flying Wing)	(…)
Primary lift-source										
Plan-view	SR	(Straight Rectangular)								
	KR	(Kinked Rectangular)								
	…	(…)								
	SD	(Straight Delta)								
	KD	(Kinked Delta)								
	…	(…)								
	J	(Joined)								
	C	(C)								
	B	(Biplane)								
	T	(Tandem)								
Aspect Ratio	HAR	(High Aspect-Ratio)								
	LAR	(Low Aspect-Ratio)								
Taper Ratio	HTR	(High Taper-Ratio)								
	LTR	(Low Taper-Ratio)								
Sweep	ZS	(Zero Sweep)								
	PS	(Positive-Sweep)								
	NS	(Negative-Sweep)								
	VS	(Variable-Sweep)								
	AS	(Asymmetric Sweep)								

(continued)

(continued)

Configurations			Symmetric					Asymmetric		
			TAC	TFC	TSC	FWC	…	OWC	OFWC	…
Concepts			(Tail Aft)	(Tail first)	(Three Surface)	(Flying Wing)	(…)	(Oblique Wing)	(Oblique Flying Wing)	(…)
Vertical integration	HW	(High-Wing)								
	MW	(Mid-Wing)								
	LW	(Low-Wing)								
Dihedral	PD	(Positive Dihedral)								
	ZD	(Zero Dihedral)								
	ND	(Negative Dihedral)								
Integration	SL	(Span-Loader)								
	BWB	(Blended-Wing-Body)								
	LB	(Lifting Body)								
	…	(…)								
Primary volume-supply										
Type	SB	(Single-Body)								
	MB	(Multi-Body)								
	BWB	(Blended-Wing-Body)								
	SL	(Span-Loader)								
	LB	(Lifting Body)								
	…	(…)								

(continued)

(continued)

Configurations			Symmetric					Asymmetric		
			TAC	TFC	TSC	FWC	…	OWC	OFWC	…
Concepts			(Tail Aft)	(Tail first)	(Three Surface)	(Flying Wing)	(…)	(Oblique Wing)	(Oblique Flying Wing)	(…)
Cross section	CS	(Circular Section)								
	ES	(Elliptical Section)								
	DS	(Diamond Section)								
	RS	(Rectangular Section)								
	ShCS	(Sharp Chined Section)								
	SmCS	(Smooth Chined Section)								
	…	(…)								
Primary control-effector										
Longitudinal (LoCE)	FS+F	(Fixed Stabilizer + Flap)								
	TS	(Trimmable Stabilizer)								
	TS+F	(Trimmable Stabilizer + Flap)								
	FFS	(Free-Floating Stabilizer)								
	HS	(High-Stabilizer)								
	MS	(Mid-Stabilizer)								
	LS	(Low-Stabilizer)								
	B	(Trimmable Butterfly)								
	B+F	(Trimmable Butterfly + Flap)								
	E	(Elevon)								
	BF	(Body Flap)								
	…	(…)								

(continued)

(continued)

Configurations			Symmetric					Asymmetric		
			TAC	TFC	TSC	FWC	…	OWC	OFWC	…
Concepts			(Tail Aft)	(Tail first)	(Three Surface)	(Flying Wing)	(…)	(Oblique Wing)	(Oblique Flying Wing)	(…)
Lateral (LaCE)	A	(Aileron)								
	S	(Spoiler)								
	E	(Elevon)								
	T	(Taileron)								
	…	(…)								
Directional (DiCE)	FF+F	(Fixed Fin + Flap)								
	FF +SF	(Fixed Fin + Split Flap)								
	TF	(Trimmable Fin)								
	TF+F	(Trimmable Fin + Flap)								
	FW	(Fixed Winglet)								
	FW +F	(Fixed Winglet + Flap)								
	FW +SF	(Fixed Winglet + Split Flap)								
	E-SF	(Elevon + Split Flap)								
	AF-D	(Auxiliary Fin - Dorsal)								
	AF-V	(Auxiliary Fin - Ventral)								
	B	(Trimable Butterfly)								
	B+F	(Trimmable Butterfly + Flap)								
	…	(…)								

(continued)

(continued)

Configurations			Symmetric					Asymmetric		
			TAC	TFC	TSC	FWC	…	OWC	OFWC	…
Concepts			(Tail Aft)	(Tail first)	(Three Surface)	(Flying Wing)	(…)	(Oblique Wing)	(Oblique Flying Wing)	(…)
Secondary control-effector (force motivator)										
Lift motivator	LES	(Slats/Krueger)								
	TEF	(Flaps)								
	A	(Aileron)								
	S	(Spoiler)								
Drag motivator	S	(Spoiler)								
	…	(…)								
Sideforce motivator	…	(…)								
Propulsion										
Type	TJ	(Turbo Jet)								
	TF	(Turbo Fan)								
	TP	(Turbo Prop)								
	R	(Rocket)								
	…	(…)								
Installation	T	(Tractor)								
	P	(Pusher)								
	AB	(Aft Body)								
	HW	(High Wing)								
	MW	(Mid Wing)								
	LW	(Low Wing)								
	…	(…)								

A.5 Final Evaluation Sequence of Computer-Based Aerodynamic Prediction Codes

Final evaluation of non-linear aerodynamic prediction codes				
Classification	Item	Singularity methods—vortex lattice methods		
		Subsonic	Subsonic/Supersonic	
Code name	Acronym	*LINAIR PRO*	*HASC95*	*VORSTAB*
	Full name	Linear aerodynamics professional	High angle of attack stability and control	Vortex-lattice stability and control
Objective		"… to obtain a quick look at new design concepts …"	"… non-linear aerodynamic prediction method suitable for conceptual/preliminary design level application…"	"… subsonic and supersonic configuration analysis … developed for vortex dominated configurations …"
		Conceptual design level subsonic and/or supersonic aerodynamic prediction codes including non-linear augmentation are used to estimate the effects of linear and non-linear (vortex lift and breakdown) flow effects		
Primary application		LinAir is intended to bridge the gap between the big aerodynamic estimation codes and the simple, approximate methods often used in advanced design	Being used primarily in high angle-of-attack applications (fighters; missiles in manoeuvering flight; low-speed portion of lifting bodies and further space applications; HSCT studies); analytical/semi-empirical prediction method	All necessary aerodynamic data is generated for flight simulation
History	Approximate date introduced	1982	1995	1975
	Generation	2nd generation (LinAir, LinAir Pro)	3rd generation (VORLAX, HASC, HASC95)	1st generation (VORSTAB)
	Responsibility for implementation	By I. Kroo at Stanford University	Lockheed Martin Tactical Aircraft Systems; Nielsen Engineering & Research, Inc.; Consulting Aviation Services	By C. E. Lan, University of Kansas
	Software rights	Desktop Aeronautics, Inc.	By M. J. Logan, Langley Research Center; by A. E. Albright at Lockheed Martin	Unicom Technology Systems (initially developed for NASA)

(continued)

(continued)

Final evaluation of non-linear aerodynamic prediction codes				
Classification	Item	Singularity methods—vortex lattice methods		
		Subsonic	Subsonic/Supersonic	
Availability	Source code	Yes (FORTRAN 77)	Yes (FORTRAN 77)	Yes (FORTRAN 77)
	Executable program	Yes (+ MS Windows environment)	Yes	Yes (+ MS Windows environment)
	Documentation	Yes	Yes	Yes
Mathematical/physical modelling	Governing equation	Prandtl-Glauert equation (potential flow theory; linear partial differential equation describing inviscid, irrotational, linearized subsonic flow)	Prandtl-Glauert equation (potential flow theory)	Prandtl-Glauert equation (potential subsonic flow, thin wing approximation, with corrections for vortex flow effects and boundary layer separation)
	Evaluation method	Discrete vortex Weissinger method (or extended lifting line theory)	Generalized vortex lattice method (GVLM)	Quasi-vortex-lattice method (QVLM)
	Boundary condition	Tangency condition on lifting surface		Tangency condition on lifting surface
	Flow singularities	Lifting surfaces represented with discrete vortex lines forming skewed horseshoe vortices	VTXCHN: the volume of the body in the transformed plane is represented by discrete point sources and sinks, and the strength of the individual singularities is determined to satisfy a flow tangency condition on the body	Horseshoe vortices (quasi-vortec-lattice method) for lifting surfaces and G. N. Ward's vortex multiplets for the fuselage (general cross-sectional shape); leading-edge singularity of pressure loading in linear theory (hence leading-edge thrust) uses the QVLM
			2D doublets account for angles of incidence and roll	
	Leading-edge suction	No	TBD	QVLM
	Pressure distribution and total force and moment calculation	The Kutta-Joukowski relation is applied to compute the forces and moments acting on the configuration; use of near-field force calculation; viscous drag is added to the forces and moments by integration over each of the elements	TBD	Thin wing theory; body of revolution

(continued)

(continued)

Final evaluation of non-linear aerodynamic prediction codes				
Classification	Item	Singularity methods—vortex lattice methods		
		Subsonic	Subsonic/Supersonic	
	Vortex Flow	No	Vortex lift - burst, … using semi-empirical methods	Vortex flow corrections: vortex lift, vortex action point, vortex breakdown
			Separated vortex flow on lifting surfaces is modelled using a combination of a VLM (modified VORLAX) and a vortex analysis code (VORLIF); the effect of airfoil thickness on LE separation is included	Method of suction analogy (POLHAMUS) to compute leading- and side-edge suction forces generated by the vortex flow
			VORLIF uses flow data (pressure and velocities) to empirically predict the transition of the core from laminar to turbulent flow and to predict the breakdown position; the influence of one vortex on another's velocity is included	The effect of leading-edge radius on vortex separation is based on the KULFAN assumption (LE vortex separation starts at aoa at which the LE drag equals the LE thrust)
			VTXCHN is an engineering prediction code to predict aerodynamic loads acting on a body including effects of nose vortex shedding. The bodies may have circular and noncircular cross sections which may include sharp chine edges	Vortex action point predicted by applying linear momentum principle of fluid mechanics; vortex-breakdown angles of attack and progression rates are obtained from semi-empirical formulas derived from analysis of experimental data
			Flow conditions include subsonic flow, high aoa, non-zero angles of sideslip, and steady-state rotational flow	Fuselage vortex lift is predicted using the method of suction analogy; asymmetric forebody vortex separation; extensive forebody vortex data; lateral-directional aerodynamics in vortex flow estimated by 'displacement-type' vortex lift
	Boundary layer	No	TBD	Boundary layer separation effects (matching the nonlinear section data (2D) with the 3D lift characteristics iteratively; nonlinear sectional data used as nearfield solution to be matched iteratively with far-field solutions from lifting surface theory)

(continued)

(continued)

Final evaluation of non-linear aerodynamic prediction codes				
Classification	Item	Singularity methods—vortex lattice methods		
		Subsonic	Subsonic/Supersonic	
				Sectional nonlinear data are used iteratively to account for the viscous effect on the lifting surfaces; classical lifting line theory not applicable to fighter type lifting surfaces (aspect ratio, sweep, stall, post-stall limitations)
	Vacuum effects, buffet onset, separation	No vacuum effects; specification of $C_{l\ max}$ (computation of local lift coefficient of the section and reduces the circulation is necessary so that the section lift coefficient never exceeds this value—crude method of modelling separated flow)	TBD	TBD
	Wake modelling	Direction of the flat wake when leaving the TE can be specified (symmetric and asymmetric wake placement)	TBD	Wake of the lifting surfaces assumed to be flat; wake deflection angle
	Subsonic speed range	Usually $M < 0.7$ or so, only at M numbers at which shock waves are not present (solves linearized potential equation)	TBD	Prandtl-Glauert compressibility corrections (stretching of the x co-ordinate system)
	Transonic speed range	Forces and moments are computed using Kutta-Joukowski relation which accounts for compressibility effects in a manner compatible with second order pressure rule; just low speed method; no vacuum limits, isentropic pressure rule corrections, or transonic eff	For low aoa (<10 deg) and thin airfoils the HASC results should be good for Mach numbers up to 0.9; for higher aoa and thin airfoils the results should be good for Mach numbers up to 0.5	Predictions of limited quality (due to true supersonic linearized code), as the methods cannot take into account transonic compressibility effects
	Supersonic speed range	No	VORLAX originally a generalized VLM for subsonic and supersonic flow applications	True linearized supersonic code
	Upwash/downwash	TBD	Multiple runs are required to obtain the necessary upwash/downwash corrections	TBD

(continued)

(continued)

Final evaluation of non-linear aerodynamic prediction codes				
Classification	Item	Singularity methods—vortex lattice methods		
		Subsonic	Subsonic/Supersonic	
	Compressibility	Prandtl-Glauert factor (in x-direction)	VTXCHN: compressibility effects on the body are included by a Gothert transformation which keeps the cross section shape unchanged but stretches the axial body coordinate	TBD
Configuration modelling	Configuration	Arbitrary 3D configuration; multi-element nonplanar lifting surfaces	Arbitrary 3D configuration; multi-element nonplanar lifting surfaces (no more than two tandem lifting surfaces can be run with vortex lift and breakdown (in VORLIF); swept fwd. wings can not be analyzed in VORLIF	Arbitrary 3D configuration; multi-element nonplanar lifting surfaces; curved LEX
		Highly generic modelling character because able to model as well the asymmetric AD-1		Lifting surfaces: wings, vertical tails, horizontal tails, LE flaps, strakes, ailerons, TE flaps, winglets; definition of curved LE
				Fuselage: body of revolution, irregular cross-section shapes; user defined wing camber, twist, thickness distribution; only one fuselage; extensive fuselage definition; only asymmetric wings
	Thickness	No	No	Optional definition of lifting surface thickness
	Camber	(a) specification of an airfoil section's zero lift pitching moment; (b) specification of aoa and incidence angles rather than the geometric incidence; (c) modelling via multi-element arrangement (elements stacked LE to TE)	Yes	Yes; definition of conical camber
	Twist	Linearly-tapered lifting elements	Yes	Definition of lifting surface incidence and twist
	Elastic axis			Definition of lifting surface elastic axis

(continued)

(continued)

Final evaluation of non-linear aerodynamic prediction codes				
Classification	Item	Singularity methods—vortex lattice methods		
		Subsonic	Subsonic/Supersonic	
	High lift/control effector deflection	Wing with flaps or control surfaces may be modelled in two pieces: an element representing the fixed part of the wing, and a smaller element representing the control surface with its LE at the TE of the first element	Accurate accounting for small LE and TE deflections (control increments), bevels and other camber devices; modelling of high-lift devices questionable; computation of the zero aoa more rigorous making the camber effects more accurate	LE and TE control effectors; vortex flaps; max. number of TE segments is 5
			A correction for the loss in spanwise lift carry-over is available for the conditions where the lifting surface unports from the fuselage when deflected	Definition of flap hinge location
	Aerodynamic derivatives	Computation of damping terms	Longitudinal/lateral stability derivatives; damping derivatives	β derivatives ($C_{y\,\beta}$, $C_{n\,\beta}$, $C_{l\,\beta}$); p-derivatives ($C_{y\,p}$, $C_{n\,p}$, $C_{l\,p}$); r-derivatives ($C_{y\,r}$, $C_{n\,r}$, $C_{l\,r}$); pitch damping derivatives at pre & post-stall conditions; longitudinal aerodynamics; control derivatives
	Hinge moments	Yes	No	Yes
	Torsional moment	No	No	Yes
	Bending moment	No	No	Yes
	Airfoil	The 2D profile drag of the sections comprising each element is represented with a parabolic fit and max. lift coefficient	Advanced airfoil definition input file (optional)	Incorporation of nonlinear airfoil (2D) section data or otherwise; extensive options for lifting surface camber definition; optional definition of lifting surface LE and TE shape
	Ground effect	Simulating the ground with an image wing (with the opposite incidence) 'flying underground'	No	In- & out ground effects
	Trim	No automatic trimming function; manual trim using approximate linearity	VORLAX: modified to trim without forebody vortex effects (VTXCHN) nor vortex lift (VORLIF) effects; thus, HASC95 in total has no automatic trimming function	No automatic trimming function; must be done manually

(continued)

(continued)

Final evaluation of non-linear aerodynamic prediction codes				
Classification	Item	Singularity methods—vortex lattice methods		
		Subsonic	Subsonic/Supersonic	
Validation	Objects	Applied to the analysis from model aircraft to manned research aircraft to VSTOL fighters	HASC95: wing; tailless fighter; Falcon 21; F-16XL; F-16; Convair Model 200 VTXCHN: circular cross section (forebody); diamond cross section neither chined nor smooth cross section; sharp chined cross section	Fighter type configurations: F/A-18; generic fighter model; F-16; f-16XL; F-5; X-29A; F-106B
	Effort	Less structured effort by I. Kroo, effort by M. Rakowitz into longitudinal evaluation (induced drag)	Highly structured validation effort for fighter type aircraft configurations	Highly structured validation effort for fighter type aircraft configurations; only very limited BAe validation done on proprietary configurations
	Summary	TBD	VTXCHN: best long. results when aoa <30 deg; for lat. cases best results when aoa <20 deg and aos <10 deg	TBD
			HASC: VORLIF predictions are sensitive to the input modelling (paneling); VORLIF predictions often result in 'noisy' trends in the middle to high aoa regions; VORLAX predictions often produced good agreement with test data;	
			lat./dir. effects poorly predicted for F-16; high aoa results (stall & post stall) will not be accurate for configurations that have TE stall before the LE stalls (i.e., wings having airfoils with large LE radius and camber);	

(continued)

(continued)

Final evaluation of non-linear aerodynamic prediction codes

Classification	Item	Singularity methods—vortex lattice methods		
		Subsonic	Subsonic/Supersonic	
			For low aoa (<10 deg) and thin airfoils the HASC results should be good for Mach numbers up to 0.9; for higher aoa and thin airfoils the results should be good for Mach numbers up to 0.5	
			Camber and airfoil shape must be represented accurately because the vortex breakdown is very sensitive to these terms; canards with their vertical placement below the wing level will cause abnormal effects at the aoa where the wake passes through the wing	
			Future validation of HASC has to cover a variety of aircraft configurations	
Computing requirement	Computer hardware requirements	Execution time depends on the size of the input deck, analysis mode utilized, and processor speed	Execution time depends on the size of the input deck, analysis mode utilized, and processor speed	Execution time depends on the size of the input deck, analysis mode utilized, and processor speed
	Graphical viewing	Advanced graphical viewing feature (allows to translate, rotate and zoom the drawing in 3 dimensions)	TBD	Build-in graphical viewing of geometries and computed results
	Input files	Reference values, global parameters and individual elements (max number of elements is 20)	Minimum of two input files required: (1) main input file containing program options, flow conditions, and geometry; (2) empirical factors for VORLIF, this file is typically not changed by a user optional files: (3) airfoil LE parameters; (4) input file for VTXCHN	Interactive input preparation

(continued)

(continued)

Final evaluation of non-linear aerodynamic prediction codes				
Classification	Item	Singularity methods—vortex lattice methods		
		Subsonic	Subsonic/Supersonic	
	Output files	Element forces; total forces; lift distribution; plot results	Provide information on the input geometry, total configuration loads, component loads, and vortex characteristics	Provide information on the input geometry, total configuration loads, component loads, and vortex characteristics
	Execution features	alpha and beta sweep; execution as a single case and 'batch' mode calculations	HASC can be run either as a complete HASC code utilizing any or all of the analysis modules	alpha and beta sweep; execution as a single case and 'batch' mode calculations
	Turn-around time	Dependent on size of input deck (number of horseshoe vortices and number of flow conditions), analysis modules utilized, and processor speed	Dependent on size of input deck (number of horseshoe vortices and number of flow conditions), analysis modules utilized, and processor speed	Dependent on size of input deck (number of horseshoe vortices and number of flow conditions), analysis modules utilized, and processor speed
Recommended improvement		Implement LE suction analogy	Activate VORLAX supersonic logic	Extend generic character to be able to model asymmetric arbitrary aircraft configurations; write subroutine for drag controls and spoilers; elasticity
Remarks		Theory valid for slender a/c (perturbations u, v, w are small)	HASC95 integrates three modules: VORLAX, VORLIF, VTXCHN (replacement for VTXCLD)	QVLM similar to conventional VLM but far superior in accuracy
		Planar surface (linearised boundary conditions and grid of horseshoe vortices)	VORLAX: generalized Vortex lattice program by L. R. Miranda (VORLAX is subsonic/supersonic code but HASC95 designed and validated for subsonic regime)	
		Proper modelling of the geometry is of critical importance to obtaining accurate answers, especially when the configuration is complex	VORLIF: semi-empirical strake/wing analysis code by Dixon	

(continued)

(continued)

Final evaluation of non-linear aerodynamic prediction codes				
Classification	Item	Singularity methods—vortex lattice methods		
		Subsonic	Subsonic/Supersonic	
			Engineering method VTXCHN (Vortex Chine) for predicting nose vortex shedding (replacement for VTXCLD; VTXCLD: 2D, unsteady, separated flow analogy for analyzing smooth forebody shapes); analysis of forebodies with chined cross-sections	
			VORLIF: semi-empirical strake/wing analysis code	
Evaluation result		Highly linear with minimum non-linear corrections just basic validation effort; additional validation by Rakowitz, Wishart (astonishing accuracy for its simplicity)	Extensive modelling of non-linear aerodynamics highly structured validation (only subsonic regime)	Extensive modelling of non-linear aerodynamics less structured but sufficient validation effort
		Only subsonic code	Limited supersonic (linear) capability (HASC95 defined as subsonic code)	Subsonic and supersonic (linear) capability (VORSTAB defined as subsonic/supersonic code)
		Limited derivative estimation capability	Derivative estimation suitable for *AeroMech* and simulation	Derivative estimation (most comprehensive) suitable for *AeroMech* and simulation
		Simple input files and easy to use	Complexity of input files depending on modelling detail	Complexity of input files depending on modelling detail
		Highly generic character	Limitations of generic modelling character (restrictions with multi-fuselages and asymmetric configurations)	Limitations of generic modelling character (restrictions with multi-fuselages and asymmetric configurations)
		Less extensive documentation of theory; sufficient user's guide	Less extensive documentation of theory; sufficient user's guide	Very good documentation of theory (highly transparent); basic user's guide
		Commercially available (DOS & Windows)	Rights by NASA; possibly not available for EC companies	Commercially available (DOS & Windows)
		CAPABLE, but RULED OUT	CAPABLE, SUITABLE but RULED OUT (similar performance to VORSTAB)	CAPABLE, SUITABLE and SELECTED

A.6 Control Allocation Case Studies

The following summarises control allocation schemes for selected case studies taken from the KBS:

TAC: *MD-91/92*

The MD-91/92 control definition includes a pitch compensation system (PCS). The PCS consists of controllers that improve flying qualities by (a) neutralising thrust induced moments (thrust moment compensation), (b) optimising elevator control power with flap deflection (flap compensation), and (c) eliminating Mach tuck (Mach trim compensation).

TFC: *SAAB JAS 39 Gripen*

The SAAB JAS 39 Gripen control definition includes a movable canard surface in combination with four elevons, rudder and leading-edge flaps. It is possible to obtain a choice of either max. L/D-ratio or max. lift, depending on what is needed at a specific flight condition, using the canard and the trailing-edge control surfaces in combination. At cruise and manoeuvre conditions, the canard and elevons are optimised for low trim drag. At low speed, high lift is desired for short field performance rather than low drag. The emphasis is then to carry a high load on the canard to provide a substantial lift increase by deflecting the wing elevon trailing-edge down for trim. Thus, to obtain a high trim load on the canard, good canard high-lift characteristics were important in the choice of the planform for the close-coupled canard layout.

TFC: *Airbus ESCT* (*European Supersonic Commercial Transport*)

The ESCT control definition includes the delta wing with the capability of changing the longitudinal equilibrium by any one or any combination of the following means: (a) elevons, (b) leading-edge and trailing-edge flaps, (c) foreplane, and (d) fuel transfer. The combinations possible become rather complex regarding the schedule of relative deflections to be used. For low speed (minimum noise), subsonic, and supersonic cruise, the allocation of these four means has to result in the highest possible L/D-ratio.

TFC: *BeechStarship*

The Starship control definition includes (a) wing trailing-edge flaps, (b) foreplane sweep, and (c) foreplane trailing-edge flaps. The control system installed prevents, in the event of a failure, any unintentional combination of flap and foreplane position. The pilot simply has a two position switch which simultaneously controls flap extension and foreplane sweep.

TFC: *North American XB-70 Valkyrie*

The XB-70 control definition includes (a) a variable incidence foreplane with trailing-edge flaps, (b) redundant wing trailing-edge flaps (elevons), and (c) variable wing tip dihedral. The canard configuration has been selected for minimum trim drag. The canard provides primary pitch control and has a flap for use during take-off and landing only. The schedule at take-off and landing requires additional down canard-elevator to trim out the pitching moments generated by deflecting the elevons down for increased lift. In cruise, the canard is geared to the elevator (elevon) action and the shift of the aerodynamic centre is controlled with the position of the wing tips.

TFC: *Concorde B*[3]

The Concorde B control definition includes (a) wing trailing-edge elevons, (b) wing leading-edge high-lift devices, (c) fuel transfer system, and (d) a canard. The low subsonic speed foreplane schedule is to give optimum performance at all higher speeds, the canard is scheduled to help recovery from high incidences. At high subsonic and transonic speeds, the canard is schedule as a function of Mach number or angle-of-attack. Above M0.8, the cruise foreplane schedule locks the canard in its optimum performance (trim drag) setting.

TFC: *HOTOL J5*

The HOTOL J5 control definition includes (a) active foreplanes and (b) full-span elevons. It is assumed that the foreplanes would be used for pitch control, supplemented when required by retaining the full-span flaperons of the datum vehicle.

TSC: *Piaggio P-180 Avanti*

The P-180 Avanti control definition includes (a) a fixed canard with trailing-edge flaps, (b) wing high-lift devices, and (c) a tailplane. The longitudinal controls are scheduled to trim the aircraft in level flight for minimum trim drag at all centre of gravity positions. The forces needed to stabilise the normal nose-down pitching moment of the wing are shared between the canard (upward balancing force) and the tail (downward balancing force). The canard-flaps are geared with the wing high-lift flaps. With this schedule, total pitch control authority is sufficient for max high-lift deflection, which is reflected in very good field performance.

TSC: *Airbus A340600 TSA*

The A340-600 TSA control definition includes (a) a canard, (b) wing high-lift devices, (c) a fuel transfer system, and (d) the trimmable tailplane. The degrees of longitudinal freedom are scheduled for maximum lift at TO for this highly geometry-limited aircraft, and minimum trim drag at cruise. The tailplane is

[3]Concorde investigation up to mid 1979, to demonstrate confidence in the foreplane as a means of improving take-off performance and noise.

retained to provide pitch control and to keep its stabilising influence. The canard generates lift and reduces the pitching moment especially during the high-lift >configuration setting. Canard control authority is dependent on the control law structure (normal law and direct law) during canard failure conditions. The trimming schedule targets 'no lift-loss' due to trimming.

TSC: *Airbus A3XX TSA*

The A3XX TSA control definition includes (a) a canard, (b) wing high-lift devices, (c) a fuel transfer system, and (d) the trimmable tailplane. The controls are scheduled for (a) reduction of the pitching moment during high-lift configuration setting, (b) trim by movable canard and tailplane, (c) manoeuvres scheduled between canard and tailplane, (d) stability augmentation superimposed on the canard and tailplane trim function, and (e) gust control with canard elevator.

FWC: *Horten HIV*

The Horten HIV control definition includes (a) high-lift device, (b) elevon, and (c) trimming surfaces. The schedule selected has to reduce unfavorable yawing moments due to aileron by making use of differential aileron movement while avoiding changes in longitudinal trim using the inner flap pair. The high speed nose down trim is generated scheduling all trailing-edge flaps while deflecting the inner elevon sections most. Summarising, the outer flaps work principally as up going ailerons, whereas the 'climbing elevator' action comes mainly from the middle flap and the 'diving elevator' action is provided by the inner flaps.

FWC: *Northrop XB-35 and YB-49*

The Northrop XB-35 and YB-49 longitudinal controls are scheduled during the low speed flight condition by deflecting the outer pitch flaps up to trim, the inner split flaps down as high-lift device and the middle elevons neutral. Deflection of the split flaps also increases the longitudinal stability at high lift coefficients.

FWC: *Northrop B-2 Spirit*

The B-2 control definition includes four pairs of control surfaces on the wing trailing-edge: (a) split drag rudders on outer wing, (b) one elevon on outer wing, (c) two elevons on inner wing, and (d) the beaver tail. The outer elevons provide primary pitch and roll control. The two inner elevons are considered secondary control surfaces used only at low-speed. The beaver-tail is working constantly to alleviate gust loads. During the beginning stages of the take-off roll, all three elevon sets are drooped until sufficient airspeed is attained. This has been done to prevent surface damage in the event of hydraulic system failure during take-off and to provide a built-in nose-down pitching moment. In low speed flight, the drag rudders are biased open. During final approach configuration, the split drag rudders are the most prominent trailing-edge control surfaces. The other surfaces remain essentially faired (providing pitch control) with the exception of the pitch trimming beaver tail itself providing a slight nose down pitching moment.

FWC: *Avro Vulcan*

The Vulcan control definition includes four elevons on each wing half. In a straight climb all operate as elevators, in a level turn all act as ailerons. In climbing turns the inboard two on each side act as elevators and the smaller outboard pair as ailerons.

FWC: *Boeing BWB*

The BWB (year 2000 project) control definition includes a variety of multi-purpose (redundant) control surfaces. Per semi-span (a) six elevons for pitch control, (b) five elevons and two spoilers for roll control, and (c) five slats and two elevons for high lift.

FWC: *BAC and Sud Aviation Concorde*

The Concorde control definition includes three elevons on each side of the aircraft: (a) outer and middle elevons always synchronised and (b) the inner elevons where the deflection in roll is less than that of the outer elevons. The reduced deflection of the inner elevons acts to minimise the aerodynamic interference of the elevons on the fin and rudder and hence the yaw moment in roll.

FWC: *Lockheed A-12, SR-71*

The A-12/SR-71 control definition includes trailing-edge elevons which provide pitch and roll control, with the outboard elevons rigged trailing-edge up relative to the inboard surfaces to reduce root bending moments at high speed. The outboard elevon units are slaved to the inboard units. Above M0.5, differential elevon travel prevents excessive aerodynamic loads from being applied to the control surfaces. The solution to reduce the static margin and excessive trim drag was to install an integrated canard surface in the form of fuselage chines ahead of the centre of gravity.

FWC: *Martin Marietta X-24B*

The X-24B control definition includes biasing of the upper/lower flaps and upper/lower rudders, providing the aerodynamic configuration's stability, longitudinal trim, and gliding performance throughout the flight envelope. The 'wedge-open' control configuration is required for stability considerations. The other extreme schedule for improved L/D for landing results in a faired longitudinal controls setting with the rudders 'toed in'. An intermediate controls setting served as the speed brake for energy management in the landing pattern.

FWC: *Rockwell International Space Shuttle Orbiter*

The Space Shuttle Orbiter control definition includes the (a) elevons, (b) body flap, (c) thrust vectoring, (d) thrusters, (e) rudder, and (f) speed brake. The elevons and body flap work together during entry to perform the trim function. The elevon is scheduled to fly a specific profile in the flight software. The body flap trims the vehicle to keep the elevons on schedule. Since the body flap moves slowly, the elevons move off-schedule to trim pitch transients and to perform longitudinal manoeuvres. The range of available CEs are phased in and out by the FCS to utilise them where their effectiveness is predictable and significant. The Orbiter's speed-brake is used below M10.0 to induce a more positive downward elevator trim deflection. The longitudinal aerodynamic CEs are phased to provide optimum control power between the c.g. limits.

A.7 Concorde Simulator Test Procedure Cards[4]

Flight no.	Subject
1a-b	Rotation on take-off
2	Ground handling/3-engine take-off
3a-c	Transonic and supersonic climb with random reverse thrust selection on 1 engine; thrust reduction
4	Aft CG clearance—CG changes at constant speed and height
5a	Longitudinal control power (high incidence)/Aft CG
5b	Longitudinal control power (high incidence)/Fwd CG
6a-b	Transonic and supersonic straight sideslips
7a-e	Response to engine power changes (symmetric)
7f	Response to engine power changes (asymmetric/engine failure)
7g	Response to engine power changes (symmetric/reverse)
8a-b	Engine-out trim
9a-b	Directional control, two engines inoperative
10	Lateral control, two engines inoperative
11	Supersonic cruise with 1 engine out failure and loss of B+Y or G+Y hydraulic systems
12	Supersonic cruise with 2 engines out failure and loss of B+Y or G+Y hydraulic systems
13	Supersonic cruise with failure of 4 engines
14a	High incidence
14b	High incidence (Fwd CG)
14c	High incidence (Aft CG)
14d	High incidence (Aft CG, Spin)
15a-b	Speed recovery
16	Longitudinal handling at V_{Ref}
17a-b	Longitudinal control (turbulence)
18a-e	Longitudinal stability (SPO)/Aft CG
19a-b	Longitudinal stability (Phugoid)/Aft CG

(continued)

[4]Chudoba, B., "Investigation of Inherent Slender-Body Characteristics Using the CONCORDE Simulator," CoA Report NFP0104, Department of Aerospace Technology, College of Aeronautics, Cranfield University, 27 February 1997.

(continued)

Flight no.	Subject
20a-f	Lateral dynamic stability (dutch roll)/rapid rudder displacement
21	Dynamic stability on all axis
22	Constant attitude deceleration/decent
23	Emergency descent
24	Emergency descent with partial loss of forward CG transfer facility
25a	Ground effect in landing
25b	Ground effect in landing (Fwd CG)
25c	Ground effect in landing (Aft CG)
26	Demonstration of max cross-wind on landing (3 engines)
27	Approach and flare with all engines stopped

A.8 Table of Contents of 'Stability and Control Design and Test Condition Matrix'[5]

[5]Chudoba, B., "Stability & Control Aircraft Design and Test Condition Matrix," Technical Report EF-039/96, Daimler-Benz Aerospace Airbus, September 1996.

CONTENTS

A.9 Euler Equations of Motion

The derivation of the coupled 6-DOF small perturbation EOM from first principles is based on Newton's laws. The approach by Etkin and Reid [1] has been selected and followed for its exemplariness. However, the inception presented here avoids the simplifications typically adopted for the symmetric type of aircraft. As a consequence, the following derivation deviates notably from Etkin and Reid's implementation. The identical first part of the derivation, until consideration of the inertia matrix I_B, has been reproduced in short for completeness.

Basic Assumptions

The aircraft is treated as a single rigid body with six degrees of freedom. The body is free to move in the atmosphere under the actions of gravity, aerodynamic forces, thrust forces, and gyroscopic effects of spinning rotors. The Earth is treated flat and stationary in inertial space, thus the rotational velocity is neglected.[6] With respect to the flat Earth assumption, Newton's laws of motion are valid.

Vector Equations of Motion of the Aircraft

The rigid-body equations are derived from first principles. Newton's laws are applied to an element dm of the aircraft, and an integration over all elements follows, see Fig. A.9.1. All velocities and accelerations are taken relative to a Newtonian (inertial) frame of reference, indicated by the fixed frame F_E. Since the aerodynamic forces depend on the velocity relative to the surrounding air mass, it is assumed in the present context that the wind velocity vector is taken zero, making the airspeed the same as the inertial velocity.

The airspeed vector of the aircraft mass center (centre of gravity) in the body frame is given by

$$\vec{V}_B = \begin{bmatrix} u & v & w \end{bmatrix}^T \tag{A.9.1}$$

The position vector of dm in the frame F_E and the frame F_B is

$$\vec{r}_{c.g.E} = \begin{bmatrix} x_E & y_E & z_E \end{bmatrix}^T \tag{A.9.2a}$$

$$\vec{r}_B = \begin{bmatrix} x & y & z \end{bmatrix}^T \tag{A.9.2b}$$

The inertial velocity of dm in the Earth frame F_E is given with Eqs. (A.9.2a) and (A.9.2b)

[6]The rotational velocity of the Earth must be taken into account for hypersonic operation.

Fig. A.9.1 Inertial frame and body frame

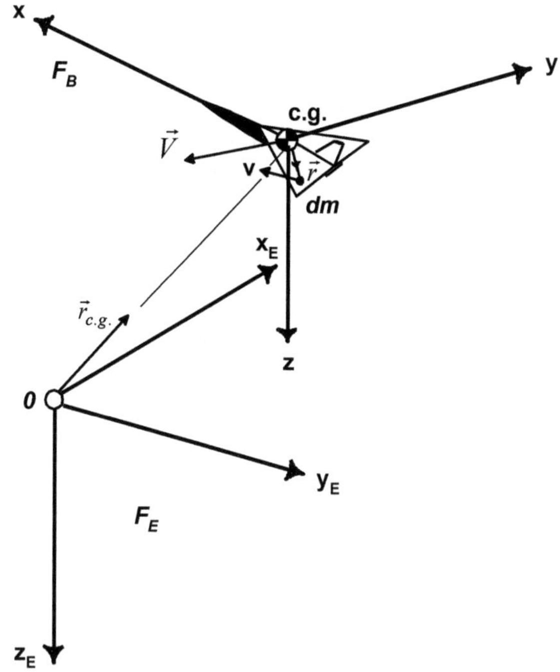

$$\vec{v}_E = \left(\dot{\vec{r}}_{c.g._E} + \dot{\vec{r}}_E\right) = \vec{V}_E + \dot{\vec{r}}_E \quad \text{(A.9.3)}$$

The momentum of the whole aircraft is

$$\int \vec{v}_E dm = \int \left(\vec{V}_E + \dot{\vec{r}}_E\right) dm = \vec{V}_E \int dm + \int \dot{\vec{r}}_E dm \quad \text{(A.9.4)}$$

With the c.g. being the mass centre and *m* the total mass of the aircraft, it follows that

$$\int \vec{v}_E dm = m\vec{V}_E \quad \text{(A.9.5)}$$

Applying Newton's second law to *dm* yields for the resultant of all forces acting on *dm*

$$d\vec{f}_E = \dot{\vec{v}}_E dm \quad \text{(A.9.6)}$$

The integral of (A.9.6) is the vector equation of motion of the aircraft

$$\vec{f}_E = \int d\vec{f}_E = \int \dot{\vec{v}}_E dm = m\dot{\vec{V}}_E \tag{A.9.7}$$

This resultant external force acts upon the aircraft and relates the force to the motion of the aircraft c.g. The relation between the external moment and the rotation of the aircraft is obtained from considering the moment of momentum for *dm* with respect to c.g.

$$d\vec{h} = \vec{r} \times \vec{v} dm \tag{A.9.8}$$

Using the matrix form of the cross product yields

$$d\vec{h}_E = \tilde{r}_E \vec{v}_E dm \tag{A.9.9}$$

With

$$\frac{d}{dt}(d\vec{h}_E) = \dot{\tilde{r}}_E \vec{v}_E dm + \tilde{r}_E \dot{\vec{v}}_E dm \tag{A.9.10}$$

From (A.9.3) follows

$$\dot{\tilde{r}}_E = \tilde{v}_E - \widetilde{V}_E \tag{A.9.11}$$

and the moment of the resultant of all forces acting on *dm* about the c.g. is

$$d\vec{G} = \vec{r} \times d\vec{f} \tag{A.9.12}$$

With (A.9.6) follows

$$d\vec{G}_E = \tilde{r}_E d\vec{f}_E = \tilde{r}_E \dot{\vec{v}}_E dm \tag{A.9.13}$$

Equation (A.9.10) becomes

$$\begin{aligned} d\vec{G}_E &= \frac{d}{dt}\left(d\vec{h}_E\right) - \dot{\tilde{r}}_E \vec{v}_E dm \\ d\vec{G}_E &= \frac{d}{dt}\left(d\vec{h}_E\right) - \left(\tilde{v}_E - \widetilde{V}_E\right)\vec{v}_E dm \\ d\vec{G}_E &= \frac{d}{dt}\left(d\vec{h}_E\right) + \widetilde{V}_E \vec{v}_E dm \end{aligned} \tag{A.9.14}$$

Integration yields the resultant moment about the c.g., denoted

$$\vec{G}_E = \int d\vec{G}_E = \frac{d}{dt}\int d\vec{h}_E + \widetilde{V}_E \int \vec{v}_E dm \tag{A.9.15}$$

Using (A.9.5), (A.9.15) reduces to the vector equation of motion of the aircraft

$$\vec{G}_E = \dot{\vec{h}}_E \tag{A.9.16}$$

where

$$\vec{h}_E = \int \tilde{r}_E \vec{v}_E dm \tag{A.9.17}$$

Equation (A.9.16) relates to and is valid only for the moving point, the c.g., as reference point. Summarising, the two vector equations of motion of the aircraft [Eqs. (A.9.7) and (A.9.16)] are equivalent to six scalar equations. Those two vector equations are written in the most general form when taking the wind vector into account as

$$\vec{f}_E = \int d\vec{f}_E = \int \dot{\vec{v}}_E^E dm = m\dot{\vec{V}}_E^E \tag{A.9.18}$$

$$\vec{G}_E = \dot{\vec{h}}_E \tag{A.9.19}$$

where velocities relative to F_E are identified by a superscript E. It is required to evaluate the components of the angular momentum in F_B, leading to

$$\vec{h}_B = \int d\vec{h}_B = \int \vec{r} \times \vec{v} dm = \int \tilde{r}_B \vec{v}_B dm \tag{A.9.20}$$

with

$$\tilde{r}_B = \begin{bmatrix} 0 & -z & y \\ z & 0 & -x \\ -y & x & 0 \end{bmatrix} \tag{A.9.21}$$

The angular velocity of the aircraft relative to inertial space is given with

$$\vec{\omega}_B = [p \quad q \quad r]^T \tag{A.9.22}$$

The velocity of a point in a rigid rotating body is given with

$$\vec{v}_B = \vec{V}_B + \tilde{\omega}_B \vec{r}_B \tag{A.9.23}$$

with the matrix

$$\tilde{\omega}_B = \begin{bmatrix} 0 & -r & q \\ r & 0 & -p \\ -q & p & 0 \end{bmatrix} \tag{A.9.24}$$

leading to

$$\begin{aligned} \vec{h}_B &= \int \tilde{r}_B \left(\vec{V}_B + \tilde{\omega}_B \vec{r}_B \right) dm \\ \vec{h}_B &= \int \tilde{r}_B \vec{V}_B dm + \int \tilde{r}_B \tilde{\omega}_B \vec{r}_B dm \\ \vec{h}_B &= \underbrace{\left(\int \tilde{r}_B dm \right)}_{0} \vec{V}_B + \int \tilde{r}_B \tilde{\omega}_B \vec{r}_B dm \end{aligned} \tag{A.9.25}$$

Since the c.g. is the origin of the position matrix, the first integral vanishes. The expansion of the triple matrix product of the second integral results with the inertia matrix

$$\vec{h}_B = I_B \vec{\omega}_B = \begin{bmatrix} I_x & -I_{xy} & -I_{xz} \\ -I_{yx} & I_y & -I_{yz} \\ -I_{zx} & -I_{zy} & I_z \end{bmatrix} \vec{\omega}_B \tag{A.9.26}$$

The elements in the inertia matrix are the moments and products of inertia of the aircraft, given by

$$\begin{aligned} I_x &= \int (y^2 + z^2) dm; \quad I_y = \int (x^2 + z^2) dm; \quad I_z = \int (x^2 + y^2) dm \\ I_{xy} &= I_{yx} = \int xy\, dm; \quad I_{xz} = I_{zx} = \int xz\, dm; \quad I_{yz} = I_{zy} = \int yz\, dm \end{aligned} \tag{A.9.27}$$

The usual assumption for symmetric aircraft is that the xz-plane is a plane of symmetry, leading to $I_{xy} = I_{yz} = 0$. If the axes are chosen to be principal axes, then is $I_{xz} = 0$. The following summarises the three principal inertia matrices of relevance:

$$I_B = \begin{bmatrix} I_x & -I_{xy} & -I_{xz} \\ -I_{yx} & I_y & -I_{yz} \\ -I_{zx} & -I_{zy} & I_z \end{bmatrix} \tag{A.9.28a}$$

$$I'_B = \begin{bmatrix} I_x & 0 & -I_{xz} \\ 0 & I_y & 0 \\ -I_{zx} & 0 & I_z \end{bmatrix} \tag{A.9.28b}$$

$$I''_B = \begin{bmatrix} I_x & 0 & 0 \\ 0 & I_y & 0 \\ 0 & 0 & I_z \end{bmatrix} \tag{A.9.28c}$$

The follow-on derivation of the EOM has taken into account the most general formulation of the inertia matrix, I_B, given with (A.9.28a), relevant for modelling asymmetric and symmetric aircraft configurations.

Flight Path of the Aircraft

Reference for the flight path of the aircraft is the Earth-fixed frame F_E with the c.g. coordinates (x_E, y_E, z_E). The Euler angles, (ψ, θ, ϕ), describe the orientation of the aircraft.[7] The determination of the flight path of the vehicle relative to F_E requires to express the velocity components in the directions of the axes of F_E.

$$\vec{V}_E = L_{EB}\vec{V}_B \tag{A.9.29}$$

The matrix of the direction cosines is given with L_{EB}, that corresponds to the transformation from F_E to F_B,

$$L_{EB} = L_z(-\psi)L_y(-\theta)L_x(-\phi) \tag{A.9.30}$$

yielding

$$L_{EB} = \begin{bmatrix} \cos\theta\cos\psi & \sin\phi\sin\theta\cos\psi - \cos\phi\sin\psi & \cos\phi\sin\theta\cos\psi + \sin\phi\sin\psi \\ \cos\theta\sin\psi & \sin\phi\sin\theta\sin\psi + \cos\phi\cos\psi & \cos\phi\sin\theta\sin\psi - \sin\phi\cos\psi \\ -\sin\theta & \sin\phi\cos\theta & \cos\phi\cos\theta \end{bmatrix} \tag{A.9.31}$$

The differential equations for the coordinates of theflight path are given with

$$\begin{bmatrix} \dot{x}_E \\ \dot{y}_E \\ \dot{z}_E \end{bmatrix} = L_{EB}\vec{V}_B \tag{A.9.32}$$

and the position of the vehicle c.g. requires integration of (A.9.32).

Orientation of the Aircraft

The following derives a set of differential equations from which the Euler angles are calculated. If the aircraft experiences an infinitesimal rotation in a time span Δt from the reference position (ψ, θ, ϕ) leading to $(\psi + \Delta\psi, \theta + \Delta\theta + \phi + \Delta\phi)$, the

[7]The angular range of the Euler angles has been limited to avoid ambiguities, see Ref. [1].

corresponding vector representing this rotation using unit vectors is approximately given with

$$\Delta\vec{n} \cong \vec{i}_3\Delta\phi + \vec{j}_2\Delta\theta + \vec{k}_1\Delta\psi \tag{A.9.33}$$

leading to the exact expression of the angular velocity

$$\vec{\omega} = \lim_{\Delta t \to 0}\frac{\Delta\vec{n}}{\Delta t} = \vec{i}_3\dot{\phi} + \vec{j}_2\dot{\theta} + \vec{k}_1\dot{\psi} \tag{A.9.34}$$

With the definition of the unit vectors (linear algebra) and the application of coordinate transformations, the unit vectors expressed in the frame F_B are

$$\vec{i}_{3_B} = \begin{bmatrix} 1 \\ 0 \\ 0 \end{bmatrix} \vec{j}_{2_B} = \begin{bmatrix} 0 \\ \cos\phi \\ -\sin\phi \end{bmatrix} \vec{k}_{1_B} = \begin{bmatrix} -\sin\theta \\ \cos\theta\sin\phi \\ \cos\theta\cos\phi \end{bmatrix} \tag{A.9.35}$$

We receive for the angular velocity

$$\vec{\omega}_B = \begin{bmatrix} p \\ q \\ r \end{bmatrix} = R\begin{bmatrix} \dot{\phi} \\ \dot{\theta} \\ \dot{\psi} \end{bmatrix} = \begin{bmatrix} 1 & 0 & -\sin\theta \\ 0 & \cos\phi & \sin\phi\cos\theta \\ 0 & -\sin\phi & \cos\phi\cos\theta \end{bmatrix}\begin{bmatrix} \dot{\phi} \\ \dot{\theta} \\ \dot{\psi} \end{bmatrix} \tag{A.9.36}$$

Inverting (A.9.36) yields the Euler angle rates as

$$\begin{bmatrix} \dot{\phi} \\ \dot{\theta} \\ \dot{\psi} \end{bmatrix} = T\begin{bmatrix} p \\ q \\ r \end{bmatrix} = \begin{bmatrix} 1 & \sin\phi\tan\theta & \cos\phi\tan\theta \\ 0 & \cos\phi & -\sin\phi \\ 0 & \sin\phi\sec\theta & \cos\phi\sec\theta \end{bmatrix}\begin{bmatrix} p \\ q \\ r \end{bmatrix} \tag{A.9.37}$$

Euler's Equations of Motion

The general form of Equation (A.9.19) contains the time derivative of the angular momentum, which contains the moments and products of inertia with reference to the axes of choice. To avoid that the inertias are variables, the equations are written in the frame F_B.

Transformation of the force equation (A.9.18) from frame F_E to frame F_B yields

$$L_{EB}\vec{f}_B = m\frac{d}{dt}\left(L_{EB}\vec{V}_B^E\right) = m\left(\dot{L}_{EB}\vec{V}_B^E + L_{EB}\dot{\vec{V}}_B^E\right) = m\left(L_{EB}\tilde{\omega}_B\vec{V}_B^E + L_{EB}\dot{\vec{V}}_B^E\right) \tag{A.9.38}$$

with the derivative of the transformation matrix given as

$$\dot{L}_{EB} = L_{EB}\tilde{\omega}_B \tag{A.9.39}$$

The transformed force- and moment[8] equations can be written as

$$\vec{f}_B = m\left(\dot{\vec{V}}_B^E + \tilde{\omega}_B \vec{V}_B^E\right) \tag{A.9.40}$$

$$\vec{G}_B = \dot{\vec{h}}_B + \tilde{\omega}_B \vec{h}_B \tag{A.9.41}$$

The force vector is the sum of gravitational-, aerodynamic-, and thrust forces:

$$\vec{f} = m\vec{g} + \vec{A} + \vec{T} \tag{A.9.42a}$$

$$\vec{f}_B = m\vec{g}_B + \vec{A}_B + \vec{T}_B \tag{A.9.42b}$$

where

$$m\vec{g}_B = mL_{BE}\vec{g}_E = mL_{BE}[0 \quad 0 \quad g]^T \tag{A.9.43}$$

$$\vec{A}_B = [X_A \quad Y_A \quad Z_A]_B^T \tag{A.9.44}$$

$$\vec{T}_B = [X_T \quad Y_T \quad Z_T]_B^T \tag{A.9.45}$$

The airspeed vector of the aircraft mass center is given with

$$\vec{V}_B^E = \left[u^E \quad v^E \quad w^E\right]^T \tag{A.9.46}$$

Expansion of the force equation (A.9.40) with (A.9.24) and (A.9.42b) yields:

$$\vec{f}_B = m\left(\dot{\vec{V}}_B^E + \tilde{\omega}_B \vec{V}_B^E\right)$$

$$m\vec{g}_B + \vec{A}_B + \vec{T}_B = m\left(\dot{\vec{V}}_B^E + \tilde{\omega}_B \vec{V}_B^E\right)$$

$$mL_{BE}\vec{g}_B + \vec{A}_B + \vec{T}_B = m\left(\dot{\vec{V}}_B^E + \tilde{\omega}_B \vec{V}_B^E\right)$$

---

[8] A similar transformation procedure yields Eq. (A.9.41).

$$m\begin{bmatrix} \cos\theta\cos\psi & \cos\theta\sin\psi & -\sin\theta \\ \sin\phi\sin\theta\cos\psi & \sin\phi\sin\theta\sin\psi & \sin\phi\cos\theta \\ -\cos\phi\sin\psi & +\cos\phi\cos\psi & \\ \cos\phi\sin\theta\cos\psi & \cos\phi\sin\theta\sin\psi & \cos\phi\cos\theta \\ +\sin\phi\sin\psi & -\sin\phi\cos\psi & \end{bmatrix}\begin{bmatrix} 0 \\ 0 \\ g \end{bmatrix} + \begin{bmatrix} X_A \\ Y_A \\ Z_A \end{bmatrix}_B + \begin{bmatrix} X_T \\ Y_T \\ Z_T \end{bmatrix}_B$$

$$= m\left(\begin{bmatrix} \dot{u}^E \\ \dot{v}^E \\ \dot{w}^E \end{bmatrix} + \begin{bmatrix} 0 & -r & q \\ r & 0 & -p \\ -q & p & 0 \end{bmatrix}\begin{bmatrix} u^E \\ v^E \\ w^E \end{bmatrix}\right)$$

$$m\begin{bmatrix} g(-\sin\theta) \\ g\sin\phi\cos\theta \\ g\cos\phi\cos\theta \end{bmatrix} + \begin{bmatrix} X_A \\ Y_A \\ Z_A \end{bmatrix}_B + \begin{bmatrix} X_T \\ Y_T \\ Z_T \end{bmatrix}_B = m\left(\begin{bmatrix} \dot{u}^E \\ \dot{v}^E \\ \dot{w}^E \end{bmatrix} + \begin{bmatrix} -rv^E + qw^E \\ ru^E - pw^E \\ -qu^E + pv^E \end{bmatrix}\right)$$

leading to the Euler force-equations of motion

$$X_A + X_T - mg\sin\theta = m\left(\dot{u}^E + qw^E - rv^E\right) \tag{A.9.47a}$$

$$Y_A + Y_T + mg\cos\theta\sin\phi = m\left(\dot{v}^E + ru^E - pw^E\right) \tag{A.9.47b}$$

$$Z_A + Z_T + mg\cos\theta\cos\phi = m\left(\dot{w}^E + pv^E - qu^E\right) \tag{A.9.47c}$$

The moment vector is the sum of aerodynamic- and thrust moments:

$$\vec{G}_B = \vec{G}_A + \vec{G}_T \tag{A.9.48}$$

where

$$\vec{G}_A = \left[L_A \quad M_A \quad N_A\right]_B^T \tag{A.9.49}$$

$$\vec{G}_T = \left[L_T \quad M_T \quad N_T\right]_B^T \tag{A.9.50}$$

Expansion of the moment equation (A.9.41) with (A.9.24), (A.9.26), and the assumption the aircraft is rigid, $(d/dt)\, I_B = 0$, yields:

$$\vec{G}_B = \vec{G}_A + \vec{G}_T = \dot{\vec{h}}_B + \tilde{\omega}_B \vec{h}_B$$

$$\vec{G}_A + \vec{G}_T = \frac{d}{dt}\left(I_B\vec{\omega}_B\right) + \tilde{\omega}_B I_B \vec{\omega}_B = \left(\dot{I}_B\vec{\omega}_B + I_B\dot{\vec{\omega}}_B\right) + \tilde{\omega}_B I_B \vec{\omega}_B = I_B\dot{\vec{\omega}}_B + \tilde{\omega}_B I_B \vec{\omega}_B$$

$$\begin{bmatrix} L_A \\ M_A \\ N_A \end{bmatrix} + \begin{bmatrix} L_T \\ M_T \\ N_T \end{bmatrix} = \begin{bmatrix} I_x & -I_{xy} & -I_{xz} \\ -I_{yx} & I_y & -I_{yz} \\ -I_{zx} & -I_{zy} & I_z \end{bmatrix} \begin{bmatrix} \dot{p} \\ \dot{q} \\ \dot{r} \end{bmatrix} + \begin{bmatrix} 0 & -r & q \\ r & 0 & -p \\ -q & p & 0 \end{bmatrix} \begin{bmatrix} I_x & -I_{xy} & -I_{xz} \\ -I_{yx} & I_y & -I_{yz} \\ -I_{zx} & -I_{zy} & I_z \end{bmatrix} \begin{bmatrix} p \\ q \\ r \end{bmatrix}$$

$$\begin{bmatrix} L_A + L_T \\ M_A + M_T \\ N_A + N_T \end{bmatrix} = \begin{bmatrix} \dot{p}I_x & -\dot{q}I_{xy} & -\dot{r}I_{xz} \\ -\dot{p}I_{yx} & \dot{q}I_y & -\dot{r}I_{yz} \\ -\dot{p}I_{zx} & -\dot{q}I_{zy} & \dot{r}I_z \end{bmatrix} + \begin{bmatrix} 0 & -r & q \\ r & 0 & -p \\ -q & p & 0 \end{bmatrix} \begin{bmatrix} pI_x & -qI_{xy} & -rI_{xz} \\ -pI_{yx} & qI_y & -rI_{yz} \\ -pI_{zx} & -qI_{zy} & rI_z \end{bmatrix}$$

$$\begin{bmatrix} L_A + L_T \\ M_A + M_T \\ N_A + N_T \end{bmatrix} = \begin{bmatrix} \dot{p}I_x & -\dot{q}I_{xy} & -\dot{r}I_{xz} \\ -\dot{p}I_{yx} & \dot{q}I_y & -\dot{r}I_{yz} \\ -\dot{p}I_{zx} & -\dot{q}I_{zy} & \dot{r}I_z \end{bmatrix} + \begin{bmatrix} rpI_{yx} - qpI_{zx} - rqI_y - q^2I_{zy} + r^2I_{yz} + qrI_z \\ rpI_x + p^2I_{zx} - rqI_{xy} + pqI_{zy} - r^2I_{xz} - prI_z \\ -qpI_x - p^2I_{yx} + q^2I_{xy} + pqI_y + qrI_{xz} - prI_{yz} \end{bmatrix}$$

leading to the Euler moment-equations of motion

$$\begin{aligned} L_A + L_T = {} & I_x\dot{p} - I_{yz}\left(q^2 - r^2\right) - I_{zx}(\dot{r} + pq) - I_{xy}(\dot{q} - rp) \\ & - \left(I_y - I_z\right)qr \end{aligned} \tag{A.9.51a}$$

$$\begin{aligned} M_A + M_T = {} & I_y\dot{q} - I_{zx}\left(r^2 - p^2\right) - I_{xy}(\dot{p} + qr) - I_{yz}(\dot{r} - pq) \\ & - (I_z - I_x)rp \end{aligned} \tag{A.9.51b}$$

$$\begin{aligned} N_A + N_T = {} & I_z\dot{r} - I_{xy}\left(p^2 - q^2\right) - I_{yz}(\dot{q} + rp) - I_{zx}(\dot{p} - qr) \\ & - \left(I_x - I_y\right)pq \end{aligned} \tag{A.9.51c}$$

Effect of Spinning Rotors

Until now, the aircraft has been assumed to be a single rigid body. However, the engine rotors are spinning relative to the body axes, thus contributing to the total angular momentum of the aircraft, see (A.9.26). The constant resultant relative angular momentum of all rotors with respect to axes parallel to C_{xyz} and to the origin of the rotor mass center is given in F_B with

$$\vec{h}' = \begin{bmatrix} h'_x & h'_y & h'_z \end{bmatrix}_B^T \tag{A.9.52}$$

The total angular momentum of the aircraft with spinning rotors is obtained in analogy to (A.9.26), where the inertias of the rotors are also included in I_B

$$\vec{h}_B = I_B\vec{\omega}_B + \vec{h}'_B \tag{A.9.53}$$

The Euler moment-equations of motion (A.9.51a–c) modify to

$$\begin{aligned}
\vec{G}_B = \vec{G}_A + \vec{G}_T &= \dot{\vec{h}}_B + \tilde{\omega}_B\vec{h}_B \\
\vec{G}_A + \vec{G}_T &= \frac{d}{dt}\left(I_B\vec{\omega}_B + \vec{h}'_B\right) + \tilde{\omega}_B\left(I_B\vec{\omega}_B + \vec{h}'_B\right) \\
&= \left(\dot{I}_B\vec{\omega}_B + I_B\dot{\vec{\omega}}_B + \dot{\vec{h}}'_B\right) + \tilde{\omega}_B I_B\vec{\omega}_B + \tilde{\omega}_B\vec{h}'_B \\
&= I_B\dot{\vec{\omega}}_B + \tilde{\omega}_B I_B\vec{\omega}_B + \tilde{\omega}_B\vec{h}'_B
\end{aligned}$$

while assuming constant moments of inertia for the aircraft and rotors.

$$\begin{aligned}
\begin{bmatrix} L_A \\ M_A \\ N_A \end{bmatrix} + \begin{bmatrix} L_T \\ M_T \\ N_T \end{bmatrix} &= \begin{bmatrix} I_x & -I_{xy} & -I_{xz} \\ -I_{yx} & I_y & -I_{yz} \\ -I_{zx} & -I_{zy} & I_z \end{bmatrix} \begin{bmatrix} \dot{p} \\ \dot{q} \\ \dot{r} \end{bmatrix} \\
&+ \begin{bmatrix} 0 & -r & q \\ r & 0 & -p \\ -q & p & 0 \end{bmatrix} \begin{bmatrix} I_x & -I_{xy} & -I_{xz} \\ -I_{yx} & I_y & -I_{yz} \\ -I_{zx} & -I_{zy} & I_z \end{bmatrix} \begin{bmatrix} p \\ q \\ r \end{bmatrix} \\
&+ \begin{bmatrix} 0 & -r & q \\ r & 0 & -p \\ -q & p & 0 \end{bmatrix} \begin{bmatrix} h'_x \\ h'_y \\ h'_z \end{bmatrix}
\end{aligned}$$

$$\begin{aligned}
\begin{bmatrix} L_A + L_T \\ M_A + M_T \\ N_A + N_T \end{bmatrix} &= \begin{bmatrix} \dot{p}I_x & -\dot{q}I_{xy} & -\dot{r}I_{xz} \\ -\dot{p}I_{yx} & \dot{q}I_y & -\dot{r}I_{yz} \\ -\dot{p}I_{zx} & -\dot{q}I_{zy} & \dot{r}I_z \end{bmatrix} \\
&+ \begin{bmatrix} rpI_{yx} - qpI_{zx} - rqI_y - q^2I_{zy} + r^2I_{yz} + qrI_z \\ rpI_x + p^2I_{zx} - rqI_{xy} + pqI_{zy} - r^2I_{xz} - prI_z \\ -qpI_x - p^2I_{yx} + q^2I_{xy} + pqI_y + qrI_{xz} - prI_{yz} \end{bmatrix} \\
&+ \begin{bmatrix} -rh'_y + qh'_z \\ rh'_x - ph'_z \\ -qh'_x + ph'_y \end{bmatrix}
\end{aligned}$$

leading to the Euler moment-equations of motion with spinning rotors

$$
\begin{aligned}
L_A + L_T = {} & I_x\dot{p} - I_{yz}\left(q^2 - r^2\right) - I_{zx}(\dot{r} + pq) - I_{xy}(\dot{q} - rp) \\
& - \left(I_y - I_z\right)qr + qh'_z - rh'_y
\end{aligned}
\tag{A.9.54a}
$$

$$
\begin{aligned}
M_A + M_T = {} & I_y\dot{q} - I_{zx}\left(r^2 - p^2\right) - I_{xy}(\dot{p} + qr) - I_{yz}(\dot{r} - pq) \\
& - (I_z - I_x)rp + rh'_x - ph'_z
\end{aligned}
\tag{A.9.54b}
$$

$$
\begin{aligned}
N_A + N_T = {} & I_z\dot{r} - I_{xy}\left(p^2 - q^2\right) - I_{yz}(\dot{q} + rp) - I_{zx}(\dot{p} - qr) \\
& - \left(I_x - I_y\right)pq + ph'_y - qh'_x
\end{aligned}
\tag{A.9.54c}
$$

The Equations Collected

According to Etkin and Reid's style, the kinematical and dynamical equations valid for symmetric and asymmetric aircraft types are collected below for convenience.

$$X_A + X_T - mg\sin\theta = m\left(\dot{u}^E + qw^E - rv^E\right) \tag{A.9.55a}$$

$$Y_A + Y_T + mg\cos\theta\sin\phi = m\left(\dot{v}^E + ru^E - pw^E\right) \tag{A.9.55b}$$

$$Z_A + Z_T + mg\cos\theta\cos\phi = m\left(\dot{w}^E + pv^E - qu^E\right) \tag{A.9.55c}$$

$$
\begin{aligned}
L_A + L_T = {} & I_x\dot{p} - I_{yz}\left(q^2 - r^2\right) - I_{zx}(\dot{r} + pq) - I_{xy}(\dot{q} - rp) \\
& - \left(I_y - I_z\right)qr + qh'_z - rh'_y
\end{aligned}
\tag{A.9.56a}
$$

$$
\begin{aligned}
M_A + M_T = {} & I_y\dot{q} - I_{zx}\left(r^2 - p^2\right) - I_{xy}(\dot{p} + qr) - I_{yz}(\dot{r} - pq) \\
& - (I_z - I_x)rp + rh'_x - ph'_z
\end{aligned}
\tag{A.9.56b}
$$

$$
\begin{aligned}
N_A + N_T = {} & I_z\dot{r} - I_{xy}\left(p^2 - q^2\right) - I_{yz}(\dot{q} + rp) - I_{zx}(\dot{p} - qr) \\
& - \left(I_x - I_y\right)pq + ph'_y - qh'_x
\end{aligned}
\tag{A.9.56c}
$$

$$p = \dot{\phi} - \dot{\psi}\sin\theta \tag{A.9.57a}$$

$$q = \dot{\theta}\cos\phi + \dot{\psi}\sin\phi\cos\theta \tag{A.9.57b}$$

$$r = -\dot{\theta}\sin\phi + \dot{\psi}\cos\phi\cos\theta \tag{A.9.57c}$$

$$\dot{\phi} = p + q(\sin\phi + r\cos\phi)\tan\theta \tag{A.9.58a}$$

$$\dot{\theta} = q \cos\phi - r \sin\phi \tag{A.9.58b}$$

$$\dot{\psi} = (q \sin\phi + r \cos\phi) \sec\theta \tag{A.9.58c}$$

$$\begin{aligned} \dot{x}_E = {} & u^E \cos\theta \cos\psi + v^E(\sin\phi \sin\theta \cos\psi - \cos\phi \sin\psi) \\ & + w^E(\cos\phi \sin\theta \cos\psi + \sin\phi \sin\psi) \end{aligned} \tag{A.9.59a}$$

$$\begin{aligned} \dot{y}_E = {} & u^E \cos\theta \sin\psi + v^E(\sin\phi \sin\theta \sin\psi + \cos\phi \cos\psi) \\ & + w^E(\cos\phi \sin\theta \sin\psi - \sin\phi \cos\psi) \end{aligned} \tag{A.9.59b}$$

$$\dot{z}_E = -u^E \sin\theta + v^E \sin\phi \cos\theta + w^E \cos\phi \cos\theta \tag{A.9.59c}$$

In summary, the above equations contain the following assumptions:

1. The Earth is treated flat and stationary in inertial space, thus rotational velocity is neglected.
2. The equations are valid for any orthogonal axes system fixed at the c.g. of the aircraft (body axes).
3. The aircraft is a rigid body ($\dot{I}_B = 0$), having attached to it any number of rigid spinning rotors.
4. The spinning rotors have constant angular speed relative to the body axes ($\dot{\vec{h}}'_B = 0$). The axes of any spinning rotor is fixed in direction relative to the body axes. This assumption is valid for thrust vectoring with a movable nozzle (usual), where the thrust vector alters direction but the axes of the spinning rotors stay constant.[9]
5. The wind velocity is zero, so that $\vec{V}^E = \vec{V}$.

The usual assumptions like, (i) the existence of a plane of symmetry (C_{xz}), (ii) the neglect of aerodynamic cross-coupling, (iii) the absence of rotor gyroscopic effects, have not been accepted in the derivation above.

Reference

1. Etkin, B. and Reid, L.D., "*Dynamics of Flight – Stability and Control,*" Third Edition, John Wiley & Sons, Inc., 1996.

[9]The assumption of fixed axes of spinning rotors requires a review, when applied to the OFWC with engines which are pivoted when the wing sweep is adjusted during flight.

A.10 Small Perturbation Equations of Motion

As a consequence of the simplifying assumptions classically made in the derivation of the EOM for conceptual design applications, the small perturbation approach usually leads to a decoupled set of longitudinal- and lateral-directional equations. However, the underlying assumptions, upon which such separation depends, have not been made in the present context. There exists no pure longitudinal motion, since (i) no plane of symmetry is assumed, and (ii) rotor gyroscopic effects are included. The absence of pure lateral-directional motions is a direct result of (i) taking rotor gyroscopic effects into account, and (ii) the inclusion of aerodynamic cross-coupling effects. Clearly, the process of linearising the 6-DOF EOM does not imply decoupled EOM.

Small Disturbance Notation

The small perturbation approach assumes, that the motion of the aircraft consists of small deviations from a steady flight reference condition. The following notation for small disturbances has been adopted from [1]:

1. Reference values of all variables are denoted with a subscript zero, the small perturbations by prefix Δ.
2. In case of a reference value being zero, the Δ is omitted (e.g., $\Delta p = p$, $\Delta q = q$, $\Delta r = r$).
3. Disturbance quantities and their derivatives are assumed to be small (small perturbation assumption), their squares and products are negligible compared to first-order quantities.
4. Only the first-order terms in disturbance quantities are kept.
5. Symmetric flight with no angular velocity is assumed to be the reference condition ($v_0 = p_0 = q_0 = r_0 = \phi_0 = 0$).
6. Stability axes are selected as standard ($w_0 = 0$); thus, u_0 is equal to the reference flight speed, and θ_0 is equal to the reference angle of climb, which is not necessarily assumed to be small.
7. No wind velocity is taken into account, so that $\vec{V}^E = \vec{V}$.
8. The DiCE (e.g., rudder) and LaCE (e.g., aileron) deflection angles are assumed zero in the reference flight condition.
9. The following transcendental functions have been used:

$$sin(\theta_0 + \Delta\theta) = \sin\theta_0 \cos\Delta\theta + \cos\theta_0 \sin\Delta\theta \cong \sin\theta_0 + \Delta\theta \cos\theta_0$$
$$cos(\theta_0 + \Delta\theta) = \cos\theta_0 \cos\Delta\theta - \sin\theta_0 \sin\Delta\theta \cong \cos\theta_0 - \Delta\theta \sin\theta_0$$

Linearisation Process

The process of linearising Eqs. (A.9.55a–c) to (A.9.59a–c) follows below:

X-Force Equation

$$\begin{aligned} &X_{A_0} + \Delta X_A + X_{T_0} + \Delta X_T - mg\ \sin(\theta_0 + \Delta\theta) \\ &\quad = m\left[\dot{u}_0^E + \Delta\dot{u}^E + (q_0 + \Delta q)\left(w_0^E + \Delta w^E\right) - (r_0 + \Delta r)\left(v_0^E + \Delta v^E\right)\right] \\ &X_{A_0} + \Delta X_A + X_{T_0} + \Delta X_T - mg(\sin\theta_0 + \Delta\theta\ \cos\theta_0) = m\Delta\dot{u} \end{aligned} \quad (A.10.1)$$

Y-Force Equation

$$\begin{aligned} &Y_{A_0} + \Delta Y_A + Y_{T_0} + \Delta Y_T + mg\,cos(\theta_0 + \Delta\theta)\,sin(\phi_0 + \Delta\phi) \\ &\quad = m\left[\dot{v}_0^E + \Delta\dot{v}^E + (r_0 + \Delta r)\left(u_0^E + \Delta u^E\right) - (p_0 + \Delta p)\left(w_0^E + \Delta w^E\right)\right] \\ &Y_{A_0} + \Delta Y_A + Y_{T_0} + \Delta Y_T + mg(\cos\theta_0 - \Delta\theta\ \sin\theta_0)\Delta\phi = m(\dot{v} + u_0 r) \\ &Y_{A_0} + \Delta Y_A + Y_{T_0} + \Delta Y_T + mg\phi\ \cos\theta_0 = m(\dot{v} + u_0 r) \end{aligned} \quad (A.10.2)$$

Z-Force Equation

$$\begin{aligned} &Z_{A_0} + \Delta Z_A + Z_{T_0} + \Delta Z_T + mg\,cos(\theta_0 + \Delta\theta)\,cos(\phi_0 + \Delta\phi) \\ &\quad = m\left[\dot{w}_0^E + \Delta\dot{w}^E + (p_0 + \Delta p)\left(v_0^E + \Delta v^E\right) - (q_0 + \Delta q)\left(u_0^E + \Delta u^E\right)\right] \\ &Z_{A_0} + \Delta Z_A + Z_{T_0} + \Delta Z_T \\ &\quad + mg(\cos\theta_0 - \Delta\theta\ \sin\theta_0)(\cos\phi_0 - \Delta\phi\ \sin\phi_0) = m(\dot{w} - u_0 q) \\ &Z_{A_0} + \Delta Z_A + Z_{T_0} + \Delta Z_T + mg(\cos\theta_0 - \Delta\theta\ \sin\theta_0) = m(\dot{w} - u_0 q) \end{aligned} \quad (A.10.3)$$

L-Moment Equation

$$\begin{aligned} &L_{A_0} + \Delta L_A + L_{T_0} + \Delta L_T = I_x(\dot{p}_0 + \Delta\dot{p}) - I_{yz}\left[(q_0 + \Delta q)^2 - (r_0 + \Delta r)^2\right] \\ &\quad - I_{zx}[\dot{r}_0 + \Delta\dot{r} + (p_0 + \Delta p)(q_0 + \Delta q)] - I_{xy}[\dot{q}_0 + \Delta\dot{q} - (r_0 + \Delta r)(p_0 + \Delta p)] \\ &\quad - \left(I_y - I_z\right)(q_0 + \Delta q)(r_0 + \Delta r) + (q_0 + \Delta q)h_z' - (r_0 + \Delta r)h_y' \\ &L_{A_0} + \Delta L_A + L_{T_0} + \Delta L_T = I_x\dot{p} - I_{zx}\dot{r} - I_{xy}\dot{q} + qh_z' - rh_y' \end{aligned}$$

$$(A.10.4)$$

M-Moment Equation

$$
\begin{aligned}
M_{A_0} + \Delta M_A + M_{T_0} + \Delta M_T &= I_y(\dot{q}_0 + \Delta\dot{q}) - I_{zx}\left[(r_0 + \Delta r)^2 - (p_0 + \Delta p)^2\right] \\
&\quad - I_{xy}[\dot{p}_0 + \Delta\dot{p} + (q_0 + \Delta q)(r_0 + \Delta r)] - I_{yz}[\dot{r}_0 + \Delta\dot{r} - (p_0 + \Delta p)(q_0 + \Delta q)] \\
&\quad - (I_z - I_x)(r_0 + \Delta r)(p_0 + \Delta p) + (r_0 + \Delta r)h'_x - (p_0 + \Delta p)h'_z \\
M_{A_0} + \Delta M_A + M_{T_0} + \Delta M_T &= I_y\dot{q} - I_{xy}\dot{p} - I_{yz}\dot{r} + rh'_x - ph'_z
\end{aligned}
\tag{A.10.5}
$$

N-Moment Equation

$$
\begin{aligned}
N_{A_0} + \Delta N_A + N_{T_0} + \Delta N_T &= I_z(\dot{r}_0 + \Delta\dot{r}) - I_{xy}\left[(p_0 + \Delta p)^2 - (q_0 + \Delta q)^2\right] \\
&\quad - I_{yz}[\dot{q}_0 + \Delta\dot{q} + (r_0 + \Delta r)(p_0 + \Delta p)] - I_{zx}[\dot{p}_0 + \Delta\dot{p} - (q_0 + \Delta q)(r_0 + \Delta r)] \\
&\quad - \left(I_x - I_y\right)(p_0 + \Delta p)(q_0 + \Delta q) + (p_0 + \Delta p)h'_y - (q_0 + \Delta q)h'_x \\
N_{A_0} + \Delta N_A + N_{T_0} + \Delta N_T &= I_z\dot{r} - I_{yz}\dot{q} - I_{zx}\dot{p} + ph'_y - qh'_x
\end{aligned}
\tag{A.10.6}
$$

p-Rate Equation

$$
\begin{aligned}
p_0 + \Delta p &= \dot{\phi}_0 + \Delta\dot{\phi} - \left(\dot{\psi}_0 + \Delta\dot{\psi}\right)\sin(\theta_0 + \Delta\theta) \\
p &= \dot{\phi} - \dot{\psi}(\sin\theta_0 + \Delta\theta\cos\theta_0) \\
p &= \dot{\phi} - \dot{\psi}\sin\theta_0
\end{aligned}
\tag{A.10.7}
$$

ϕ-Change of Rate Equation

$$
\begin{aligned}
\dot{\phi}_0 + \Delta\dot{\phi} &= p_0 + \Delta p + (q_0 + \Delta q)[\sin(\phi_0 + \Delta\phi) \\
&\quad + (r_0 + \Delta r)\cos(\phi_0 + \Delta\phi)]\tan(\theta_0 + \Delta\theta) \\
\dot{\phi} &= p + q[\sin\phi_0 + \Delta\phi\cos\phi_0 + r(\cos\phi_0 - \Delta\phi\sin\phi_0)]\frac{\tan\theta_0 + \tan\Delta\theta}{1 - \tan\theta_0\tan\Delta\theta} \\
\dot{\phi} &= p + r\tan\theta_0
\end{aligned}
\tag{A.10.8}
$$

θ-Change of Rate Equation

$$
\begin{aligned}
\dot{\theta}_0 + \Delta\dot{\theta} &= (q_0 + \Delta q)\cos(\phi_0 + \Delta\phi) - (r_0 + \Delta r)\sin(\phi_0 + \Delta\phi) \\
\dot{\theta} &= q(\cos\phi_0 - \Delta\phi\sin\phi_0) - r(\sin\phi_0 + \Delta\phi\cos\phi_0) \\
\dot{\theta} &= q
\end{aligned}
\tag{A.10.9}
$$

ψ-Change of Rate Equation

$$\dot{\psi}_0 + \varDelta\dot{\psi} = [(q_0 + \varDelta q)\ \sin(\phi_0 + \varDelta\phi) + (r_0 + \varDelta r)\ \cos(\phi_0 + \varDelta\phi)]\frac{1}{\cos(\theta_0 + \varDelta\theta)}$$

$$\dot{\psi} = [q(\sin\phi_0 + \varDelta\phi\ \cos\phi_0) + r(\cos\phi_0 - \varDelta\phi\ \sin\phi_0)]\frac{1}{\cos\theta_0 - \varDelta\theta\ \sin\theta_0}$$

$$\dot{\psi} = r\ \sec\theta_0$$

(A.10.10)

x-Velocity-Component Equation

$$\begin{aligned}
\dot{x}_{E0} + \Delta\dot{x}_E &= \left(u_0^E + \Delta u^E\right)\cos(\theta_0 + \Delta\theta)\cos(\psi_0 + \Delta\psi) \\
&\quad + \left(v_0^E + \Delta v^E\right)[\sin(\phi_0 + \Delta\phi)\sin(\theta_0 + \Delta\theta)\cos(\psi_0 + \Delta\psi) \\
&\quad - \cos(\phi_0 + \Delta\phi)\sin(\psi_0 + \Delta\psi)] \\
&\quad + \left(w_0^E + \Delta w^E\right)[(\cos\phi_0 - \Delta\phi\ \sin\phi_0)(\sin\theta_0 + \Delta\theta\ \cos\theta_0)(\cos\psi_0 - \Delta\psi\ \sin\psi_0) \\
&\quad + (\sin\phi_0 + \Delta\phi\ \cos\phi_0)(\sin\psi_0 + \Delta\psi\ \cos\psi_0)] \\
\dot{x}_E &= (u_0 + \Delta u)\cos\theta_0 - u_0\Delta\theta\ \sin\theta_0 + v[(\phi\ \sin\theta_0 + \phi\Delta\theta\ \cos\theta_0) - \psi] \\
&\quad + w[\sin\theta_0 + \Delta\theta\ \cos\theta_0 + \phi\psi] \\
\dot{x}_E &= (u_0 + \Delta u)\cos\theta_0 - u_0\Delta\theta\ \sin\theta_0 + w\ \sin\theta_0
\end{aligned}$$

(A.10.11)

y-Velocity-Component Equation

$$\begin{aligned}
\dot{y}_{E0} + \varDelta\dot{y}_E &= \left(u_0^E + \varDelta u^E\right)\ \cos(\theta_0 + \varDelta\theta)\ \sin(\psi_0 + \varDelta\psi) \\
&\quad + \left(v_0^E + \varDelta v^E\right)[\sin(\phi_0 + \varDelta\phi)\ sin(\theta_0 + \varDelta\theta)\ \sin(\psi_0 + \varDelta\psi) \\
&\quad + \cos(\phi_0 + \varDelta\phi)\ \cos(\psi_0 + \varDelta\psi)] \\
&\quad + \left(w_0^E + \varDelta w^E\right)[\cos(\phi_0 + \varDelta\phi)\ \sin(\theta_0 + \varDelta\theta)\ \sin(\psi_0 + \varDelta\psi) \\
&\quad - \sin(\phi_0 + \varDelta\phi)\ \cos(\psi_0 + \varDelta\psi)] \\
\dot{y}_E &= \left[u_0^E\ \cos\theta_0 - u_0^E\varDelta\theta\ \sin\theta_0 + \varDelta u^E\ \cos\theta_0 - \varDelta u^E\varDelta\theta\ \sin\theta_0\right]\psi \\
&\quad + v^E[\phi(\sin\theta_0 + \varDelta\theta\ \cos\theta_0)\psi + 1] \\
&\quad + w^E[(\sin\theta_0 + \varDelta\theta\ \cos\theta_0)\psi - \phi] \\
\dot{y}_E &= u_0\psi\ \cos\theta_0 + v
\end{aligned}$$

(A.10.12)

z-Velocity-Component Equation

$$\begin{aligned}
\dot{z}_{E0} + \Delta\dot{z}_E &= -\left(u_0^E + \Delta u^E\right)\,\sin(\theta_0 + \Delta\theta) + \left(v_0^E + \Delta v^E\right)\,\sin(\phi_0 + \Delta\phi)\,\cos(\theta_0 + \Delta\theta) \\
&\quad + \left(w_0^E + \Delta w^E\right)\,\cos(\phi_0 + \Delta\phi)\,\cos(\theta_0 + \Delta\theta) \\
\dot{z}_E &= -\left(u_0^E + \Delta u^E\right)(\sin\theta_0 + \Delta\theta\,\cos\theta_0) \\
&\quad + \left(v_0^E + \Delta v^E\right)(\sin\phi_0 + \Delta\phi\,\cos\phi_0)(\cos\theta_0 - \Delta\theta\,\sin\theta_0) \\
&\quad + \left(w_0^E + \Delta w^E\right)(\cos\phi_0 - \Delta\phi\,\sin\phi_0)(\cos\theta_0 - \Delta\theta\,\sin\theta_0) \\
\dot{z}_E &= -(u_0 + \Delta u)\,\sin\theta_0 - u_0\Delta\theta\,\cos\theta_0 + w\,\cos\theta_0
\end{aligned} \tag{A.10.13}$$

The Equations Collected

Taking the small disturbance notation into account we receive the coupled 6-DOF small perturbation EOM:

$$X_{A_0} + \Delta X_A + X_{T_0} + \Delta X_T - mg(\sin\theta_0 + \Delta\theta\,\cos\theta_0) = m\Delta\dot{u} \tag{A.10.14a}$$

$$Y_{A_0} + \Delta Y_A + Y_{T_0} + \Delta Y_T + mg\phi\,\cos\theta_0 = m(\dot{v} + u_0 r) \tag{A.10.14b}$$

$$Z_{A_0} + \Delta Z_A + Z_{T_0} + \Delta Z_T + mg(\cos\theta_0 - \Delta\theta\,\sin\theta_0) = m(\dot{w} - u_0 q) \tag{A.10.14c}$$

$$L_{A_0} + \Delta L_A + L_{T_0} + \Delta L_T = I_x\dot{p} - I_{zx}\dot{r} - I_{xy}\dot{q} + qh'_z - rh'_y \tag{A.10.15a}$$

$$M_{A_0} + \Delta M_A + M_{T_0} + \Delta M_T = I_y\dot{q} - I_{xy}\dot{p} - I_{yz}\dot{r} + rh'_x - ph'_z \tag{A.10.15b}$$

$$N_{A_0} + \Delta N_A + N_{T_0} + \Delta N_T = I_z\dot{r} - I_{yz}\dot{q} - I_{zx}\dot{p} + ph'_y - qh'_x \tag{A.10.15c}$$

$$p = \dot{\phi} - \dot{\psi}\,\sin\theta_0 \tag{A.10.16}$$

$$\dot{\phi} = p + r\,\tan\theta_0 \tag{A.10.17a}$$

$$\dot{\theta} = q \tag{A.10.17b}$$

$$\dot{\psi} = r\,sec\,\theta_0 \tag{A.10.17c}$$

$$\dot{x}_E = (u_0 + \Delta u)\,\cos\theta_0 - u_0\Delta\theta\,\sin\theta_0 + w\,\sin\theta_0 \tag{A.10.18a}$$

$$\dot{y}_E = u_0\psi\,\cos\theta_0 + v \tag{A.10.18b}$$

$$\dot{z}_E = -(u_0 + \Delta u)\,\sin\theta_0 - u_0\Delta\theta\,\cos\theta_0 + w\,\cos\theta_0 \tag{A.10.18c}$$

Reference Steady State

All disturbance quantities are set equal zero in Eqs. (A.10.14a–c) to (A.10.18a–c), to obtain the reference forces and moments which apply to the reference flight condition. The following set of equations represents the trim state of the reference flight condition.

$$X_{A_0} + X_{T_0} - mg\sin\theta_0 = 0 \tag{A.10.19a}$$

$$Y_{A_0} + Y_{T_0} = 0 \tag{A.10.19b}$$

$$Z_{A_0} + Z_{T_0} + mg\cos\theta_0 = 0 \tag{A.10.19c}$$

$$L_{A_0} + L_{T_0} = 0 \tag{A.10.20a}$$

$$M_{A_0} + M_{T_0} = 0 \tag{A.10.20b}$$

$$N_{A_0} + N_{T_0} = 0 \tag{A.10.20c}$$

$$\dot{x}_E = u_0\cos\theta_0 \tag{A.10.21a}$$

$$\dot{y}_E = 0 \tag{A.10.21b}$$

$$\dot{z}_E = -u_0\sin\theta_0 \tag{A.10.21c}$$

Equations (A.10.19a–c) to (A.10.21a–c) are substituted into Eqs. (A.10.14a–c) to (A.10.18a–c), yielding

$$\Delta X_A + \Delta X_T - mg\Delta\theta\cos\theta_0 = m\Delta\dot{u} \tag{A.10.22a}$$

$$\Delta Y_A + \Delta Y_T + mg\phi\cos\theta_0 = m(\dot{v} + u_0 r) \tag{A.10.22b}$$

$$\Delta Z_A + \Delta Z_T - mg\Delta\theta\sin\theta_0 = m(\dot{w} - u_0 q) \tag{A.10.22c}$$

$$\Delta L_A + \Delta L_T = I_x\dot{p} - I_{zx}\dot{r} - I_{xy}\dot{q} + qh'_z - rh'_y \tag{A.10.23a}$$

$$\Delta M_A + \Delta M_T = I_y\dot{q} - I_{xy}\dot{p} - I_{yz}\dot{r} + rh'_x - ph'_z \tag{A.10.23b}$$

$$\Delta N_A + \Delta N_T = I_z\dot{r} - I_{yz}\dot{q} - I_{zx}\dot{p} + ph'_y - qh'_x \tag{A.10.23c}$$

$$p = \dot{\phi} - \dot{\psi}\sin\theta_0 \tag{A.10.24}$$

$$\dot{\phi} = p + r \tan \theta_0 \tag{A.10.25a}$$

$$\dot{\theta} = q \tag{A.10.25b}$$

$$\dot{\psi} = r \sec \theta_0 \tag{A.10.25c}$$

$$\Delta \dot{x}_E = \Delta u \cos \theta_0 - u_0 \Delta\theta \sin \theta_0 + w \sin \theta_0 \tag{A.10.26a}$$

$$\Delta \dot{y}_E = u_0 \psi \cos \theta_0 + v \tag{A.10.26b}$$

$$\Delta \dot{z}_E = -\Delta u \sin \theta_0 - u_0 \Delta\theta \cos \theta_0 + w \cos \theta_0 \tag{A.10.26c}$$

Equations (A.10.22a–c) to (A.10.26a–c) form the basis for aircraft stability and response discussions, and for automatic flight control theory and applications. It is required to rewrite and solve Eqs. (A.10.22–c) to (A.10.26a–c) for the desired form to have the first derivatives of the dependent variables,$\Delta\dot{u}$, $\dot{v}$, $\dot{w}$, $\dot{p}$, $\dot{q}$, $\dot{r}$, $\dot{\phi}$, p, $\dot{\theta}$, $\dot{\psi}$, $\Delta\dot{x}_E$, $\Delta\dot{y}_E$, $\Delta\dot{z}_E$, on the left. Equation (A.10.23a–c) have to be rewritten to form a linear system of equations solved simultaneously for $\dot{p}$, $\dot{q}$, and $\dot{r}$.

$$\begin{aligned} &I_x\dot{p} - I_{zx}\dot{r} - I_{xy}\dot{q} = \left(\Delta L_A + \Delta L_T - qh'_z + rh'_y\right) \\ &I_x\dot{p} - I_{xy}\dot{q} - I_{zx}\dot{r} = (A) \end{aligned} \tag{A.10.27a}$$

$$\begin{aligned} &I_y\dot{q} - I_{xy}\dot{p} - I_{yz}\dot{r} = \left(\Delta M_A + \Delta M_T - rh'_x + ph'_z\right) \\ &-I_{xy}\dot{p} + I_y\dot{q} - I_{yz}\dot{r} = (B) \end{aligned} \tag{A.10.27b}$$

$$\begin{aligned} &I_z\dot{r} - I_{yz}\dot{q} - I_{zx}\dot{p} = \left(\Delta N_A + \Delta N_T - ph'_y + qh'_x\right) \\ &-I_{zx}\dot{p} - I_{yz}\dot{q} + I_z\dot{r} = (C) \end{aligned} \tag{A.10.27c}$$

The linear system written in matrix format

$$\begin{bmatrix} I_x & -I_{xy} & -I_{zx} \\ -I_{xy} & I_y & -I_{yz} \\ -I_{zx} & -I_{yz} & I_z \end{bmatrix} \begin{bmatrix} \dot{p} \\ \dot{q} \\ \dot{r} \end{bmatrix} = \begin{bmatrix} A \\ B \\ C \end{bmatrix} \tag{A.10.28}$$

The linear system (A.10.28) has been solved using Maple V [2, 3], the result is given with (A.10.29a–c).

$$\dot{p} = \frac{A\left(I_yI_z - I_{yz}^2\right) + B\left(I_{xy}I_z + I_{zx}I_{yz}\right) + C\left(I_{xy}I_{yz} + I_{zx}I_y\right)}{I_xI_yI_z - I_xI_{yz}^2 - I_{xy}^2I_z - 2I_{yz}I_{xy}I_{zx} - I_yI_{zx}^2} \tag{A.10.29a}$$

$$\dot{q} = \frac{A\left(I_{yz}I_{zx} + I_{xy}I_z\right) + B\left(I_xI_z - I_{zx}^2\right) + C\left(I_xI_{yz} + I_{zx}I_{xy}\right)}{I_xI_yI_z - I_xI_{yz}^2 - I_{xy}^2I_z - 2I_{yz}I_{xy}I_{zx} - I_yI_{zx}^2} \tag{A.10.29b}$$

$$\dot{r} = \frac{A\left(I_{yz}I_{xy} + I_yI_{zx}\right) + B\left(I_xI_{yz} + I_{zx}I_{xy}\right) + C\left(I_xI_y - I_{xy}^2\right)}{I_xI_yI_z - I_xI_{yz}^2 - I_{xy}^2I_z - 2I_{yz}I_{xy}I_{zx} - I_yI_{zx}^2} \tag{A.10.29c}$$

with A, B, and C given in (A.10.27a–c). Note, that Eq. (A.10.29a–c) differ only by the moments of inertia terms in the numerator. The equations are summarised for convenience:

$$\Delta\dot{u} = m^{-1}(\Delta X_A + \Delta X_T) - g\Delta\theta\cos\theta_0 \tag{A.10.30a}$$

$$\dot{v} = m^{-1}(\Delta Y_A + \Delta Y_T) + g\phi\cos\theta_0 - u_0 r \tag{A.10.30b}$$

$$\dot{w} = m^{-1}(\Delta Z_A + \Delta Z_T) - g\Delta\theta\sin\theta_0 + u_0 q \tag{A.10.30c}$$

$$\begin{aligned}\dot{p} = {} & \left(I_xI_yI_z - I_xI_{yz}^2 - I_{xy}^2I_z - 2I_{yz}I_{xy}I_{zx} - I_yI_{zx}^2\right)^{-1} \\ & \times \left[\begin{array}{l}\left(\Delta L_A + \Delta L_T - qh_z' + rh_y'\right)\left(I_yI_z - I_{yz}^2\right) \\ + \left(\Delta M_A + \Delta M_T - rh_x' + ph_z'\right)\left(I_{xy}I_z + I_{zx}I_{yz}\right) \\ + \left(\Delta N_A + \Delta N_T - ph_y' + qh_x'\right)\left(I_{xy}I_{yz} + I_{zx}I_y\right)\end{array}\right]\end{aligned} \tag{A.10.31a}$$

$$\begin{aligned}\dot{q} = {} & \left(I_xI_yI_z - I_xI_{yz}^2 - I_{xy}^2I_z - 2I_{yz}I_{xy}I_{zx} - I_yI_{zx}^2\right)^{-1} \\ & \times \left[\begin{array}{l}\left(\Delta L_A + \Delta L_T - qh_z' + rh_y'\right)\left(I_{yz}I_{zx} + I_{xy}I_z\right) \\ + \left(\Delta M_A + \Delta M_T - rh_x' + ph_z'\right)\left(I_xI_z - I_{zx}^2\right) \\ + \left(\Delta N_A + \Delta N_T - ph_y' + qh_x'\right)\left(I_xI_{yz} + I_{zx}I_{xy}\right)\end{array}\right]\end{aligned} \tag{A.10.31b}$$

$$\begin{aligned}\dot{r} = {} & \left(I_xI_yI_z - I_xI_{yz}^2 - I_{xy}^2I_z - 2I_{yz}I_{xy}I_{zx} - I_yI_{zx}^2\right)^{-1} \\ & \times \left[\begin{array}{l}\left(\Delta L_A + \Delta L_T - qh_z' + rh_y'\right)\left(I_{yz}I_{xy} + I_yI_{zx}\right) \\ + \left(\Delta M_A + \Delta M_T - rh_x' + ph_z'\right)\left(I_xI_{yz} + I_{zx}I_{xy}\right) \\ + \left(\Delta N_A + \Delta N_T - ph_y' + qh_x'\right)\left(I_xI_y - I_{xy}^2\right)\end{array}\right]\end{aligned} \tag{A.10.31c}$$

$$p = \dot{\phi} - \dot{\psi}\sin\theta_0 \tag{A.10.32}$$

$$\dot{\phi} = p + r \tan\theta_0 \quad \text{(A.10.33a)}$$

$$\dot{\theta} = q \quad \text{(A.10.33b)}$$

$$\dot{\psi} = r\, sec\, \theta_0 \quad \text{(A.10.33c)}$$

$$\Delta \dot{x}_E = \Delta u \cos\theta_0 - u_0 \Delta\theta \sin\theta_0 + w \sin\theta_0 \quad \text{(A.10.34a)}$$

$$\Delta \dot{y}_E = u_0 \psi \cos\theta_0 + v \quad \text{(A.10.34b)}$$

$$\Delta \dot{z}_E = -\Delta u \sin\theta_0 - u_0 \Delta\theta \cos\theta_0 + w \cos\theta_0 \quad \text{(A.10.34c)}$$

Linear Air Reactions

When considering the general case of asymmetric aircraft configurations and concepts, the usual assumption to neglect cross-coupling derivatives is not valid. In writing out the complete linear expression for the aerodynamic forces and moments, the following is assumed:

1. The derivatives with respect to rates of change of motion variables are neglected except for $Z_{\dot{w}}$ and $M_{\dot{w}}$.
2. The density of the atmosphere is assumed not to vary with small altitude variations.

The linear forces and moments considered basic for asymmetric type of aircraft are given below:

$$\Delta X_A + \Delta X_T = \Delta X = X_u \Delta u + X_v v + X_w w + X_p p + X_q q + X_r r + \Delta X_{CE} \quad \text{(A.10.35a)}$$

$$\Delta Y_A + \Delta Y_T = \Delta Y = Y_u \Delta u + Y_v v + Y_w w + Y_p p + Y_q q + Y_r r + \Delta Y_{CE} \quad \text{(A.10.35b)}$$

$$\Delta Z_A + \Delta Z_T = \Delta Z = Z_u \Delta u + Z_v v + Z_w w + Z_{\dot{w}} \dot{w} + Z_p p + Z_q q + Z_r r + \Delta Z_{CE} \quad \text{(A.10.35c)}$$

$$\Delta L_A + \Delta L_T = \Delta L = L_u \Delta u + L_v v + L_w w + L_p p + L_q q + L_r r + \Delta L_{CE} \quad \text{(A.10.36a)}$$

$$\Delta M_A + \Delta M_T = \Delta M = M_u \Delta u + M_v v + M_w w + M_{\dot{w}} \dot{w} + M_p p + M_q q + M_r r + \Delta M_{CE} \quad \text{(A.10.36b)}$$

$$\Delta N_A + \Delta N_T = \Delta N = N_u \Delta u + N_v v + N_w w + N_p p + N_q q + N_r r + \Delta N_{CE} \quad \text{(A.10.36c)}$$

The terms with subscript CE are control forces and moments described by the control vector

$$\vec{c} = \left[\, \delta_{LoCE_i} \quad \delta_{DiCE_i} \quad \delta_{LaCE_i} \quad \delta_{T_i} \,\right] \tag{A.10.37}$$

Linear Small Perturbation Equations of Motion

It is required to substitute (A.10.35a–c) and (A.10.36a–c) into (A.10.30a–c) to (A.10.31a–c).

$$\Delta\dot{u} = m^{-1}\left(X_u \Delta u + X_v v + X_w w + X_p p + X_q q + X_r r + \Delta X_{CE}\right) - g\Delta\theta \cos\theta_0 \tag{A.10.38a}$$

$$\dot{v} = m^{-1}\left(Y_u \Delta u + Y_v v + Y_w w + Y_p p + Y_q q + Y_r r + \Delta Y_{CE}\right) + g\phi \cos\theta_0 - u_0 r \tag{A.10.38b}$$

$$\begin{aligned}\dot{w} = m^{-1}&\left(Z_u \Delta u + Z_v v + Z_w w + Z_{\dot{w}}\dot{w} + Z_p p + Z_q q + Z_r r + \Delta Z_{CE}\right)\\ &- g\Delta\theta \sin\theta_0 + u_0 q\end{aligned} \tag{A.10.38c}$$

$$\begin{aligned}\dot{p} = &\left(I_x I_y I_z - I_x I_{yz}^2 - I_{xy}^2 I_z - 2I_{yz}I_{xy}I_{zx} - I_y I_{zx}^2\right)^{-1}\\ &\begin{bmatrix}\left(L_u \Delta u + L_v v + L_w w + L_p p + L_q q + L_r r + \Delta L_{CE} - qh'_z + rh'_y\right)\left(I_y I_z - I_{yz}^2\right)\\ +\left(M_u \Delta u + M_v v + M_w w + M_{\dot{w}}\dot{w} + M_p p + M_q q + M_r r + \Delta M_{CE} - rh'_x + ph'_z\right)\left(I_{xy}I_z + I_{zx}I_{yz}\right)\\ +\left(N_u \Delta u + N_v v + N_w w + N_p p + N_q q + N_r r + \Delta N_{CE} - ph'_y + qh'_x\right)\left(I_{xy}I_{yz} + I_{zx}I_y\right)\end{bmatrix}\end{aligned} \tag{A.10.39a}$$

$$\begin{aligned}\dot{q} = &\left(I_x I_y I_z - I_x I_{yz}^2 - I_{xy}^2 I_z - 2I_{yz}I_{xy}I_{zx} - I_y I_{zx}^2\right)^{-1}\\ &\begin{bmatrix}\left(L_u \Delta u + L_v v + L_w w + L_p p + L_q q + L_r r + \Delta L_{CE} - qh'_z + rh'_y\right)\left(I_{yz}I_{zx} + I_{xy}I_z\right)\\ +\left(M_u \Delta u + M_v v + M_w w + M_{\dot{w}}\dot{w} + M_p p + M_q q + M_r r + \Delta M_{CE} - rh'_x + ph'_z\right)\left(I_x I_z - I_{zx}^2\right)\\ +\left(N_u \Delta u + N_v v + N_w w + N_p p + N_q q + N_r r + \Delta N_{CE} - ph'_y + qh'_x\right)\left(I_x I_{yz} + I_{zx}I_{xy}\right)\end{bmatrix}\end{aligned} \tag{A.10.39b}$$

$$\dot{r} = \left(I_xI_yI_z - I_xI_{yz}^2 - I_{xy}^2I_z - 2I_{yz}I_{xy}I_{zx} - I_yI_{zx}^2\right)^{-1}$$
$$\begin{bmatrix} \left(L_u\Delta u + L_v v + L_w w + L_p p + L_q q + L_r r + \Delta L_{CE} - qh_z' + rh_y'\right)\left(I_{yz}I_{xy} + I_yI_{zx}\right) \\ + \left(M_u\Delta u + M_v v + M_w w + M_{\dot{w}}\dot{w} + M_p p + M_q q + M_r r + \Delta M_{CE} - rh_x' + ph_z'\right)\left(I_xI_{yz} + I_{zx}I_{xy}\right) \\ + \left(N_u\Delta u + N_v v + N_w w + N_p p + N_q q + N_r r + \Delta N_{CE} - ph_y' + qh_x'\right)\left(I_xI_y - I_{xy}^2\right) \end{bmatrix} \tag{A.10.39c}$$

Four of Eqs. (A.10.38c) and (A.10.39a–c), contain $\dot{w}$-terms on the right-hand side. Since it is desirable to retain the form of having first derivatives of the dependent variables on the left, it is required to substitute $\dot{w}$ from (A.10.38c) into the three equations given by (A.10.39a–c).

It is required to manipulate Eq. (A.10.38c) algebraically to obtain $\dot{w}$ only on the left-hand side:

$$\begin{aligned} \dot{w} &= m^{-1}\left(Z_u\Delta u + Z_v v + Z_w w + Z_{\dot{w}}\dot{w} + Z_p p + Z_q q + Z_r r + \Delta Z_{CE}\right) \\ &\quad - g\Delta\theta \sin\theta_0 + u_0 q \\ \dot{w} - m^{-1}Z_{\dot{w}}\dot{w} &= \dot{w}\left(1 - m^{-1}Z_{\dot{w}}\right) = m^{-1}\left(Z_u\Delta u + Z_v v + Z_w w + Z_p p + Z_q q + Z_r r + \Delta Z_{CE}\right) \\ &\quad - g\Delta\theta \sin\theta_0 + u_0 q \\ \dot{w} &= \left(1 - m^{-1}Z_{\dot{w}}\right)^{-1}\left[m^{-1}\left(Z_u\Delta u + Z_v v + Z_w w + Z_p p + Z_q q + Z_r r + \Delta Z_{CE}\right)\right. \\ &\quad \left. - g\Delta\theta \sin\theta_0 + u_0 q\right] \\ \dot{w} &= \left(m - Z_{\dot{w}}\right)^{-1}\left(Z_u\Delta u + Z_v v + Z_w w + Z_p p + Z_q q + Z_r r + \Delta Z_{CE}\right) \\ &\quad - \left(m - Z_{\dot{w}}\right)^{-1} mg\Delta\theta \sin\theta_0 + \left(m - Z_{\dot{w}}\right)^{-1} m u_0 q \end{aligned} \tag{A.10.40}$$

Equation (A.10.40) is substituted into (A.10.39a), where terms are rearranged and conveniently collected.

$$\dot{p} = \left(I_xI_yI_z - I_xI_{yz}^2 - I_{xy}^2I_z - 2I_{yz}I_{xy}I_{zx} - I_yI_{zx}^2\right)^{-1}$$
$$\begin{bmatrix} \left(L_u\Delta u + L_v v + L_w w + L_p p + L_q q + L_r r + \Delta L_{CE} - qh_z' + rh_y'\right)\left(I_yI_z - I_{yz}^2\right) \\ + \begin{pmatrix} M_u\Delta u + M_v v + M_w w \\ + M_{\dot{w}} \begin{bmatrix} \left(m - Z_{\dot{w}}\right)^{-1}\left(Z_u\Delta u + Z_v v + Z_w w + Z_p p + Z_q q + Z_r r + \Delta Z_{CE}\right) \\ -\left(m - Z_{\dot{w}}\right)^{-1} mg\Delta\theta \sin\theta_0 + \left(m - Z_{\dot{w}}\right)^{-1} m u_0 q \end{bmatrix} \\ + M_p p + M_q q + M_r r + \Delta M_{CE} - rh_x' + ph_z' \end{pmatrix} \left(I_{xy}I_z + I_{zx}I_{yz}\right) \\ + \left(N_u\Delta u + N_v v + N_w w + N_p p + N_q q + N_r r + \Delta N_{CE} - ph_y' + qh_x'\right)\left(I_{xy}I_{yz} + I_{zx}I_y\right) \end{bmatrix}$$

$$
\begin{aligned}
\dot{p} = {} & \left(I_x I_y I_z - I_x I_{yz}^2 - I_{xy}^2 I_z - 2I_{yz} I_{xy} I_{zx} - I_y I_{zx}^2\right)^{-1} \\
& \left[\begin{array}{l}
\left(\Delta u L_u + v L_v + w L_w + p L_p + q\left(L_q - h_z'\right) + r\left(L_r + h_y'\right) + \Delta L_{CE}\right)\left(I_y I_z - I_{yz}^2\right) \\
+ \left(\begin{array}{l}
\Delta u M_u + v M_v + w M_w \\
+ M_{\dot{w}}(m - Z_{\dot{w}})^{-1}\left(Z_u \Delta u + Z_v v + Z_w w + Z_p p + Z_q q + Z_r r + \Delta Z_{CE}\right) \\
- M_{\dot{w}}(m - Z_{\dot{w}})^{-1} mg \Delta\theta \sin\theta_0 \\
+ M_{\dot{w}}(m - Z_{\dot{w}})^{-1} m u_0 q \\
+ M_p p + M_q q + M_r r + \Delta M_{CE} - r h_x' + p h_z'
\end{array} \right) \left(I_{xy} I_z + I_{zx} I_{yz}\right) \\
+ \left(\Delta u N_u + v N_v + w N_w + p\left(N_p - h_y'\right) + q\left(N_q + h_x'\right) + r N_r + \Delta N_{CE}\right)\left(I_{xy} I_{yz} + I_{zx} I_y\right)
\end{array} \right]
\end{aligned}
$$

$$
\begin{aligned}
\dot{p} = {} & \left(I_x I_y I_z - I_x I_{yz}^2 - I_{xy}^2 I_z - 2I_{yz} I_{xy} I_{zx} - I_y I_{zx}^2\right)^{-1}\left(I_y I_z - I_{yz}^2\right) \quad (\equiv I_L^*) \\
& \left[\Delta u L_u + v L_v + w L_w + p L_p + q\left(L_q - h_z'\right) + r\left(L_r + h_y'\right) + \Delta L_{CE}\right] \\
& + \left(I_x I_y I_z - I_x I_{yz}^2 - I_{xy}^2 I_z - 2I_{yz} I_{xy} I_{zx} - I_y I_{zx}^2\right)^{-1}\left(I_{xy} I_z + I_{zx} I_{yz}\right) \quad (\equiv I_M^*) \\
& \left[\begin{array}{l}
\Delta u M_u + v M_v + w M_w + \Delta u M_{\dot{w}} Z_u (m - Z_{\dot{w}})^{-1} + v M_{\dot{w}} Z_v (m - Z_{\dot{w}})^{-1} + w M_{\dot{w}} Z_w (m - Z_{\dot{w}})^{-1} \\
+ p M_{\dot{w}} Z_p (m - Z_{\dot{w}})^{-1} + q M_{\dot{w}} Z_q (m - Z_{\dot{w}})^{-1} + r M_{\dot{w}} Z_r (m - Z_{\dot{w}})^{-1} + M_{\dot{w}} \Delta Z_{CE} (m - Z_{\dot{w}})^{-1} \\
- \Delta\theta M_{\dot{w}} mg \sin\theta_0 (m - Z_{\dot{w}})^{-1} + q M_{\dot{w}} m u_0 (m - Z_{\dot{w}})^{-1} + p\left(M_p + h_z'\right) + q M_q + r\left(M_r - h_x'\right) + \Delta M_{CE}
\end{array} \right] \\
& + \left(I_x I_y I_z - I_x I_{yz}^2 - I_{xy}^2 I_z - 2I_{yz} I_{xy} I_{zx} - I_y I_{zx}^2\right)^{-1}\left(I_{xy} I_{yz} + I_{zx} I_y\right) \quad (\equiv I_N^*) \\
& \left[\Delta u N_u + v N_v + w N_w + p\left(N_p - h_y'\right) + q\left(N_q + h_x'\right) + r N_r + \Delta N_{CE}\right]
\end{aligned}
$$

$$
\begin{aligned}
\dot{p} = {} & I_L^*\left[\Delta u L_u + v L_v + w L_w + p L_p + q\left(L_q - h_z'\right) + r\left(L_r + h_y'\right) + \Delta L_{CE}\right] \\
& + I_M^* \left[\begin{array}{l}
\Delta u\left(M_u + M_{\dot{w}} Z_u (m - Z_{\dot{w}})^{-1}\right) + v\left(M_v + M_{\dot{w}} Z_v (m - Z_{\dot{w}})^{-1}\right) + w\left(M_w + M_{\dot{w}} Z_w (m - Z_{\dot{w}})^{-1}\right) \\
+ p\left(M_{\dot{w}} Z_p (m - Z_{\dot{w}})^{-1} + M_p + h_z'\right) + q\left(M_{\dot{w}} Z_q (m - Z_{\dot{w}})^{-1} + M_w m u_0 (m \quad Z_{\dot{w}})^{-1} + M_q\right) \\
+ r\left(M_{\dot{w}} Z_r (m - Z_{\dot{w}})^{-1} + M_r - h_x'\right) - \Delta\theta M_{\dot{w}} mg \sin\theta_0 (m - Z_{\dot{w}})^{-1} + \Delta M_{CE} + M_{\dot{w}} \Delta Z_{CE} (m - Z_{\dot{w}})^{-1}
\end{array} \right] \\
& + I_N^*\left[\Delta u N_u + v N_v + w N_w + p\left(N_p - h_y'\right) + q\left(N_q + h_x'\right) + r N_r + \Delta N_{CE}\right]
\end{aligned}
$$

$$
\begin{aligned}
\dot{p} = {} & \Delta u L_u I_L^* + v L_v I_L^* + w L_w I_L^* + p L_p I_L^* + q\left(L_q - h_z'\right) I_L^* \\
& + r\left(L_r + h_y'\right) I_L^* + \Delta L_{CE} I_L^* \\
& + \Delta u I_M^* \left(M_u + M_{\dot{w}} Z_u (m - Z_{\dot{w}})^{-1}\right) + v I_M^* \left(M_v + M_{\dot{w}} Z_v (m - Z_{\dot{w}})^{-1}\right) \\
& + w I_M^* \left(M_w + M_{\dot{w}} Z_w (m - Z_{\dot{w}})^{-1}\right) \\
& + p I_M^* \left(M_{\dot{w}} Z_p (m - Z_{\dot{w}})^{-1} + M_p + h_z'\right) + q I_M^* \left(M_{\dot{w}} Z_q (m - Z_{\dot{w}})^{-1}\right. \\
& \left. + M_{\dot{w}} m u_0 (m - Z_{\dot{w}})^{-1} + M_q\right) \\
& + r I_M^* \left(M_{\dot{w}} Z_r (m - Z_{\dot{w}})^{-1} + M_r - h_x'\right) - \Delta\theta I_M^* M_{\dot{w}} m g \sin\theta_0 (m - Z_{\dot{w}})^{-1} \\
& + \Delta M_{CE} I_M^* + M_{\dot{w}} I_M^* \Delta Z_{CE} (m - Z_{\dot{w}})^{-1} \\
& + \Delta u N_u I_N^* + v N_v I_N^* + w N_w I_N^* + p I_N^* \left(N_p - h_y'\right) + q I_N^* \left(N_q + h_x'\right) \\
& + r N_r I_N^* + \Delta N_{CE} I_N^*
\end{aligned}
$$

$$
\begin{aligned}
\dot{p} = {} & \Delta u\left[L_u I_L^* + \left(M_u + M_{\dot{w}} Z_u (m - Z_{\dot{w}})^{-1}\right) I_M^* + N_u I_N^*\right] \\
& v\left[L_v I_L^* + \left(M_v + M_{\dot{w}} Z_v (m - Z_{\dot{w}})^{-1}\right) I_M^* + N_v I_N^*\right] \\
& w\left[L_w I_L^* + \left(M_w + M_{\dot{w}} Z_w (m - Z_{\dot{w}})^{-1}\right) I_M^* + N_w I_N^*\right] \\
& p\left[L_p I_L^* + \left(M_{\dot{w}} Z_p (m - Z_{\dot{w}})^{-1} + M_p + h_z'\right) I_M^* + \left(N_p - h_y'\right) I_N^*\right] \\
& q\left[\left(L_q - h_z'\right) I_L^* + \left(M_{\dot{w}} Z_q (m - Z_{\dot{w}})^{-1} + M_{\dot{w}} m u_0 (m - Z_{\dot{w}})^{-1} + M_q\right) I_M^*\right. \\
& \left. + \left(N_q + h_x'\right) I_N^*\right] \\
& r\left[\left(L_r + h_y'\right) I_L^* + \left(M_{\dot{w}} Z_r (m - Z_{\dot{w}})^{-1} + M_r - h_x'\right) I_M^* + N_r I_N^*\right] \\
& - \Delta\theta I_M^* M_{\dot{w}} m g \sin\theta_0 (m - Z_{\dot{w}})^{-1} \\
& + \Delta L_{CE} I_L^* + \Delta M_{CE} I_M^* + \Delta N_{CE} I_N^* + \Delta Z_{CE} I_M^* M_{\dot{w}} (m - Z_{\dot{w}})^{-1}
\end{aligned}
\tag{A.10.41}
$$

with the moments of inertia given with

$$
I_L^* = \left(I_x I_y I_z - I_x I_{yz}^2 - I_{xy}^2 I_z - 2 I_{yz} I_{xy} I_{zx} - I_y I_{zx}^2\right)^{-1} \left(I_y I_z - I_{yz}^2\right) \tag{A.10.42a}
$$

$$
I_M^* = \left(I_x I_y I_z - I_x I_{yz}^2 - I_{xy}^2 I_z - 2 I_{yz} I_{xy} I_{zx} - I_y I_{zx}^2\right)^{-1} \left(I_{xy} I_z + I_{zx} I_{yz}\right) \tag{A.10.42b}
$$

$$
I_N^* = \left(I_x I_y I_z - I_x I_{yz}^2 - I_{xy}^2 I_z - 2 I_{yz} I_{xy} I_{zx} - I_y I_{zx}^2\right)^{-1} \left(I_{xy} I_{yz} + I_{zx} I_y\right) \tag{A.10.42c}
$$

Equation (A.10.40) is substituted into (A.10.39b), where terms are rearranged and conveniently collected.

$$\dot{q} = \left(I_x I_y I_z - I_x I_{yz}^2 - I_{xy}^2 I_z - 2I_{yz}I_{xy}I_{zx} - I_y I_{zx}^2\right)^{-1}$$

$$\left[\begin{array}{l} \left(L_u \Delta u + L_v v + L_w w + L_p p + L_q q + L_r r + \Delta L_{CE} - qh'_z + rh'_y\right)\left(I_{yz}I_{zx} + I_{xy}I_z\right) \\ + \left(\begin{array}{l} M_u \Delta u + M_v v + M_w w \\ + M_{\dot{w}} \left[\begin{array}{l} (m - Z_{\dot{w}})^{-1}(Z_u \Delta u + Z_v v + Z_w w + Z_p p + Z_q q + Z_r r + \Delta Z_{CE}) \\ -(m - Z_{\dot{w}})^{-1} mg\Delta\theta \sin\theta_0 + (m - Z_{\dot{w}})^{-1} mu_0 q \end{array}\right] \\ + M_p p + M_q q + M_r r + \Delta M_{CE} - rh'_x + ph'_z \end{array}\right)\left(I_x I_z - I_{zx}^2\right) \\ + \left(N_u \Delta u + N_v v + N_w w + N_p p + N_q q + N_r r + \Delta N_{CE} - ph'_y + qh'_x\right)\left(I_x I_{yz} + I_{zx}I_{xy}\right) \end{array}\right]$$

$$\begin{aligned} \dot{q} = {} & \left(I_x I_y I_z - I_x I_{yz}^2 - I_{xy}^2 I_z - 2I_{yz}I_{xy}I_{zx} - I_y I_{zx}^2\right)^{-1}\left(I_{yz}I_{zx} + I_{xy}I_z\right) \quad (\equiv I_L^{**}) \\ & \left[\Delta u L_u + vL_v + wL_w + pL_p + q(L_q - h'_z) + r\left(L_r + h'_y\right) + \Delta L_{CE}\right] \\ & + \left(I_x I_y I_z - I_x I_{yz}^2 - I_{xy}^2 I_z - 2I_{yz}I_{xy}I_{zx} - I_y I_{zx}^2\right)^{-1}\left(I_x I_z - I_{zx}^2\right) \quad (\equiv I_M^{**}) \\ & \left[\begin{array}{l} \Delta u M_u + vM_v + wM_w + \Delta u M_{\dot{w}} Z_u (m - Z_{\dot{w}})^{-1} + vM_{\dot{w}} Z_v (m - Z_{\dot{w}})^{-1} + wM_{\dot{w}} Z_w (m - Z_{\dot{w}})^{-1} \\ + pM_{\dot{w}} Z_p (m - Z_{\dot{w}})^{-1} + qM_{\dot{w}} Z_q (m - Z_{\dot{w}})^{-1} + rM_{\dot{w}} Z_r (m - Z_{\dot{w}})^{-1} + M_{\dot{w}} \Delta Z_{CE} (m - Z_{\dot{w}})^{-1} \\ - \Delta\theta M_{\dot{w}} mg \sin\theta_0 (m - Z_{\dot{w}})^{-1} + qM_{\dot{w}} mu_0 (m - Z_{\dot{w}})^{-1} + p(M_p + h'_z) + qM_q + r(M_r - h'_x) + \Delta M_{CE} \end{array}\right] \\ & + \left(I_x I_y I_z - I_x I_{yz}^2 - I_{xy}^2 I_z - 2I_{yz}I_{xy}I_{zx} - I_y I_{zx}^2\right)^{-1}\left(I_x I_{yz} + I_{zx}I_{xy}\right) \quad (\equiv I_N^{**}) \\ & \left[\Delta u N_u + vN_v + wN_w + p\left(N_p - h'_y\right) + q(N_q + h'_x) + rN_r + \Delta N_{CE}\right] \end{aligned}$$

$$\begin{aligned} \dot{q} = {} & \Delta u\left[L_u I_L^{**} + \left(M_u + M_{\dot{w}} Z_u (m - Z_{\dot{w}})^{-1}\right) I_M^{**} + N_u I_N^{**}\right] \\ & v\left[L_v I_L^{**} + \left(M_v + M_{\dot{w}} Z_v (m - Z_{\dot{w}})^{-1}\right) I_M^{**} + N_v I_N^{**}\right] \\ & w\left[L_w I_L^{**} + \left(M_w + M_{\dot{w}} Z_w (m - Z_{\dot{w}})^{-1}\right) I_M^{**} + N_w I_N^{**}\right] \\ & p\left[L_p I_L^{**} + \left(M_{\dot{w}} Z_p (m - Z_{\dot{w}})^{-1} + M_p + h'_z\right) I_M^{**} + \left(N_p - h'_y\right) I_N^{**}\right] \\ & q\left[(L_q - h'_z) I_L^{**} + \left(M_{\dot{w}} Z_q (m - Z_{\dot{w}})^{-1} + M_{\dot{w}} mu_0 (m - Z_{\dot{w}})^{-1} + M_q\right) I_M^{**} \right. \\ & \left. + (N_q + h'_x) I_N^{**}\right] \\ & r\left[\left(L_r + h'_y\right) I_L^{**} + \left(M_{\dot{w}} Z_r (m - Z_{\dot{w}})^{-1} + M_r - h'_x\right) I_M^{**} + N_r I_N^{**}\right] \\ & - \Delta\theta I_M^{**} M_{\dot{w}} mg \sin\theta_0 (m - Z_{\dot{w}})^{-1} \\ & + \Delta L_{CE} I_L^{**} + \Delta M_{CE} I_M^{**} + \Delta N_{CE} I_N^{**} + \Delta Z_{CE} I_M^{**} M_{\dot{w}} (m - Z_{\dot{w}})^{-1} \end{aligned} \tag{A.10.43}$$

with the moments of inertia given with

$$I_L^{**} = \left(I_xI_yI_z - I_xI_{yz}^2 - I_{xy}^2I_z - 2I_{yz}I_{xy}I_{zx} - I_yI_{zx}^2\right)^{-1}\left(I_{yz}I_{zx} + I_{xy}I_z\right) \quad \text{(A.10.44a)}$$

$$I_M^{**} = \left(I_xI_yI_z - I_xI_{yz}^2 - I_{xy}^2I_z - 2I_{yz}I_{xy}I_{zx} - I_yI_{zx}^2\right)^{-1}\left(I_xI_z - I_{zx}^2\right) \quad \text{(A.10.44b)}$$

$$I_N^{**} = \left(I_xI_yI_z - I_xI_{yz}^2 - I_{xy}^2I_z - 2I_{yz}I_{xy}I_{zx} - I_yI_{zx}^2\right)^{-1}\left(I_xI_{yz} + I_{zx}I_{xy}\right) \quad \text{(A.10.44c)}$$

Equation (A.10.40) is substituted into (A.10.39c), where terms are rearranged and conveniently collected.

$$\dot{r} = \left(I_xI_yI_z - I_xI_{yz}^2 - I_{xy}^2I_z - 2I_{yz}I_{xy}I_{zx} - I_yI_{zx}^2\right)^{-1}$$

$$\left[\begin{array}{l} \left(L_u\Delta u + L_v v + L_w w + L_p p + L_q q + L_r r + \Delta L_{CE} - qh'_z + rh'_y\right)\left(I_{yz}I_{xy} + I_yI_{zx}\right) \\ + \left(\begin{array}{l} M_u\Delta u + M_v v + M_w w \\ + M_{\dot{w}}\left[\begin{array}{l} (m - Z_{\dot{w}})^{-1}\left(Z_u\Delta u + Z_v v + Z_w w + Z_p p + Z_q q + Z_r r + \Delta Z_{CE}\right) \\ -(m - Z_{\dot{w}})^{-1} mg\Delta\theta \sin\theta_0 + (m - Z_{\dot{w}})^{-1} mu_0 q \end{array}\right] \\ + M_p p + M_q q + M_r r + \Delta M_{CE} - rh'_x + ph'_z \end{array}\right)\left(I_xI_{yz} + I_{zx}I_{xy}\right) \\ + \left(N_u\Delta u + N_v v + N_w w + N_p p + N_q q + N_r r + \Delta N_{CE} - ph'_y + qh'_x\right)\left(I_xI_y - I_{xy}^2\right) \end{array}\right]$$

$$\begin{aligned} \dot{r} = {} & \left(I_xI_yI_z - I_xI_{yz}^2 - I_{xy}^2I_z - 2I_{yz}I_{xy}I_{zx} - I_yI_{zx}^2\right)^{-1}\left(I_{yz}I_{xy} + I_yI_{zx}\right) \quad (\equiv I_L^{***}) \\ & \left[\Delta uL_u + vL_v + wL_w + pL_p + q\left(L_q - h'_z\right) + r\left(L_r + h'_y\right) + \Delta L_{CE}\right] \\ & + \left(I_xI_yI_z - I_xI_{yz}^2 - I_{xy}^2I_z - 2I_{yz}I_{xy}I_{zx} - I_yI_{zx}^2\right)^{-1}\left(I_xI_{yz} + I_{zx}I_{xy}\right) \quad (\equiv I_M^{***}) \\ & \left[\begin{array}{l} \Delta uM_u + vM_v + wM_w + \Delta uM_{\dot{w}}Z_u(m - Z_{\dot{w}})^{-1} + vM_{\dot{w}}Z_v(m - Z_{\dot{w}})^{-1} + wM_{\dot{w}}Z_w(m - Z_{\dot{w}})^{-1} \\ + pM_{\dot{w}}Z_p(m - Z_{\dot{w}})^{-1} + qM_{\dot{w}}Z_q(m - Z_{\dot{w}})^{-1} + rM_{\dot{w}}Z_r(m - Z_{\dot{w}})^{-1} + M_{\dot{w}}\Delta Z_{CE}(m - Z_{\dot{w}})^{-1} \\ -\Delta\theta M_{\dot{w}}mg \sin\theta_0(m - Z_{\dot{w}})^{-1} + qM_{\dot{w}}mu_0(m - Z_{\dot{w}})^{-1} + p\left(M_p + h'_z\right) + qM_q + r\left(M_r - h'_x\right) + \Delta M_{CE} \end{array}\right] \\ & + \left(I_xI_yI_z - I_xI_{yz}^2 - I_{xy}^2I_z - 2I_{yz}I_{xy}I_{zx} - I_yI_{zx}^2\right)^{-1}\left(I_xI_y - I_{xy}^2\right) \quad (\equiv I_N^{***}) \\ & \left[\Delta uN_u + vN_v + wN_w + p\left(N_p - h'_y\right) + q\left(N_q + h'_x\right) + rN_r + \Delta N_{CE}\right] \end{aligned}$$

$$\begin{aligned}\dot{r} = \Delta u\Big[&L_u I_L^{***} + \Big(M_u + M_{\dot{w}} Z_u (m - Z_{\dot{w}})^{-1}\Big) I_M^{***} + N_u I_N^{***}\Big]\\
&v\Big[L_v I_L^{***} + \Big(M_v + M_{\dot{w}} Z_v (m - Z_{\dot{w}})^{-1}\Big) I_M^{***} + N_v I_N^{***}\Big]\\
&w\Big[L_w I_L^{***} + \Big(M_w + M_{\dot{w}} Z_w (m - Z_{\dot{w}})^{-1}\Big) I_M^{***} + N_w I_N^{***}\Big]\\
&p\Big[L_p I_L^{***} + \Big(M_{\dot{w}} Z_p (m - Z_{\dot{w}})^{-1} + M_p + h_z'\Big) I_M^{***} + \Big(N_p - h_y'\Big) I_N^{***}\Big]\\
&q\Big[(L_q - h_z') I_L^{***} + \Big(M_{\dot{w}} Z_q (m - Z_{\dot{w}})^{-1} + M_{\dot{w}} m u_0 (m - Z_{\dot{w}})^{-1} + M_q\Big) I_M^{***}\\
&+ (N_q + h_x') I_N^{***}\Big]\\
&r\Big[\Big(L_r + h_y'\Big) I_L^{***} + \Big(M_{\dot{w}} Z_r (m - Z_{\dot{w}})^{-1} + M_r - h_x'\Big) I_M^{***} + N_r I_N^{***}\Big]\\
&- \Delta\theta I_M^{***} M_{\dot{w}} m g \sin\theta_0 (m - Z_{\dot{w}})^{-1}\\
&+ \Delta L_{CE} I_L^{***} + \Delta M_{CE} I_M^{***} + \Delta N_{CE} I_N^{***} + \Delta Z_{CE} I_M^{***} M_{\dot{w}} (m - Z_{\dot{w}})^{-1}\end{aligned} \tag{A.10.45}$$

with the moments of inertia given with

$$I_L^{***} = \Big(I_x I_y I_z - I_x I_{yz}^2 - I_{xy}^2 I_z - 2 I_{yz} I_{xy} I_{zx} - I_y I_{zx}^2\Big)^{-1} \big(I_{yz} I_{xy} + I_y I_{zx}\big) \tag{A.10.46a}$$

$$I_M^{***} = \Big(I_x I_y I_z - I_x I_{yz}^2 - I_{xy}^2 I_z - 2 I_{yz} I_{xy} I_{zx} - I_y I_{zx}^2\Big)^{-1} \big(I_x I_{yz} + I_{zx} I_{xy}\big) \tag{A.10.46b}$$

$$I_N^{***} = \Big(I_x I_y I_z - I_x I_{yz}^2 - I_{xy}^2 I_z - 2 I_{yz} I_{xy} I_{zx} - I_y I_{zx}^2\Big)^{-1} \Big(I_x I_y - I_{xy}^2\Big) \tag{A.10.46c}$$

The Equations Collected

The <u>6-DOF</u> <u>Small Perturbation</u> <u>EOM</u> are summarised for convenience:

$$\Delta\dot{u} = m^{-1}\big(X_u \Delta u + X_v v + X_w w + X_p p + X_q q + X_r r + \Delta X_{CE}\big) - g\Delta\theta \cos\theta_0 \tag{A.10.38a}$$

$$\dot{v} = m^{-1}\big(Y_u \Delta u + Y_v v + Y_w w + Y_p p + Y_q q + Y_r r + \Delta Y_{CE}\big) + g\phi \cos\theta_0 - u_0 r \tag{A.10.38b}$$

$$\begin{aligned}\dot{w} = {}&(m - Z_{\dot{w}})^{-1}\big(Z_u \Delta u + Z_v v + Z_w w + Z_p p + Z_q q + Z_r r + \Delta Z_{CE}\big)\\ &- (m - Z_{\dot{w}})^{-1} m g \Delta\theta \sin\theta_0 + (m - Z_{\dot{w}})^{-1} m u_0 q\end{aligned} \tag{A.10.40}$$

$$
\begin{aligned}
\dot{p} = {} & \Delta u\Big[L_u I_L^* + \Big(M_u + M_{\dot{w}} Z_u (m - Z_{\dot{w}})^{-1}\Big) I_M^* + N_u I_N^*\Big] \\
& v\Big[L_v I_L^* + \Big(M_v + M_{\dot{w}} Z_v (m - Z_{\dot{w}})^{-1}\Big) I_M^* + N_v I_N^*\Big] \\
& w\Big[L_w I_L^* + \Big(M_w + M_{\dot{w}} Z_w (m - Z_{\dot{w}})^{-1}\Big) I_M^* + N_w I_N^*\Big] \\
& p\Big[L_p I_L^* + \Big(M_{\dot{w}} Z_p (m - Z_{\dot{w}})^{-1} + M_p + h_z'\Big) I_M^* + \Big(N_p - h_y'\Big) I_N^*\Big] \\
& q\Big[\big(L_q - h_z'\big) I_L^* + \Big(M_{\dot{w}} Z_q (m - Z_{\dot{w}})^{-1} + M_{\dot{w}} m u_0 (m - Z_{\dot{w}})^{-1} + M_q\Big) I_M^* \\
& + \big(N_q + h_x'\big) I_N^*\Big] \\
& r\Big[\Big(L_r + h_y'\Big) I_L^* + \Big(M_{\dot{w}} Z_r (m - Z_{\dot{w}})^{-1} + M_r - h_x'\Big) I_M^* + N_r I_N^*\Big] \\
& - \Delta\theta I_M^* M_{\dot{w}} m g \sin\theta_0 (m - Z_{\dot{w}})^{-1} \\
& + \Delta L_{CE} I_L^* + \Delta M_{CE} I_M^* + \Delta N_{CE} I_N^* + \Delta Z_{CE} I_M^* M_{\dot{w}} (m - Z_{\dot{w}})^{-1}
\end{aligned}
\tag{A.10.41}
$$

$$
\begin{aligned}
\dot{q} = {} & \Delta u\Big[L_u I_L^{**} + \Big(M_u + M_{\dot{w}} Z_u (m - Z_{\dot{w}})^{-1}\Big) I_M^{**} + N_u I_N^{**}\Big] \\
& v\Big[L_v I_L^{**} + \Big(M_v + M_{\dot{w}} Z_v (m - Z_{\dot{w}})^{-1}\Big) I_M^{**} + N_v I_N^{**}\Big] \\
& w\Big[L_w I_L^{**} + \Big(M_w + M_{\dot{w}} Z_w (m - Z_{\dot{w}})^{-1}\Big) I_M^{**} + N_w I_N^{**}\Big] \\
& p\Big[L_p I_L^{**} + \Big(M_{\dot{w}} Z_p (m - Z_{\dot{w}})^{-1} + M_p + h_z'\Big) I_M^{**} + \Big(N_p - h_y'\Big) I_N^{**}\Big] \\
& q\Big[\big(L_q - h_z'\big) I_L^{**} + \Big(M_{\dot{w}} Z_q (m - Z_{\dot{w}})^{-1} + M_{\dot{w}} m u_0 (m - Z_{\dot{w}})^{-1} + M_q\Big) I_M^{**} \\
& + \big(N_q + h_x'\big) I_N^{**}\Big] \\
& r\Big[\Big(L_r + h_y'\Big) I_L^{**} + \Big(M_{\dot{w}} Z_r (m - Z_{\dot{w}})^{-1} + M_r - h_x'\Big) I_M^{**} + N_r I_N^{**}\Big] \\
& - \Delta\theta I_M^{**} M_{\dot{w}} m g \sin\theta_0 (m - Z_{\dot{w}})^{-1} \\
& + \Delta L_{CE} I_L^{**} + \Delta M_{CE} I_M^{**} + \Delta N_{CE} I_N^{**} + \Delta Z_{CE} I_M^{**} M_{\dot{w}} (m - Z_{\dot{w}})^{-1}
\end{aligned}
\tag{A.10.43}
$$

$$
\begin{aligned}
\dot{r} = \Delta u & \left[L_u I_L^{***} + \left(M_u + M_{\dot{w}} Z_u (m - Z_{\dot{w}})^{-1} \right) I_M^{***} + N_u I_N^{***} \right] \\
& v \left[L_v I_L^{***} + \left(M_v + M_{\dot{w}} Z_v (m - Z_{\dot{w}})^{-1} \right) I_M^{***} + N_v I_N^{***} \right] \\
& w \left[L_w I_L^{***} + \left(M_w + M_{\dot{w}} Z_w (m - Z_{\dot{w}})^{-1} \right) I_M^{***} + N_w I_N^{***} \right] \\
& p \left[L_p I_L^{***} + \left(M_{\dot{w}} Z_p (m - Z_{\dot{w}})^{-1} + M_p + h'_z \right) I_M^{***} + \left(N_p - h'_y \right) I_N^{***} \right] \\
& q \left[(L_q - h'_z) I_L^{***} + \left(M_{\dot{w}} Z_q (m - Z_{\dot{w}})^{-1} + M_{\dot{w}} m u_0 (m - Z_{\dot{w}})^{-1} + M_q \right) I_M^{***} \right. \\
& \left. + (N_q + h'_x) I_N^{***} \right] \\
& r \left[\left(L_r + h'_y \right) I_L^{***} + \left(M_{\dot{w}} Z_r (m - Z_{\dot{w}})^{-1} + M_r - h'_x \right) I_M^{***} + N_r I_N^{***} \right] \\
& - \Delta\theta I_M^{***} M_{\dot{w}} m g \sin\theta_0 (m - Z_{\dot{w}})^{-1} \\
& + \Delta L_{CE} I_L^{***} + \Delta M_{CE} I_M^{***} + \Delta N_{CE} I_N^{***} + \Delta Z_{CE} I_M^{***} M_{\dot{w}} (m - Z_{\dot{w}})^{-1}
\end{aligned}
\tag{A.10.45}
$$

$$p = \dot{\phi} - \dot{\psi} \sin\theta_0 \tag{A.10.32}$$

$$\dot{\phi} = p + r \tan\theta_0 \tag{A.10.33a}$$

$$\dot{\theta} = q \tag{A.10.33b}$$

$$\dot{\psi} = r \sec\theta_0 \tag{A.10.33c}$$

$$\Delta\dot{x}_E = \Delta u \cos\theta_0 - u_0 \Delta\theta \sin\theta_0 + w \sin\theta_0 \tag{A.10.34a}$$

$$\Delta\dot{y}_E = u_0 \psi \cos\theta_0 + v \tag{A.10.34b}$$

$$\Delta\dot{z}_E = -\Delta u \sin\theta_0 - u_0 \Delta\theta \cos\theta_0 + w \cos\theta_0 \tag{A.10.34c}$$

The <u>*State Vector Form of the Linear 6-DOF*</u> <u>*Small Perturbation*</u> <u>*EOM*</u>, written in vector/matrix notation, are presented in Eq. (A.10.47).

Open-Loop, Coupled 6-Degree Of Freedom Small Perturbation Equations of Motion, Eq. (A.10.47a).

$$
\underbrace{\begin{bmatrix} \Delta\dot{u} \\ \dot{v} \\ \dot{w} \\ \dot{p} \\ \dot{q} \\ \dot{r} \\ \dot{\phi} \\ \Delta\dot{\theta} \\ \dot{\psi} \end{bmatrix}}_{\dot{\bar{x}}} = \underbrace{\begin{bmatrix} \quad \end{bmatrix}}_{A} \bullet \underbrace{\begin{bmatrix} \Delta u \\ v \\ w \\ p \\ q \\ r \\ \phi \\ \Delta\theta \\ \psi \end{bmatrix}}_{\bar{x}} + \underbrace{\begin{bmatrix} \dfrac{\Delta X_{CE}}{m} \\ \dfrac{\Delta Y_{CE}}{m} \\ \dfrac{\Delta Z_{CE}}{(m - Z_{\dot{w}})} \\ \Delta L_{CE} I_L^{*} + \Delta M_{CE} I_M^{*} + \Delta N_{CE} I_N^{*} + \dfrac{\Delta Z_{CE} M_{\dot{w}} I_M^{*}}{(m - Z_{\dot{w}})} \\ \Delta L_{CE} I_L^{**} + \Delta M_{CE} I_M^{**} + \Delta N_{CE} I_N^{**} + \dfrac{\Delta Z_{CE} M_{\dot{w}} I_M^{**}}{(m - Z_{\dot{w}})} \\ \Delta L_{CE} I_L^{***} + \Delta M_{CE} I_M^{***} + \Delta N_{CE} I_N^{***} + \dfrac{\Delta Z_{CE} M_{\dot{w}} I_M^{***}}{(m - Z_{\dot{w}})} \\ 0 \\ 0 \\ 0 \end{bmatrix}}_{B\bar{c}}
$$

(A.10.47)

Open-Loop, Coupled 6-Degree Of Freedom Small Perturbation Equations of Motion, Eq. (A.10.47a), cont.

$$
\underbrace{\begin{bmatrix}
\frac{X_u}{m} & \frac{X_v}{m} & \frac{X_w}{m} & \frac{X_p}{m} & \frac{X_q}{m} & \frac{X_r}{m} & 0 & -g\cos\theta_0 & 0 \\
\frac{Y_u}{m} & \frac{Y_v}{m} & \frac{Y_w}{m} & \frac{Y_p}{m} & \frac{Y_q}{m} & \frac{Y_r}{m}-u_0 & g\cos\theta_0 & 0 & 0 \\
\frac{Z_u}{(m-Z_{\dot{w}})} & \frac{Z_v}{(m-Z_{\dot{w}})} & \frac{Z_w}{(m-Z_{\dot{w}})} & \frac{Z_p}{(m-Z_{\dot{w}})} & \frac{Z_q+mu_0}{(m-Z_{\dot{w}})} & \frac{Z_r}{(m-Z_{\dot{w}})} & 0 & \frac{-mg\sin\theta_0}{(m-Z_{\dot{w}})} & 0 \\
\begin{array}{l} L_uI_L^{*} \\ +\left(M_u+\frac{M_{\dot{w}}Z_u}{(m-Z_{\dot{w}})}\right)I_M^{*} \\ +N_uI_N^{*} \end{array} &
\begin{array}{l} L_vI_L^{*} \\ +\left(M_v+\frac{M_{\dot{w}}Z_v}{(m-Z_{\dot{w}})}\right)I_M^{*} \\ +N_vI_N^{*} \end{array} &
\begin{array}{l} L_wI_L^{*} \\ +\left(M_w+\frac{M_{\dot{w}}Z_w}{(m-Z_{\dot{w}})}\right)I_M^{*} \\ +N_wI_N^{*} \end{array} &
\begin{array}{l} L_pI_L^{*} \\ +\left(M_p+h_z'+\frac{M_{\dot{w}}Z_p}{(m-Z_{\dot{w}})}\right)I_M^{*} \\ +(N_p-h_y')I_N^{*} \end{array} &
\begin{array}{l} (L_q-h_z')I_L^{*} \\ +\left(M_q+\frac{M_{\dot{w}}Z_q+M_{\dot{w}}mu_0}{(m-Z_{\dot{w}})}\right)I_M^{*} \\ +(N_q+h_x')I_N^{*} \end{array} &
\begin{array}{l} (L_r+h_y')I_L^{*} \\ +\left(M_r-h_x'+\frac{M_{\dot{w}}Z_r}{(m-Z_{\dot{w}})}\right)I_M^{*} \\ +N_rI_N^{*} \end{array} &
0 & \frac{-mgM_{\dot{w}}I_M^{*}\sin\theta_0}{(m-Z_{\dot{w}})} & 0 \\
\begin{array}{l} L_uI_L^{**} \\ +\left(M_u+\frac{M_{\dot{w}}Z_u}{(m-Z_{\dot{w}})}\right)I_M^{**} \\ +N_uI_N^{**} \end{array} &
\begin{array}{l} L_vI_L^{**} \\ +\left(M_v+\frac{M_{\dot{w}}Z_v}{(m-Z_{\dot{w}})}\right)I_M^{**} \\ +N_vI_N^{**} \end{array} &
\begin{array}{l} L_wI_L^{**} \\ +\left(M_w+\frac{M_{\dot{w}}Z_w}{(m-Z_{\dot{w}})}\right)I_M^{**} \\ +N_wI_N^{**} \end{array} &
\begin{array}{l} L_pI_L^{**} \\ +\left(M_p+h_z'+\frac{M_{\dot{w}}Z_p}{(m-Z_{\dot{w}})}\right)I_M^{**} \\ +(N_p-h_y')I_N^{**} \end{array} &
\begin{array}{l} (L_q-h_z')I_L^{**} \\ +\left(M_q+\frac{M_{\dot{w}}Z_q+M_{\dot{w}}mu_0}{(m-Z_{\dot{w}})}\right)I_M^{**} \\ +(N_q+h_x')I_N^{**} \end{array} &
\begin{array}{l} (L_r+h_y')I_L^{**} \\ +\left(M_r-h_x'+\frac{M_{\dot{w}}Z_r}{(m-Z_{\dot{w}})}\right)I_M^{**} \\ +N_rI_N^{**} \end{array} &
0 & \frac{-mgM_{\dot{w}}I_M^{**}\sin\theta_0}{(m-Z_{\dot{w}})} & 0 \\
\begin{array}{l} L_uI_L^{***} \\ +\left(M_u+\frac{M_{\dot{w}}Z_u}{(m-Z_{\dot{w}})}\right)I_M^{***} \\ +N_uI_N^{***} \end{array} &
\begin{array}{l} L_vI_L^{***} \\ +\left(M_v+\frac{M_{\dot{w}}Z_v}{(m-Z_{\dot{w}})}\right)I_M^{***} \\ +N_vI_N^{***} \end{array} &
\begin{array}{l} L_wI_L^{***} \\ +\left(M_w+\frac{M_{\dot{w}}Z_w}{(m-Z_{\dot{w}})}\right)I_M^{***} \\ +N_wI_N^{***} \end{array} &
\begin{array}{l} L_pI_L^{***} \\ +\left(M_p+h_z'+\frac{M_{\dot{w}}Z_p}{(m-Z_{\dot{w}})}\right)I_M^{***} \\ +(N_p-h_y')I_N^{***} \end{array} &
\begin{array}{l} (L_q-h_z')I_L^{***} \\ +\left(M_q+\frac{M_{\dot{w}}Z_q+M_{\dot{w}}mu_0}{(m-Z_{\dot{w}})}\right)I_M^{***} \\ +(N_q+h_x')I_N^{***} \end{array} &
\begin{array}{l} (L_r+h_y')I_L^{***} \\ +\left(M_r-h_x'+\frac{M_{\dot{w}}Z_r}{(m-Z_{\dot{w}})}\right)I_M^{***} \\ +N_rI_N^{***} \end{array} &
0 & \frac{-mgM_{\dot{w}}I_M^{***}\sin\theta_0}{(m-Z_{\dot{w}})} & 0 \\
0 & 0 & 0 & 1 & 0 & \tan\theta_0 & 0 & 0 & 0 \\
0 & 0 & 0 & 0 & 1 & 0 & 0 & 0 & 0 \\
0 & 0 & 0 & 0 & 0 & \sec\theta_0 & 0 & 0 & 0
\end{bmatrix}}_{A}
\tag{A.10.47a}
$$

Expanding the system matrix $\boldsymbol{B}$ and the control vector $\vec{c}$ in (A.10.47a) yields (A.10.47b). Note, that the Δ in the derivative of the state vector, $\dot{\vec{x}}$, and the state vector itself, $\vec{x}$, indicates, that the reference value is not zero.

Open-Loop, Coupled 6-Degree of Freedom Small Perturbation Equations of Motion, Eq. (A.10.47b)

$$\underbrace{\begin{bmatrix} \Delta\dot{u} \\ \dot{v} \\ \dot{w} \\ \dot{p} \\ \dot{q} \\ \dot{r} \\ \dot{\phi} \\ \Delta\dot{\theta} \\ \dot{\psi} \end{bmatrix}}_{\dot{\bar{x}}} = \underbrace{\begin{bmatrix} \quad \end{bmatrix}}_{A} \bullet \underbrace{\begin{bmatrix} \Delta u \\ v \\ w \\ p \\ q \\ r \\ \phi \\ \Delta\theta \\ \psi \end{bmatrix}}_{\bar{x}} + \underbrace{\begin{bmatrix} \frac{X_{\delta_{LoCE}}}{m} & \frac{X_{\delta_{DiCE}}}{m} & \frac{X_{\delta_{LaCE}}}{m} & \frac{X_{\tau}}{m} \\ \frac{Y_{\delta_{LoCE}}}{m} & \frac{Y_{\delta_{DiCE}}}{m} & \frac{Y_{\delta_{LaCE}}}{m} & \frac{Y_{\tau}}{m} \\ \frac{Z_{\delta_{LoCE}}}{(m-Z_{\dot{w}})} & \frac{Z_{\delta_{DiCE}}}{(m-Z_{\dot{w}})} & \frac{Z_{\delta_{LaCE}}}{(m-Z_{\dot{w}})} & \frac{Z_{\tau}}{(m-Z_{\dot{w}})} \\ L_{\delta_{LoCE}}I_L^{*} + M_{\delta_{LoCE}}I_M^{*} + N_{\delta_{LoCE}}I_N^{*} + \frac{Z_{\delta_{LoCE}}M_{\dot{w}}I_M^{*}}{(m-Z_{\dot{w}})} & L_{\delta_{DiCE}}I_L^{*} + M_{\delta_{DiCE}}I_M^{*} + N_{\delta_{DiCE}}I_N^{*} + \frac{Z_{\delta_{DiCE}}M_{\dot{w}}I_M^{*}}{(m-Z_{\dot{w}})} & L_{\delta_{LaCE}}I_L^{*} + M_{\delta_{LaCE}}I_M^{*} + N_{\delta_{LaCE}}I_N^{*} + \frac{Z_{\delta_{LaCE}}M_{\dot{w}}I_M^{*}}{(m-Z_{\dot{w}})} & L_{\tau}I_L^{*} + M_{\tau}I_M^{*} + N_{\tau}I_N^{*} \\ L_{\delta_{LoCE}}I_L^{**} + M_{\delta_{LoCE}}I_M^{**} + N_{\delta_{LoCE}}I_N^{**} + \frac{Z_{\delta_{LoCE}}M_{\dot{w}}I_M^{**}}{(m-Z_{\dot{w}})} & L_{\delta_{DiCE}}I_L^{**} + M_{\delta_{DiCE}}I_M^{**} + N_{\delta_{DiCE}}I_N^{**} + \frac{Z_{\delta_{DiCE}}M_{\dot{w}}I_M^{**}}{(m-Z_{\dot{w}})} & L_{\delta_{LaCE}}I_L^{**} + M_{\delta_{LaCE}}I_M^{**} + N_{\delta_{LaCE}}I_N^{**} + \frac{Z_{\delta_{LaCE}}M_{\dot{w}}I_M^{**}}{(m-Z_{\dot{w}})} & L_{\tau}I_L^{**} + M_{\tau}I_M^{**} + N_{\tau}I_N^{**} \\ L_{\delta_{LoCE}}I_L^{***} + M_{\delta_{LoCE}}I_M^{***} + N_{\delta_{LoCE}}I_N^{***} + \frac{Z_{\delta_{LoCE}}M_{\dot{w}}I_M^{***}}{(m-Z_{\dot{w}})} & L_{\delta_{DiCE}}I_L^{***} + M_{\delta_{DiCE}}I_M^{***} + N_{\delta_{DiCE}}I_N^{***} + \frac{Z_{\delta_{DiCE}}M_{\dot{w}}I_M^{***}}{(m-Z_{\dot{w}})} & L_{\delta_{LaCE}}I_L^{***} + M_{\delta_{LaCE}}I_M^{***} + N_{\delta_{LaCE}}I_N^{***} + \frac{Z_{\delta_{LaCE}}M_{\dot{w}}I_M^{***}}{(m-Z_{\dot{w}})} & L_{\tau}I_L^{***} + M_{\tau}I_M^{***} + N_{\tau}I_N^{***} \\ 0 & 0 & 0 & 0 \\ 0 & 0 & 0 & 0 \\ 0 & 0 & 0 & 0 \end{bmatrix}}_{B} \bullet \underbrace{\begin{bmatrix} \delta_{LoCE_i} \\ \delta_{DiCE_i} \\ \delta_{LaCE_i} \\ \tau_i \end{bmatrix}}_{\bar{c}} \tag{A.10.47b}$$

Concise Form and Unit Check of the State Space Form of the EOM

Equation (A.10.48) presents the state space form of the coupled dynamic EOM in concise form:

$$\underbrace{\begin{bmatrix} \Delta\dot{u} \\ \dot{v} \\ \dot{w} \\ \dot{p} \\ \dot{q} \\ \dot{r} \\ \dot{\phi} \\ \Delta\dot{\theta} \\ \dot{\psi} \end{bmatrix}}_{\vec{\dot{x}}} = \underbrace{\begin{bmatrix} x_u & x_v & x_w & x_p & x_q & x_r & 0 & x_\theta & 0 \\ y_u & y_v & y_w & y_p & y_q & y_r & y_\phi & 0 & 0 \\ z_u & z_v & z_w & z_p & z_q & z_r & 0 & z_\theta & 0 \\ l_u & l_v & l_w & l_p & l_q & l_r & 0 & l_\theta & 0 \\ m_u & m_v & m_w & m_p & m_q & m_r & 0 & m_\theta & 0 \\ n_u & n_v & n_w & n_p & n_q & n_r & 0 & n_\theta & 0 \\ 0 & 0 & 0 & 1 & 0 & \tan\theta_0 & 0 & 0 & 0 \\ 0 & 0 & 0 & 0 & 1 & 0 & 0 & 0 & 0 \\ 0 & 0 & 0 & 0 & 0 & sec\,\theta_0 & 0 & 0 & 0 \end{bmatrix}}_{A} \underbrace{\begin{bmatrix} \Delta u \\ v \\ w \\ p \\ q \\ r \\ \phi \\ \theta \\ \psi \end{bmatrix}}_{\vec{x}}$$
$$+ \underbrace{\begin{bmatrix} x_{\delta_{LoCE}} & x_{\delta_{DiCE}} & x_{\delta_{LaCE}} & x_\tau \\ y_{\delta_{LoCE}} & y_{\delta_{DiCE}} & y_{\delta_{LaCE}} & y_\tau \\ z_{\delta_{LoCE}} & z_{\delta_{DiCE}} & z_{\delta_{LaCE}} & z_\tau \\ l_{\delta_{LoCE}} & l_{\delta_{DiCE}} & l_{\delta_{LaCE}} & l_\tau \\ m_{\delta_{LoCE}} & m_{\delta_{DiCE}} & m_{\delta_{LaCE}} & m_\tau \\ n_{\delta_{LoCE}} & n_{\delta_{DiCE}} & n_{\delta_{LaCE}} & n_\tau \\ 0 & 0 & 0 & 0 \\ 0 & 0 & 0 & 0 \\ 0 & 0 & 0 & 0 \end{bmatrix}}_{B} \underbrace{\begin{bmatrix} \delta_{LoCE} \\ \delta_{DiCE} \\ \delta_{LaCE} \\ \tau \end{bmatrix}}_{\vec{c}} \tag{A.10.48}$$

The definitions of the concise derivatives in (A.10.48) are given within (A.10.47). However, some more detailed information is required with view to the system matrix $\boldsymbol{B}$ and the control vector$\vec{c}$

$$x_{\delta_i} = \frac{\Delta X_{CE}}{m} = \sum_i \frac{X_{\delta_i}}{m}\delta_i \tag{A.10.49a}$$

$$y_{\delta_i} = \frac{\Delta Y_{CE}}{m} = \sum_i \frac{Y_{\delta_i}}{m}\delta_i \tag{A.10.49b}$$

$$z_{\delta_i} = \frac{\Delta Z_{CE}}{(m - Z_{\dot{w}})} = \sum_i \frac{Z_{\delta_i}}{(m - Z_{\dot{w}})}\delta_i \tag{A.10.49c}$$

$$\begin{aligned} l_{\delta_i} &= \Delta L_{CE} I_L^* + \Delta M_{CE} I_M^* + \Delta N_{CE} I_N^* + \frac{\Delta Z_{CE} M_{\dot{w}} I_M^*}{(m - Z_{\dot{w}})} \\ &= \sum L_{\delta_i} \delta_i I_L^* + \sum M_{\delta_i} \delta_i I_M^* + \sum N_{\delta_i} \delta_i I_N^* + \sum \frac{Z_{\delta_i} \delta_i M_{\dot{w}} I_M^*}{(m - Z_{\dot{w}})} \end{aligned} \tag{A.10.50a}$$

$$\begin{aligned} m_{\delta_i} &= \Delta L_{CE} I_L^{**} + \Delta M_{CE} I_M^{**} + \Delta N_{CE} I_N^{**} + \frac{\Delta Z_{CE} M_{\dot{w}} I_M^{**}}{(m - Z_{\dot{w}})} \\ &= \sum L_{\delta_i} \delta_i I_L^{**} + \sum M_{\delta_i} \delta_i I_M^{**} + \sum N_{\delta_i} \delta_i I_N^{**} + \sum \frac{Z_{\delta_i} \delta_i M_{\dot{w}} I_M^{**}}{(m - Z_{\dot{w}})} \end{aligned} \tag{A.10.50b}$$

$$\begin{aligned} n_{\delta_i} &= \Delta L_{CE} I_L^{***} + \Delta M_{CE} I_M^{***} + \Delta N_{CE} I_N^{***} + \frac{\Delta Z_{CE} M_{\dot{w}} I_M^{***}}{(m - Z_{\dot{w}})} \\ &= \sum L_{\delta_i} \delta_i I_L^{***} + \sum M_{\delta_i} \delta_i I_M^{***} + \sum N_{\delta_i} \delta_i I_N^{***} + \sum \frac{Z_{\delta_i} \delta_i M_{\dot{w}} I_M^{***}}{(m - Z_{\dot{w}})} \end{aligned} \tag{A.10.50c}$$

$$x_\tau = \frac{X_\tau}{m} \tau \tag{A.10.51a}$$

$$y_\tau = \frac{Y_\tau}{m} \tau \tag{A.10.51b}$$

$$z_\tau = \frac{Z_\tau}{m} \tau \tag{A.10.51c}$$

$$l_\tau = L_\tau \tau I_L^* + M_\tau \tau I_M^* + N_\tau \tau I_N^* \tag{A.10.52a}$$

$$m_\tau = L_\tau \tau I_L^{**} + M_\tau \tau I_M^{**} + N_\tau \tau I_N^{**} \tag{A.10.52b}$$

$$n_\tau = L_\tau \tau I_L^{***} + M_\tau \tau I_M^{***} + N_\tau \tau I_N^{***} \tag{A.10.52c}$$

A thorough check of the units of the derivatives and definitions has been performed. ‘Equation’ (A.10.53) confirms the units of the terms in Eqs. (A.10.47) and (A.10.48).

$$\begin{bmatrix} m/s^2 \\ m/s^2 \\ m/s^2 \\ 1/s^2 \\ 1/s^2 \\ 1/s^2 \\ rad/s \\ rad/s \\ rad/s \end{bmatrix} = \begin{bmatrix} 1/s & 1/s & 1/s & m/s & m/s & m/s & 0 & m/s^2 & 0 \\ 1/s & 1/s & 1/s & m/s & m/s & m/s & m/s^2 & 0 & 0 \\ 1/s & 1/s & 1/s & m/s & m/s & m/s & 0 & m/s^2 & 0 \\ 1/sm & 1/sm & 1/sm & 1/s & 1/s & 1/s & 0 & 1/s^2 & 0 \\ 1/sm & 1/sm & 1/sm & 1/s & 1/s & 1/s & 0 & 1/s^2 & 0 \\ 1/sm & 1/sm & 1/sm & 1/s & 1/s & 1/s & 0 & 1/s^2 & 0 \\ 0 & 0 & 0 & 1 & 0 & \tan\theta_0 & 0 & 0 & 0 \\ 0 & 0 & 0 & 0 & 1 & 0 & 0 & 0 & 0 \\ 0 & 0 & 0 & 0 & 0 & sec\theta_0 & 0 & 0 & 0 \end{bmatrix} \begin{bmatrix} m/s \\ m/s \\ m/s \\ rad/s \\ rad/s \\ rad/s \\ rad \\ rad \\ rad \end{bmatrix}$$

$$+ \begin{bmatrix} m/s^2rad & m/s^2rad & m/s^2rad & m/s^2rad \\ m/s^2rad & m/s^2rad & m/s^2rad & m/s^2rad \\ m/s^2rad & m/s^2rad & m/s^2rad & m/s^2rad \\ 1/s^2rad & 1/s^2rad & 1/s^2rad & 1/s^2rad \\ 1/s^2rad & 1/s^2rad & 1/s^2rad & 1/s^2rad \\ 1/s^2rad & 1/s^2rad & 1/s^2rad & 1/s^2rad \\ 0 & 0 & 0 & 0 \\ 0 & 0 & 0 & 0 \\ 0 & 0 & 0 & 0 \end{bmatrix} \begin{bmatrix} rad \\ rad \\ rad \\ rad \end{bmatrix} \tag{A.10.53}$$

References

1. Etkin, B. and Reid, L.D., *"Dynamics of Flight – Stability and Control,"* Third Edition, John Wiley & Sons, Inc., 1996.
2. Anon., Maple V Release 5, Version 5.00, Student Version, Waterloo Maple Inc., 1998.
3. Kreyszig, E. and Norminton, E.J., *"Advanced Engineering Mathematics - Maple Computer Manual,"* Seventh Edition, John Wiley & Sons, Inc., 1994.

A.11 Steady State Equations of Motion

A.11.1 Thrust Forces and Moments

At first it is required to model the thrust forces and moments acting on the airframe.

The thrust vector is given as

$$\vec{T}_i = \begin{bmatrix} T_{i_x} & T_{i_y} & T_{i_y} \end{bmatrix}^T \tag{A.11.1}$$

having the following components in body axes

$$T_{i_{x_b}} = T_i \cos\phi_{T_i} \cos\psi_{T_i} \tag{A.11.2a}$$

$$T_{i_{y_b}} = T_i \cos\phi_{T_i} \sin\psi_{T_i} \tag{A.11.2b}$$

$$T_{i_{z_b}} = T_i \sin\phi_{T_i} \tag{A.11.2c}$$

The total thrust-moments around the c.g. are given with

$$L_b = -T_{i_{y_b}} z_T + T_{i_{z_b}} y_T \tag{A.11.3a}$$

$$M_b = T_{i_{x_b}} z_T - T_{i_{z_b}} x_T \tag{A.11.3b}$$

$$N_b = T_{i_{y_b}} x_T - T_{i_{x_b}} y_T \tag{A.11.3c}$$

The transformation from body axes to stability axes (b → s) is

$$\begin{pmatrix} T_{i_{x_s}} \\ T_{i_{y_s}} \\ T_{i_{z_s}} \end{pmatrix} = \begin{bmatrix} \cos\alpha & 0 & \sin\alpha \\ 0 & 1 & 0 \\ -\sin\alpha & 0 & \cos\alpha \end{bmatrix} \begin{pmatrix} T_{i_{x_b}} \\ T_{i_{y_b}} \\ T_{i_{z_b}} \end{pmatrix} \tag{A.11.4}$$

yielding

$$\begin{aligned} T_{i_{x_s}} &= T_{i_{x_b}} \cos\alpha + T_{i_{z_b}} \sin\alpha \\ T_{i_{x_s}} &= T_i \cos\phi_{T_i} \cos\psi_{T_i} \cos\alpha + T_i \sin\phi_{T_i} \sin\alpha \\ T_{i_{x_s}} &= T_i\left(\cos\phi_{T_i} \cos\psi_{T_i} \cos\alpha + \sin\phi_{T_i} \sin\alpha\right) \end{aligned} \tag{A.11.5a}$$

$$\begin{aligned} T_{i_{y_s}} &= T_{i_{y_b}} \\ T_{i_{y_s}} &= T_i \cos\phi_{T_i} \sin\psi_{T_i} \end{aligned} \tag{A.11.5b}$$

$$
\begin{aligned}
T_{i_{z_s}} &= -T_{i_{x_b}} \sin\alpha + T_{i_{z_b}} \cos\alpha \\
T_{i_{z_s}} &= -T_i \cos\phi_{T_i} \cos\psi_{T_i} \sin\alpha + T_i \sin\phi_{T_i} \cos\alpha \\
T_{i_{z_s}} &= T_i\left(-\cos\phi_{T_i} \cos\psi_{T_i} \sin\alpha + \sin\phi_{T_i} \cos\alpha\right)
\end{aligned}
\tag{A.11.5c}
$$

We obtain for the moments around the c.g.

$$
\begin{aligned}
L_s &= L_b \cos\alpha + N_b \sin\alpha \\
L_s &= -T_{i_{y_b}} z_T \cos\alpha + T_{i_{z_b}} y_T \cos\alpha + T_{i_{y_b}} x_T \sin\alpha - T_{i_{x_b}} y_T \sin\alpha \\
L_s &= -T_i \cos\phi_{T_i} \sin\psi_{T_i} z_T \cos\alpha + T_i \sin\phi_{T_i} y_T \cos\alpha \\
&\quad + T_i \cos\phi_{T_i} \sin\psi_{T_i} x_T \sin\alpha - T_i \cos\phi_{T_i} \cos\psi_{T_i} y_T \sin\alpha \\
L_s &= T_i \cos\alpha\left(-\cos\phi_{T_i} \sin\psi_{T_i} z_T + \sin\phi_{T_i} y_T\right) \\
&\quad + T_i \sin\alpha\left(\cos\phi_{T_i} \sin\psi_{T_i} x_T - \cos\phi_{T_i} \cos\psi_{T_i} y_T\right)
\end{aligned}
\tag{A.11.6a}
$$

$$
\begin{aligned}
M_s &= M_b \\
M_s &= T_{i_{x_b}} z_T - T_{i_{z_b}} x_T \\
M_s &= T_i \cos\phi_{T_i} \cos\psi_{T_i} z_T - T_i \sin\phi_{T_i} x_T \\
M_s &= T_i\left(\cos\phi_{T_i} \cos\psi_{T_i} z_T - \sin\phi_{T_i} x_T\right)
\end{aligned}
\tag{A.11.6b}
$$

$$
\begin{aligned}
N_s &= -L_b \sin_\alpha + N_b \cos\alpha \\
L_s &= T_{i_{y_b}} z_T \sin\alpha - T_{i_{z_b}} y_T \sin\alpha + T_{i_{y_b}} x_T \cos\alpha - T_{i_{x_b}} y_T \cos\alpha \\
L_s &= T_i \cos\phi_{T_i} \sin\psi_{T_i} z_T \sin\alpha - T_i \sin\phi_{T_i} y_T \sin\alpha \\
&\quad + T_i \cos\phi_{T_i} \sin\psi_{T_i} x_T \cos\alpha - T_i \cos\phi_{T_i} \cos\psi_{T_i} y_T \cos\alpha \\
L_s &= T_i \sin\alpha\left(\cos\phi_{T_i} \sin\psi_{T_i} z_T - \sin\phi_{T_i} y_T\right) \\
&\quad + T_i \cos\alpha\left(\cos\phi_{T_i} \sin\psi_{T_i} x_T - \cos\phi_{T_i} \cos\psi_{T_i} y_T\right)
\end{aligned}
\tag{A.11.6c}
$$

The Thrust Terms in their final form

$$
F_{T_{x_s}} = \sum_{i=1}^{n} T_i\left(\cos\phi_{T_i} \cos\psi_{T_i} \cos\alpha + \sin\phi_{T_i} \sin\alpha\right)
\tag{A.11.7a}
$$

$$
F_{T_{y_s}} = \sum_{i=1}^{n} T_i\left(\cos\phi_{T_i} \sin\psi_{T_i}\right)
\tag{A.11.7b}
$$

$$
F_{T_{z_s}} = \sum_{i=1}^{n} T_i\left(-\cos\phi_{T_i} \cos\psi_{T_i} \sin\alpha + \sin\phi_{T_i} \cos\alpha\right)
\tag{A.11.7c}
$$

$$
L_s = \sum_{i=1}^{n} \begin{bmatrix} T_i \cos\alpha\left(-\cos\phi_{T_i} \sin\psi_{T_i} z_T + \sin\phi_{T_i} y_T\right) \\ + T_i \sin\alpha\left(\cos\phi_{T_i} \sin\psi_{T_i} x_T - \cos\phi_{T_i} \cos\psi_{T_i} y_T\right) \end{bmatrix}
\tag{A.11.8a}
$$

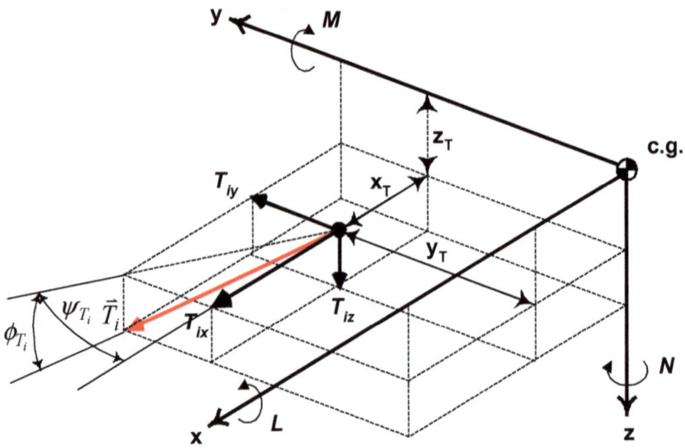

Fig. A.11.1 Thrust force component break-down

$$M_s = \sum_{i=1}^{n} \left[T_i \left(\cos\phi_{T_i} \cos\psi_{T_i} z_T - \sin\phi_{T_i} x_T \right) \right] \tag{A.11.8b}$$

$$N_s = \sum_{i=1}^{n} \begin{bmatrix} T_i \sin\alpha \left(\cos\phi_{T_i} \sin\psi_{T_i} z_T - \sin\phi_{T_i} y_T \right) \\ + T_i \cos\alpha \left(\cos\phi_{T_i} \sin\psi_{T_i} x_T - \cos\phi_{T_i} \cos\psi_{T_i} y_T \right) \end{bmatrix} \tag{A.11.8c}$$

Note, the angle ϕ_{T_i} in Fig. A.11.1 is negative.

A.11.2 *Steady State Straight Line Flight*

The form of the EOM required for steady state straight line flight is given with Eqs. (A.9.55a–c)–(A.9.58a–c). Steady state rectilinear flight is characterised by zero perturbation, $\Delta = 0$, and $\vec{\omega} = 0$, which in turn implies $p = q = r = 0$. The kinematic equations become trivial and the force and moment equations become

$$0 = -mg \sin\theta + X_A + X_T \tag{A.11.9a}$$

$$0 = mg \sin\phi \cos\theta + Y_A + Y_T \tag{A.11.9b}$$

$$0 = mg \cos\theta \cos\phi + Z_A + Z_T \tag{A.11.9c}$$

$$0 = L_A + L_T \tag{A.11.10a}$$

$$0 = M_A + M_T \tag{A.11.10b}$$

$$0 = N_A + N_T \tag{A.11.10c}$$

The corresponding aerodynamic and thrust forces and moments are substituted into (A.11.9a–c)–(A.11.10a–c), to give the following set of *Non-Linear Trim EOM for Steady State Straight Line* flight written in stability axes($\theta = \gamma$). A thorough check of the units in Eqs. (A.11.11a–c) and (A.11.12a–c) has been performed.

$$mg\,\sin\gamma = -\begin{bmatrix} C_{D_0} + C_{D_\alpha}\alpha + \sum_{j=1}^{n} C_{D_{i_{LoCE_j}}} i_{LoCE_j} + \sum_{k=1}^{m} C_{D_{\delta_{LoCE_k}}} \delta_{LoCE_k} \\ + \sum_{j=1}^{n} C_{D_{i_{DiCE_j}}} i_{DiCE_j} + \sum_{k=1}^{m} C_{D_{\delta_{DiCE_k}}} \delta_{DiCE_k} \\ + \sum_{j=1}^{n} C_{D_{i_{LaCE_j}}} i_{LaCE_j} + \sum_{k=1}^{m} C_{D_{\delta_{LaCE_k}}} \delta_{LaCE_k} \end{bmatrix} \bar{q}S + \sum_{i=1}^{n} T_i\left(\cos\phi_{T_i}\,\cos\psi_{T_i}\,\cos\alpha + \sin\phi_{T_i}\,\sin\alpha\right) - \sum_{i=1}^{n} \Delta X_{D_i} \tag{A.11.11a}$$

$$-mg\,\sin\phi\,\cos\gamma = \begin{bmatrix} C_{y_0} + C_{y_\beta}\beta + \sum_{j=1}^{n} C_{y_{i_{LoCE_j}}} i_{LoCE_j} + \sum_{k=1}^{m} C_{y_{\delta_{LoCE_k}}} \delta_{LoCE_k} \\ + \sum_{j=1}^{n} C_{y_{i_{DiCE_j}}} i_{DiCE_j} + \sum_{k=1}^{m} C_{y_{\delta_{DiCE_k}}} \delta_{DiCE_k} \\ + \sum_{j=1}^{n} C_{y_{i_{LaCE_j}}} i_{LaCE_j} + \sum_{k=1}^{m} C_{y_{\delta_{LaCE_k}}} \delta_{LaCE_k} \end{bmatrix} \bar{q}S + \sum_{i=1}^{n} T_i\,\cos\phi_{T_i}\,\sin\psi_{T_i} \tag{A.11.11b}$$

$$-mg\,\cos\gamma\,\cos\phi = -\begin{bmatrix} C_{L_0} + C_{L_\alpha}\alpha + \sum_{j=1}^{n} C_{L_{i_{LoCE_j}}} i_{LoCE_j} + \sum_{k=1}^{m} C_{L_{\delta_{LoCE_k}}} \delta_{LoCE_k} \\ + \sum_{j=1}^{n} C_{L_{i_{DiCE_j}}} i_{DiCE_j} + \sum_{k=1}^{m} C_{L_{\delta_{DiCE_k}}} \delta_{DiCE_k} \\ + \sum_{j=1}^{n} C_{L_{i_{LaCE_j}}} i_{LaCE_j} + \sum_{k=1}^{m} C_{L_{\delta_{LaCE_k}}} \delta_{LaCE_k} \end{bmatrix} \bar{q}S + \sum_{i=1}^{n} T_i\left(-\cos\phi_{T_i}\,\cos\psi_{T_i}\,\sin\alpha + \sin\phi_{T_i}\,\cos\alpha\right) \tag{A.11.11c}$$

$$0=\begin{bmatrix} C_{l_0}+C_{l_\beta}\beta+\sum_{j=1}^{n}C_{l_{i_{LoCE_j}}}i_{LoCE_j}+\sum_{k=1}^{m}C_{l_{\delta_{LoCE_k}}}\delta_{LoCE_k} \\ +\sum_{j=1}^{n}C_{l_{i_{DiCE_j}}}i_{DiCE_j}+\sum_{k=1}^{m}C_{l_{\delta_{DiCE_k}}}\delta_{DiCE_k} \\ +\sum_{j=1}^{n}C_{l_{i_{LaCE_j}}}i_{LaCE_j}+\sum_{k=1}^{m}C_{l_{\delta_{LaCE_k}}}\delta_{LaCE_k} \end{bmatrix}\bar{q}Sb$$
$$+\sum_{i=1}^{n}T_i\cos\alpha\left(-\cos\phi_{T_i}\sin\psi_{T_i}z_T+\sin\phi_{T_i}y_T\right)$$
$$+\sum_{i=1}^{n}T_i\sin\alpha\left(\cos\phi_{T_i}\sin\psi_{T_i}x_T-\cos\phi_{T_i}\cos\psi_{T_i}y_T\right) \tag{A.11.12a}$$

$$0=\begin{bmatrix} C_{m_0}+C_{m_\alpha}\alpha+\sum_{j=1}^{n}C_{m_{i_{LoCE_j}}}i_{LoCE_j}+\sum_{k=1}^{m}C_{m_{\delta_{LoCE_k}}}\delta_{LoCE_k} \\ +\sum_{j=1}^{n}C_{m_{i_{DiCE_j}}}i_{DiCE_j}+\sum_{k=1}^{m}C_{m_{\delta_{DiCE_k}}}\delta_{DiCE_k} \\ +\sum_{j=1}^{n}C_{m_{i_{LaCE_j}}}i_{LaCE_j}+\sum_{k=1}^{m}C_{m_{\delta_{LaCE_k}}}\delta_{LaCE_k} \end{bmatrix}\bar{q}Sb$$
$$+\sum_{i=1}^{n}T_i\left(\cos\phi_{T_i}\cos\psi_{T_i}z_T-\sin\phi_{T_i}x_T\right)\pm\sum_{i=1}^{n}\Delta M_{D_i} \tag{A.11.12b}$$

$$0=\begin{bmatrix} C_{n_0}+C_{n_\beta}\beta+\sum_{j=1}^{n}C_{n_{i_{LoCE_j}}}i_{LoCE_j}+\sum_{k=1}^{m}C_{n_{\delta_{LoCE_k}}}\delta_{LoCE_k} \\ +\sum_{j=1}^{n}C_{n_{i_{DiCE_j}}}i_{DiCE_j}+\sum_{k=1}^{m}C_{n_{\delta_{DiCE_k}}}\delta_{DiCE_k} \\ +\sum_{j=1}^{n}C_{n_{i_{LaCE_j}}}i_{LaCE_j}+\sum_{k=1}^{m}C_{n_{\delta_{LaCE_k}}}\delta_{LaCE_k} \end{bmatrix}\bar{q}Sb$$
$$+\sum_{i=1}^{n}T_i\sin\alpha\left(\cos\phi_{T_i}\sin\psi_{T_i}z_T-\sin\phi_{T_i}y_T\right)$$
$$+\sum_{i=1}^{n}T_i\cos\alpha\left(\cos\phi_{T_i}\sin\psi_{T_i}x_T-\cos\phi_{T_i}\cos\psi_{T_i}y_T\right)\pm\sum_{i=1}^{n}\Delta N_{D_i} \tag{A.11.12c}$$

<u>Note:</u> In stability axes we have

$$\begin{aligned} X_{aerodynamic_s} &= -D \\ Z_{aerodynamic_s} &= -L \\ M_{aerodynamic_s} &= M \end{aligned} \tag{A.11.13}$$

Furthermore is has to be recalled that $C_{D_0}, C_{Y_0}, C_{L_0}, C_{l_0}, C_{m_0}, C_{n_0}$ are all defined at $\alpha = 0$ and not at $C_L = 0$. Equations (A.11.11a–c) and (A.11.12a–c) are not presented in matrix format because of their non-linear character.

A.11.3 Steady State Turning Flight

The following derivation considers only horizontal steady turning flight, see Fig. A.11.2.

Steady turning flight is specified by the angular velocity vector of the aircraft relative to inertial space

$$\vec{\omega} = [p \quad q \quad r]^T \tag{A.11.14}$$

and expressed with Euler angles

$$\vec{\omega} = \vec{i}_3\dot{\phi} + \vec{j}_2\dot{\theta} + \vec{k}_1\dot{\psi} \tag{A.11.15}$$

Horizontal steady state turning flight[10] is characterised by a constant rate of turn only, leading to an angular velocity vector

$$\vec{\omega} = \vec{k}_1\dot{\psi} \tag{A.11.16}$$

Referring to Fig. A.11.2, the following equilibrium conditions must hold for a steady, level turn expressed in stability axes:

$$F_{CF} = m\omega^2 R = m\dot{\psi}^2 R = L \sin\phi \tag{A.11.17}$$

$$mg = L \cos\phi \tag{A.11.18}$$

Satisfying the following kinematic relationship

$$u = \omega R = \dot{\psi} R \tag{A.11.19}$$

The turn radius is obtained with

$$\dot{\psi} = \frac{u}{R} \tag{A.11.20}$$

[10]Since the turning manoeuver is assumed to take place in a horizontal plane, the stability x-axis lies consequently in the same horizontal plane. Since the rate of turn vector $\vec{\dot{\psi}}$ is perpendicular to this horizontal plane, it follows that $p = 0$.

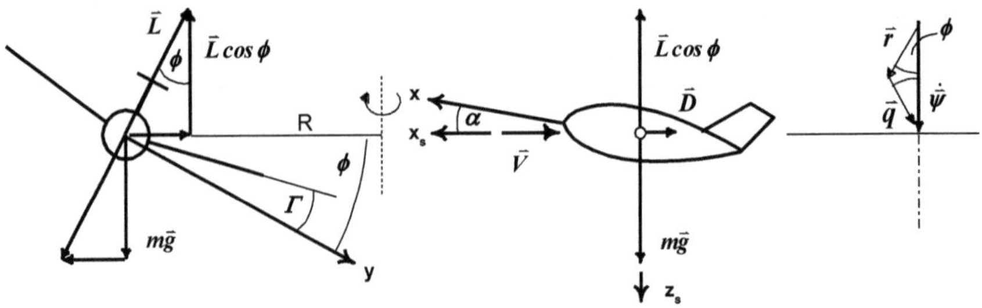

Fig. A.11.2 Horizontal steady state turning flight

substituted into (A.11.17) yields

$$m\dot{\psi}^2 R = m\frac{u^2}{R^2}R = L\sin\phi$$
$$R = \frac{mu^2}{L\sin\phi}$$

with

$$m = \frac{L\cos\phi}{g}$$

follows

$$R = \frac{L\cos\phi u^2}{gL\sin\phi}$$
$$R = \frac{u^2}{g}\tan^{-1}\phi \quad \text{(A.11.21)}$$

The corresponding turn rate is found by eliminating the turn radius from Eqs. (A.11.19) and (A.11.20)

$$u = \dot{\psi}\frac{u^2}{g}\,tan^{-1}\phi$$
$$\dot{\psi} = \frac{g}{u}\tan\phi \quad \text{(A.11.22)}$$

At this point, the concept of the load factor, n, is introduced:

$$L = nW = nmg \quad \text{(A.11.23)}$$

By referring to Eq. (A.11.18) it is seen that

$$L = n\frac{L\cos\phi}{g}g \qquad n = cos^{-1}\phi \tag{A.11.24}$$

By combining the kinematic equations (A.9.57a–c) with (A.11.22) and (A.11.23), the following expressions for the angular velocity components are derived

$$p = 0\ (\text{since}\ \gamma = 0) \tag{A.11.25a}$$

$$q = \dot{\psi}\sin\phi = \frac{g}{u}\tan\phi\sin\phi = \frac{g}{u}\frac{\sin\phi}{\cos\phi}\sin\phi \qquad q = \frac{g}{u}n\,sin^2\phi \tag{A.11.25b}$$

$$r = \dot{\psi}\cos\phi = \frac{g}{u}\tan\phi\cos\phi \qquad r = \frac{g}{u}\tan\phi\frac{1}{n} \tag{A.11.25c}$$

In the case of steady turning flight, the dotted quantities in (A.9.55a–c) and (A.9.56a–c) are zero, as in the case of steady state rectilinear flight.

The Non-Linear Trim EOM for Steady State Turning Flight written in stability axes$(\theta = \gamma)$ are derived below. A thorough check of the units in Eqs. (A.11.26a–c) and (A.11.27a–c) has been performed.

$$-mrv = -mg\frac{\tan\phi}{nu}V\sin\beta = X_A + X_T$$

$$-\frac{mg}{n}\tan\phi\sin\beta = -\begin{bmatrix} C_{D_0} + C_{D_\alpha}\alpha + C_{D_q}\frac{q\bar{c}}{2V} + C_{D_r}\frac{r\bar{c}}{2V} \\ + C_{D_{i_{LoCE}}} i_{LoCE} + C_{D_{\delta_{LoCE}}}\delta_{LoCE} \\ + C_{D_{i_{DiCE}}} i_{DiCE} + C_{D_{\delta_{DiCE}}}\delta_{DiCE} \\ + C_{D_{i_{LaCE}}} i_{LaCE} + C_{D_{\delta_{LaCE}}}\delta_{LaCE} \end{bmatrix}\bar{q}S$$

$$+\sum_{i=1}^{n} T_i\left(\cos\phi_{T_i}\cos\psi_{T_i}\cos\alpha + \sin\phi_{T_i}\sin\alpha\right) - \sum_{i=1}^{n}\Delta X_{D_i}$$

$$-\frac{mg}{n}\tan\phi\sin\beta = -\begin{bmatrix} C_{D_0} + C_{D_\alpha}\alpha + C_{D_q}\frac{gn\bar{c}}{2V^2}\sin^2\phi + C_{D_r}\frac{g\bar{c}}{2V^2 n}\tan\phi \\ + C_{D_{i_{LoCE}}} i_{LoCE} + C_{D_{\delta_{LoCE}}}\delta_{LoCE} \\ + C_{D_{i_{DiCE}}} i_{DiCE} + C_{D_{\delta_{DiCE}}}\delta_{DiCE} \\ + C_{D_{i_{LaCE}}} i_{LaCE} + C_{D_{\delta_{LaCE}}}\delta_{LaCE} \end{bmatrix}\bar{q}S$$

$$+\sum_{i=1}^{n} T_i\left(\cos\phi_{T_i}\cos\psi_{T_i}\cos\alpha + \sin\phi_{T_i}\sin\alpha\right) - \sum_{i=1}^{n}\Delta X_{D_i} \tag{A.11.26a}$$

$$mru - mg\sin\phi = Y_A + Y_T$$

$$mg\left(\frac{\tan\phi}{n} - \sin\phi\right) = \begin{bmatrix} C_{Y_0} + C_{Y_\beta}\beta + C_{Y_q}\frac{qb}{2V} + C_{Y_r}\frac{rb}{2V} \\ + C_{Y_{i_{LoCE}}} i_{LoCE} + C_{Y_{\delta_{LoCE}}}\delta_{LoCE} \\ + C_{Y_{i_{DiCE}}} i_{DiCE} + C_{Y_{\delta_{DiCE}}}\delta_{DiCE} \\ + C_{Y_{i_{LaCE}}} i_{LaCE} + C_{Y_{\delta_{LaCE}}}\delta_{LaCE} \end{bmatrix}\bar{q}S$$

$$+ \sum_{i=1}^{n} T_i\left(\cos\phi_{T_i}\sin\psi_{T_i}\right)$$

$$mg\left(\frac{\tan\phi}{n} - \sin\phi\right) = \begin{bmatrix} C_{Y_0} + C_{Y_\beta}\beta + C_{Y_q}\frac{gnb}{2V^2}\sin^2\phi + C_{Y_r}\frac{gb}{2V^2 n}\tan\phi \\ + C_{Y_{i_{LoCE}}} i_{LoCE} + C_{Y_{\delta_{LoCE}}}\delta_{LoCE} \\ + C_{Y_{i_{DiCE}}} i_{DiCE} + C_{Y_{\delta_{DiCE}}}\delta_{DiCE} \\ + C_{Y_{i_{LaCE}}} i_{LaCE} + C_{Y_{\delta_{LaCE}}}\delta_{LaCE} \end{bmatrix}\bar{q}S$$

$$+ \sum_{i=1}^{n} T_i\left(\cos\phi_{T_i}\sin\psi_{T_i}\right) \tag{A.11.26b}$$

$$-mqu - mg\cos\phi = -L_A + Z_T$$

$$-mg\left(n\,sin^2\phi - \cos\phi\right) = \begin{bmatrix} C_{L_0} + C_{L_\alpha}\alpha + C_{L_q}\frac{q\bar{c}}{2V} + C_{L_r}\frac{r\bar{c}}{2V} \\ + C_{L_{i_{LoCE}}} i_{LoCE} + C_{L_{\delta_{LoCE}}}\delta_{LoCE} \\ + C_{L_{i_{DiCE}}} i_{DiCE} + C_{L_{\delta_{DiCE}}}\delta_{DiCE} \\ + C_{L_{i_{LaCE}}} i_{LaCE} + C_{L_{\delta_{LaCE}}}\delta_{LaCE} \end{bmatrix}\bar{q}S$$

$$+ \sum_{i=1}^{n} T_i\left(-\cos\phi_{T_i}\cos\psi_{T_i}\sin\alpha + \sin\phi_{T_i}\cos\alpha\right)$$

$$-mg\left(n\sin^2\phi - \cos\phi\right) = \begin{bmatrix} C_{L_0} + C_{L_\alpha}\alpha + C_{L_q}\frac{gn\bar{c}}{2V^2}\sin^2\phi + C_{L_r}\frac{g\bar{c}}{2V^2 n}\tan\phi \\ + C_{L_{i_{LoCE}}} i_{LoCE} + C_{L_{\delta_{LoCE}}}\delta_{LoCE} \\ + C_{L_{i_{DiCE}}} i_{DiCE} + C_{L_{\delta_{DiCE}}}\delta_{DiCE} \\ + C_{L_{i_{LaCE}}} i_{LaCE} + C_{L_{\delta_{LaCE}}}\delta_{LaCE} \end{bmatrix}\bar{q}S$$

$$+ \sum_{i=1}^{n} T_i\left(-\cos\phi_{T_i}\cos\psi_{T_i}\sin\alpha + \sin\phi_{T_i}\cos\alpha\right) \tag{A.11.26c}$$

$$
\begin{aligned}
&-I_{yz}\left(q^2-r^2\right)-\left(I_y-I_z\right)qr+qh'_z-rh'_y=L_A+L_T\\
&-\frac{I_{yz}g^2\left(n^4\sin^4\phi-\tan^2\phi\right)}{u^2n^2}-\frac{\left(I_y-I_z\right)g^2\sin^3\phi}{u^2\cos\phi}+\frac{gnh'_z\sin^2\phi}{u}-\frac{gh'_y\tan\phi}{nu}=L_A+L_T\\
&-\frac{I_{yz}g^2\left(n^4\sin^4\phi-\tan^2\phi\right)}{V^2n^2}-\frac{\left(I_y-I_z\right)g^2\sin^3\phi}{V^2\cos\phi}+\frac{gnh'_z\sin^2\phi}{V}-\frac{gh'_y\tan\phi}{nV}\\
&=\begin{bmatrix}C_{l_0}+C_{l_\beta}\beta+C_{l_q}\frac{qb}{2V}+C_{l_r}\frac{rb}{2V}\\+C_{l_{i_{LoCE}}}i_{LoCE}+C_{l_{\delta_{LoCE}}}\delta_{LoCE}\\+C_{l_{i_{DiCE}}}i_{DiCE}+C_{l_{\delta_{DiCE}}}\delta_{DiCE}\\+C_{l_{i_{LaCE}}}i_{LaCE}+C_{l_{\delta_{LaCE}}}\delta_{LaCE}\end{bmatrix}\bar{q}Sb\\
&\quad+\sum_{i=1}^{n}T_i\cos\alpha\left(-\cos\phi_{T_i}\sin\psi_{T_i}z_T+\sin\phi_{T_i}y_T\right)\\
&\quad+\sum_{i=1}^{n}T_i\sin\alpha\left(\cos\phi_{T_i}\sin\psi_{T_i}x_T-\cos\phi_{T_i}\cos\psi_{T_i}y_T\right)\\
&-\frac{I_{yz}g^2\left(n^4\sin^4\phi-\tan^2\phi\right)}{V^2n^2}-\frac{\left(I_y-I_z\right)g^2\sin^3\phi}{V^2\cos\phi}+\frac{gnh'_z\sin^2\phi}{V}-\frac{gh'_y\tan\phi}{nV}\\
&=\begin{bmatrix}C_{l_0}+C_{l_\beta}\beta+C_{l_q}\frac{gnb}{2V^2}\sin^2\phi+C_{l_r}\frac{gb}{2V^2n}\tan\phi\\+C_{l_{i_{LoCE}}}i_{LoCE}+C_{l_{\delta_{LoCE}}}\delta_{LoCE}\\+C_{l_{i_{DiCE}}}i_{DiCE}+C_{l_{\delta_{DiCE}}}\delta_{DiCE}\\+C_{l_{i_{LaCE}}}i_{LaCE}+C_{l_{\delta_{LaCE}}}\delta_{LaCE}\end{bmatrix}\bar{q}Sb\\
&\quad+\sum_{i=1}^{n}T_i\cos\alpha\left(-\cos\phi_{T_i}\sin\psi_{T_i}z_T+\sin\phi_{T_i}y_T\right)\\
&\quad+\sum_{i=1}^{n}T_i\sin\alpha\left(\cos\phi_{T_i}\sin\psi_{T_i}x_T-\cos\phi_{T_i}\cos\psi_{T_i}y_T\right)
\end{aligned}
\tag{A.11.27a}
$$

$$
\begin{aligned}
&-I_{zx}r^2-I_{xy}qr+rh'_x=M_A+M_T\\
&-I_{zx}\frac{g^2\tan^2\phi}{n^2u^2}-I_{xy}\frac{g^2\sin^2\phi\tan\phi}{u^2}+\frac{g\tan\phi}{nu}h'_x=M_A+M_T\\
&-I_{zx}\frac{g^2\tan^2\phi}{n^2u^2}-I_{xy}\frac{g^2\sin^2\phi\tan\phi}{u^2}+\frac{g\tan\phi}{nu}h'_x=\begin{bmatrix}C_{m_0}+C_{m_\alpha}\alpha+C_{m_q}\frac{q\bar{c}}{2V}+C_{m_r}\frac{r\bar{c}}{2V}\\+C_{m_{i_{LoCE}}}i_{LoCE}+C_{m_{\delta_{LoCE}}}\delta_{LoCE}\\+C_{m_{i_{DiCE}}}i_{DiCE}+C_{m_{\delta_{DiCE}}}\delta_{DiCE}\\+C_{m_{i_{LaCE}}}i_{LaCE}+C_{m_{\delta_{LaCE}}}\delta_{LaCE}\end{bmatrix}\bar{q}S\bar{c}\\
&+\sum_{i=1}^{n}T_i\left(\cos\phi_{T_i}\cos\psi_{T_i}z_T-\sin\phi_{T_i}x_T\right)\pm\sum_{i=1}^{n}\Delta M_{D_i}
\end{aligned}
$$

$$
\begin{aligned}
&- I_{zx}\frac{g^2\, tan^2\phi}{n^2u^2} - I_{xy}\frac{g^2\,\sin^2\phi\,\tan\phi}{u^2} + \frac{g\,\tan\phi}{nu}h'_x \\
&= \begin{bmatrix} C_{m_0} + C_{m_\alpha}\alpha + C_{m_q}\frac{gn\bar{c}}{2V^2}\sin^2\phi + C_{m_r}\frac{g\bar{c}}{2V^2n}\tan\phi \\ + C_{m_{i_{LoCE}}} i_{LoCE} + C_{m_{\delta_{LoCE}}}\delta_{LoCE} \\ + C_{m_{i_{DiCE}}} i_{DiCE} + C_{m_{\delta_{DiCE}}}\delta_{DiCE} \\ + C_{m_{i_{LaCE}}} i_{LaCE} + C_{m_{\delta_{LaCE}}}\delta_{LaCE} \end{bmatrix}\bar{q}S\bar{c} \\
&+ \sum_{i=1}^{n} T_i\left(\cos\phi_{T_i}\,\cos\psi_{T_i}z_T - \sin\phi_{T_i}x_T\right) \pm \sum_{i=1}^{n}\Delta M_{D_i}
\end{aligned}
\tag{A.11.27b}
$$

$$
I_{xy}q^2 + I_{zx}qr - qh'_x = N_A + N_T
$$

$$
I_{xy}\frac{g^2n^2}{u^2}\sin^4\phi + I_{zx}\frac{g^2\,\sin^2\phi\,\tan\phi}{u^2} - \frac{gn\,\sin^2\phi}{u}h'_x = N_A + N_T
$$

$$
\begin{aligned}
&I_{xy}\frac{g^2n^2}{u^2}\sin^4\phi + I_{zx}\frac{g^2\,\sin^2\phi\,\tan\phi}{u^2} - \frac{gn\,sin^2\phi}{u}h'_x = \begin{bmatrix} C_{n_0} + C_{n_\beta}\beta + C_{n_q}\frac{qb}{2V} + C_{n_r}\frac{rb}{2V} \\ + C_{n_{i_{LoCE}}} i_{LoCE} + C_{n_{\delta_{LoCE}}}\delta_{LoCE} \\ + C_{n_{i_{DiCE}}} i_{DiCE} + C_{n_{\delta_{DiCE}}}\delta_{DiCE} \\ + C_{n_{i_{LaCE}}} i_{LaCE} + C_{n_{\delta_{LaCE}}}\delta_{LaCE} \end{bmatrix}\bar{q}Sb \\
&+ \sum_{i=1}^{n} T_i\,\sin\alpha\left(\cos\phi_{T_i}\,\sin\psi_{T_i}z_T - \sin\phi_{T_i}y_T\right) \\
&+ \sum_{i=1}^{n} T_i\,\cos\alpha\left(\cos\phi_{T_i}\,\sin\psi_{T_i}x_T - \cos\phi_{T_i}\,\cos\psi_{T_i}y_T\right) - \sum_{i=1}^{n}\Delta N_{D_i}
\end{aligned}
$$

$$
\begin{aligned}
&I_{xy}\frac{g^2n^2}{u^2}\sin^4\phi + I_{zx}\frac{g^2\,sin^2\phi\,\tan\phi}{u^2} - \frac{gn\,\sin^2\phi}{u}h'_x \\
&= \begin{bmatrix} C_{n_0} + C_{n_\beta}\beta + C_{n_q}\frac{gnb}{2V^2}\sin^2\phi + C_{n_r}\frac{gb}{2V^2n}\tan\phi \\ + C_{n_{i_{LoCE}}} i_{LoCE} + C_{n_{\delta_{LoCE}}}\delta_{LoCE} \\ + C_{n_{i_{DiCE}}} i_{DiCE} + C_{n_{\delta_{DiCE}}}\delta_{DiCE} \\ + C_{n_{i_{LaCE}}} i_{LaCE} + C_{n_{\delta_{LaCE}}}\delta_{LaCE} \end{bmatrix}\bar{q}Sb \\
&+ \sum_{i=1}^{n} T_i\,\cos\alpha\left(-\cos\phi_{T_i}\,\sin\psi_{T_i}z_T + \sin\phi_{T_i}y_T\right) \\
&+ \sum_{i=1}^{n} T_i\,\sin\alpha\left(\cos\phi_{T_i}\,\sin\psi_{T_i}x_T - \cos\phi_{T_i}\,\cos\psi_{T_i}y_T\right) - \sum_{i=1}^{n}\Delta N_{D_i}
\end{aligned}
\tag{A.11.27c}
$$

The steady pitch- and yaw rates in this type of turning manoeuver are functions of load factor and bank angle. It must be noted that the climbing/descending steady turning flight demands additional control power compared to the horizontal turning flight, see Etkin et al. [1] and Brüning et al. [2]. These cases are, however, not considered in the present context due to the additional complexity involved. If the turn is coordinated, then no net lateral acceleration acts on the aircraft. This condition implies, that in a steady, level turn the aerodynamic sideforce in Eq. (A.10.14b) is equal to zero. However, the coordinated turn is considered here to be a special case only, thus, the algorithm presented allows steady sideslipping flight cases including thrust asymmetry.

A.11.4 Steady State Pull-Up and Push-Over Flight

The following derivation considers the general case of the aircraft flying with $\beta \neq 0$ in the xz-plane, see Fig. A.11.3.

Since a quasi-steady flight condition is considered in stability axes at the top/bottom of the curved flight path, the following conditions apply:

$$\vec{\omega} = [0 \quad q \quad r]^T_{asymmetric\,flight} \tag{A.11.28a}$$

$$\vec{\omega} = [0 \quad q \quad 0]^T_{symmetric\,flight} \tag{A.11.28b}$$

$$(u \neq 0, v \neq 0, w = 0)_{asymmetric\,flight} \tag{A.11.29a}$$

$$(u \neq 0, v = 0, w = 0)_{symmetric\,flight} \tag{A.11.29b}$$

$$(\phi \neq 0, \theta = \gamma = 0, \psi \neq 0)_{asymmetric\,flight} \tag{A.11.30}$$

$$\left(\dot{\phi} = \dot{\psi} = 0, \dot{\theta} \neq 0\right)_{asymmetric\ and\ symmetric\,flight} \tag{A.11.31}$$

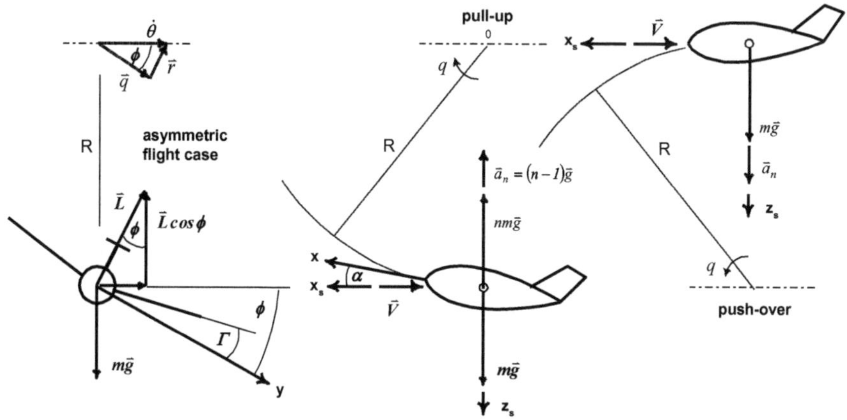

Fig. A.11.3 Symmetric steady pull-up and push-over flight

$$\left(u^E, v^E, w^E\right) = (u, v, w)(\text{no wind}) \tag{A.11.32}$$

$$v = V \sin\beta \tag{A.11.33}$$

$$u = V \tag{A.11.34}$$

The flight path tangent is horizontal at the top/bottom of the curved flight path, hence the net normal force is perpendicular relative to the *xy*-plane. It follows for the load factor:

$$\begin{aligned} n &= \frac{L}{mg} \\ \frac{L}{mg} - 1 &= n - 1 \\ \underbrace{L - mg}_{\text{net normal force } N_n} &= (n-1)mg \end{aligned} \tag{A.11.35}$$

It follows for the normal acceleration

$$\begin{aligned} N_n &= (n-1)mg \\ \frac{N_n}{m} &= (n-1)g \\ a_n &= (n-1)g \end{aligned} \tag{A.11.36}$$

The load factor$n$ in Eq. (A.11.36) covers the following flight conditions:

$n = 1$ non-manoeuvring flight
$n > 1$ pull-up
$n < 1$ push-over
$n = 0$ ballistic flight

It is desirable to express the steady state angular rates as a function of the load factor$n$. Referring to Fig. A.11.3, the following equilibrium condition must hold for a steady pull-up/push-over manoeuver.

$$\begin{aligned} mg + ma_n &= L \\ mg + m(n-1)g &= L \\ mg + mng - mg &= L \\ mng &= L \end{aligned} \tag{A.11.37}$$

with

$$q = \dot{\gamma}$$
$$L = mg + m\dot{\theta}^2 R$$
$$\dot{\theta} = \frac{V}{R} = \dot{\gamma}$$

follows

$$q = \frac{V}{R} \qquad \text{(A.11.38)}$$

with

$$L = m\left(g + \frac{V^2}{R^2}R\right)$$
$$\frac{L}{m} = g + \frac{V^2}{R} \qquad \text{(A.11.39)}$$
$$\frac{V}{R} = \frac{L}{mV} - \frac{g}{V}$$

follows for the angular velocity component q

$$q = \dot{\theta}\cos\phi = \frac{V}{R}\cos\phi$$
$$q = \left(\frac{L}{mV} - \frac{g}{V}\right)\cos\phi = \left(\frac{gn}{V} - \frac{g}{V}\right)\cos\phi \qquad \text{(A.11.40)}$$
$$q = \frac{g}{V}(n-1)\cos\phi$$

and it follows for the angular velocity component r

$$r = -\dot{\theta}\sin\phi = -\frac{V}{R}\sin\phi$$
$$r = -\left(\frac{L}{mV} - \frac{g}{V}\right)\sin\phi = -\left(\frac{gn}{V} - \frac{g}{V}\right)\sin\phi \qquad \text{(A.11.41)}$$
$$r = -\frac{g}{V}(n-1)\sin\phi$$

When considering the quasi-steady condition of the pull-up or push-over with a horizontal flight path tangent, the dotted quantities in Eqs. (A.9.55a–c) to (A.9.59a–c) are zero, as in the case of steady state rectilinear flight and steady state level turning flight. Equations (A.11.42a–c) to (A.11.44a–c) are written in stability axes.

$$X_A + X_T = -mrV \sin\beta \tag{A.11.42a}$$

$$Y_A + Y_T + mg \sin\phi = mrV \tag{A.11.42b}$$

$$Z_A + Z_T + mg \cos\phi = -mqV \tag{A.11.42c}$$

$$L_A + L_T = -I_{yz}(q^2 - r^2) - (I_y - I_z)qr + qh'_z - rh'_y \tag{A.11.43a}$$

$$M_A + M_T = -I_{zx}r^2 - I_{xy}qr + rh'_x \tag{A.11.43b}$$

$$N_A + N_T = I_{xy}q^2 + I_{zx}qr - qh'_x \tag{A.11.43c}$$

$$p = 0 \tag{A.11.44a}$$

$$q = \dot{\gamma} \cos\phi \tag{A.11.44b}$$

$$r = -\dot{\gamma} \sin\phi \tag{A.11.44c}$$

Equations (A.11.40) to (A.11.43a–c) can be assembled to the *Linear Trim EOM for Steady State Pull-Up and Push-Over Flight*. A thorough check of the units in Eqs. (A.11.45a–c) and (A.11.46a–c) has been performed.

$$-mrV \sin\beta = X_A + X_T$$

$$mg(n-1)\sin\phi\sin\beta = -\begin{bmatrix} C_{D_0} + C_{D_\alpha}\alpha + C_{D_q}\frac{q\bar{c}}{2V} + C_{D_r}\frac{r\bar{c}}{2V} \\ + C_{D_{i_{LoCE}}} i_{LoCE} + C_{D_{\delta_{LoCE}}} \delta_{LoCE} \\ + C_{D_{i_{DiCE}}} i_{DiCE} + C_{D_{\delta_{DiCE}}} \delta_{DiCE} \\ + C_{D_{i_{LaCE}}} i_{LaCE} + C_{D_{\delta_{LaCE}}} \delta_{LaCE} \end{bmatrix} \bar{q}S$$

$$+ \sum_{i=1}^{n} T_i(\cos\phi_{T_i}\cos\psi_{T_i}\cos\alpha + \sin\phi_{T_i}\sin\alpha) - \sum_{i=1}^{n} \Delta X_{D_i}$$

$$mg(n-1)\sin\phi\sin\beta = -\begin{bmatrix} C_{D_0} + C_{D_\alpha}\alpha + C_{D_q}\frac{g\bar{c}(n-1)\cos\phi}{2V^2} - C_{D_r}\frac{g\bar{c}(n-1)\sin\phi}{2V^2} \\ + C_{D_{i_{LoCE}}} i_{LoCE} + C_{D_{\delta_{LoCE}}} \delta_{LoCE} \\ + C_{D_{i_{DiCE}}} i_{DiCE} + C_{D_{\delta_{DiCE}}} \delta_{DiCE} \\ + C_{D_{i_{LaCE}}} i_{LaCE} + C_{D_{\delta_{LaCE}}} \delta_{LaCE} \end{bmatrix} \bar{q}S$$

$$+ \sum_{i=1}^{n} T_i(\cos\phi_{T_i}\cos\psi_{T_i}\cos\alpha + \sin\phi_{T_i}\sin\alpha) - \sum_{i=1}^{n} \Delta X_{D_i} \tag{A.11.45a}$$

$$mrV - mg\sin\phi = Y_A + Y_T$$

$$-mgn\sin\phi = \begin{bmatrix} C_{Y_0} + C_{Y_\beta}\beta + C_{Y_q}\frac{qb}{2V} + C_{Y_r}\frac{rb}{2V} \\ + C_{Y_{i_{LoCE}}} i_{LoCE} + C_{Y_{\delta_{LoCE}}}\delta_{LoCE} \\ + C_{Y_{i_{DiCE}}} i_{DiCE} + C_{Y_{\delta_{DiCE}}}\delta_{DiCE} \\ + C_{Y_{i_{LaCE}}} i_{LaCE} + C_{Y_{\delta_{LaCE}}}\delta_{LaCE} \end{bmatrix}\bar{q}S + \sum_{i=1}^{n} T_i\left(\cos\phi_{T_i}\sin\psi_{T_i}\right)$$

$$-mgn\sin\phi = \begin{bmatrix} C_{Y_0} + C_{Y_\beta}\beta + C_{Y_q}\frac{gb(n-1)\cos\phi}{2V^2} - C_{Y_r}\frac{gb(n-1)\sin\phi}{2V^2} \\ + C_{Y_{i_{LoCE}}} i_{LoCE} + C_{Y_{\delta_{LoCE}}}\delta_{LoCE} \\ + C_{Y_{i_{DiCE}}} i_{DiCE} + C_{Y_{\delta_{DiCE}}}\delta_{DiCE} \\ + C_{Y_{i_{LaCE}}} i_{LaCE} + C_{Y_{\delta_{LaCE}}}\delta_{LaCE} \end{bmatrix}\bar{q}S$$

$$+ \sum_{i=1}^{n} T_i\left(\cos\phi_{T_i}\sin\psi_{T_i}\right) \tag{A.11.45b}$$

$$-mqV - mg\cos\phi = -L_A + Z_T$$

$$-mgn\cos\phi = -\begin{bmatrix} C_{L_0} + C_{L_\alpha}\alpha + C_{L_q}\frac{q\bar{c}}{2V} + C_{L_r}\frac{r\bar{c}}{2V} \\ + C_{L_{i_{LoCE}}} i_{LoCE} + C_{L_{\delta_{LoCE}}}\delta_{LoCE} \\ + C_{L_{i_{DiCE}}} i_{DiCE} + C_{L_{\delta_{DiCE}}}\delta_{DiCE} \\ + C_{L_{i_{LaCE}}} i_{LaCE} + C_{L_{\delta_{LaCE}}}\delta_{LaCE} \end{bmatrix}\bar{q}S$$

$$+ \sum_{i=1}^{n} T_i\left(-\cos\phi_{T_i}\cos\psi_{T_i}\sin\alpha + \sin\phi_{T_i}\cos\alpha\right)$$

$$-mgn\cos\phi = -\begin{bmatrix} C_{L_0} + C_{L_\alpha}\alpha + C_{L_q}\frac{g\bar{c}(n-1)\cos\phi}{2V^2} - C_{L_r}\frac{g\bar{c}(n-1)\sin\phi}{2V^2} \\ + C_{L_{i_{LoCE}}} i_{LoCE} + C_{L_{\delta_{LoCE}}}\delta_{LoCE} \\ + C_{L_{i_{DiCE}}} i_{DiCE} + C_{L_{\delta_{DiCE}}}\delta_{DiCE} \\ + C_{L_{i_{LaCE}}} i_{LaCE} + C_{L_{\delta_{LaCE}}}\delta_{LaCE} \end{bmatrix}\bar{q}S$$

$$+ \sum_{i=1}^{n} T_i\left(-\cos\phi_{T_i}\cos\psi_{T_i}\sin\alpha + \sin\phi_{T_i}\cos\alpha\right) \tag{A.11.45c}$$

$$-I_{yz}\left(q^2 - r^2\right) - \left(I_y - I_z\right)qr + qh'_z - rh'_y = L_A + L_T$$

$$-I_{yz}\frac{g^2(n-1)^2\left(\cos^2\phi - \sin^2\phi\right)}{V^2} + \left(I_y - I_z\right)\frac{g^2(n-1)^2\cos\phi\sin\phi}{V^2}$$

$$+\left(h'_z\cos\phi + h'_y\sin\phi\right)\frac{g(n-1)}{V} = L_A + L_T$$

$$
\begin{aligned}
&- I_{yz}\frac{g^2(n-1)^2\left(cos^2\phi - \sin^2\phi\right)}{V^2} + \left(I_y - I_z\right)\frac{g^2(n-1)^2\cos\phi\,\sin\phi}{V^2} \\
&+ \left(h_z'\cos\phi + h_y'\sin\phi\right)\frac{g(n-1)}{V} \\
&= \begin{bmatrix} C_{l_0} + C_{l_\beta}\beta + C_{l_q}\frac{qb}{2V} + C_{l_r}\frac{rb}{2V} \\ + C_{l_{i_{LoCE}}} i_{LoCE} + C_{l_{\delta_{LoCE}}}\delta_{LoCE} \\ + C_{l_{i_{DiCE}}} i_{DiCE} + C_{l_{\delta_{DiCE}}}\delta_{DiCE} \\ + C_{l_{i_{LaCE}}} i_{LaCE} + C_{l_{\delta_{LaCE}}}\delta_{LaCE} \end{bmatrix}\bar{q}Sb + \sum_{i=1}^{n} T_i\cos\alpha\left(-\cos\phi_{T_i}\sin\psi_{T_i}z_T + \sin\phi_{T_i}y_T\right) \\
&+ \sum_{i=1}^{n} T_i\sin\alpha\left(\cos\phi_{T_i}\sin\psi_{T_i}x_T - \cos\phi_{T_i}\cos\psi_{T_i}y_T\right) \\
&- I_{yz}\frac{g^2(n-1)^2\left(\cos^2\phi - \sin^2\phi\right)}{V^2} + \left(I_y - I_z\right)\frac{g^2(n-1)^2\cos\phi\,\sin\phi}{V^2} \\
&+ \left(h_z'\cos\phi + h_y'\sin\phi\right)\frac{g(n-1)}{V} \\
&= \begin{bmatrix} C_{l_0} + C_{l_\beta}\beta + C_{l_q}\frac{gb(n-1)\cos\phi}{2V^2} - C_{l_r}\frac{gb(n-1)\sin\phi}{2V^2} \\ + C_{l_{i_{LoCE}}} i_{LoCE} + C_{l_{\delta_{LoCE}}}\delta_{LoCE} \\ + C_{l_{i_{DiCE}}} i_{DiCE} + C_{l_{\delta_{DiCE}}}\delta_{DiCE} \\ + C_{l_{i_{LaCE}}} i_{LaCE} + C_{l_{\delta_{LaCE}}}\delta_{LaCE} \end{bmatrix}\bar{q}Sb \\
&+ \sum_{i=1}^{n} T_i\cos\alpha\left(-\cos\phi_{T_i}\sin\psi_{T_i}z_T + \sin\phi_{T_i}y_T\right) \\
&+ \sum_{i=1}^{n} T_i\sin\alpha\left(\cos\phi_{T_i}\sin\psi_{T_i}x_T - \cos\phi_{T_i}\cos\psi_{T_i}y_T\right)
\end{aligned}
\tag{A.11.46a}
$$

$$
\begin{aligned}
&- I_{zx}r^2 - I_{xy}qr + rh_x' = M_A + M_T \\
&- \frac{g^2(n-1)^2\sin^2\phi}{V^2}I_{zx} + \frac{g^2(n-1)^2\cos\phi\,\sin\phi}{V^2}I_{xy} - \frac{g(n-1)\sin\phi}{V}h_x' = M_A + M_T \\
&- \frac{g^2(n-1)^2\sin^2\phi}{V^2}I_{zx} + \frac{g^2(n-1)^2\cos\phi\,\sin\phi}{V^2}I_{xy} - \frac{g(n-1)\sin\phi}{V}h_x' \\
&= \begin{bmatrix} C_{m_0} + C_{m_\alpha}\alpha + C_{m_q}\frac{q\bar{c}}{2V} + C_{m_r}\frac{r\bar{c}}{2V} \\ + C_{m_{i_{LoCE}}} i_{LoCE} + C_{m_{\delta_{LoCE}}}\delta_{LoCE} \\ + C_{m_{i_{DiCE}}} i_{DiCE} + C_{m_{\delta_{DiCE}}}\delta_{DiCE} \\ + C_{m_{i_{LaCE}}} i_{LaCE} + C_{m_{\delta_{LaCE}}}\delta_{LaCE} \end{bmatrix}\bar{q}S\bar{c} \\
&+ \sum_{i=1}^{n} T_i\left(\cos\phi_{T_i}\cos\psi_{T_i}z_T - \sin\phi_{T_i}x_T\right) \pm \sum_{i=1}^{n}\Delta M_{D_i} \\
&- \frac{g^2(n-1)^2\sin^2\phi}{V^2}I_{zx} + \frac{g^2(n-1)^2\cos\phi\,\sin\phi}{V^2}I_{xy} - \frac{g(n-1)\sin\phi}{V}h_x'
\end{aligned}
$$

$$= \begin{bmatrix} C_{m_0} + C_{m_\alpha}\alpha + C_{m_q}\frac{g\bar{c}(n-1)\cos\phi}{2V^2} - C_{m_r}\frac{g\bar{c}(n-1)\sin\phi}{2V^2} \\ + C_{m_{i_{LoCE}}} i_{LoCE} + C_{m_{\delta_{LoCE}}}\delta_{LoCE} \\ + C_{m_{i_{DiCE}}} i_{DiCE} + C_{m_{\delta_{DiCE}}}\delta_{DiCE} \\ + C_{m_{i_{LaCE}}} i_{LaCE} + C_{m_{\delta_{LaCE}}}\delta_{LaCE} \end{bmatrix}\bar{q}S\bar{c} \tag{A.11.46b}$$
$$+ \sum_{i=1}^{n} T_i\left(\cos\phi_{T_i}\cos\psi_{T_i}z_T - \sin\phi_{T_i}x_T\right) \pm \sum_{i=1}^{n}\Delta M_{D_i}$$

$$I_{xy}q^2 + I_{zx}qr - qh'_x = N_A + N_T$$

$$I_{xy}\frac{g^2(n-1)^2\cos^2\phi}{V^2} - I_{zx}\frac{g^2(n-1)^2\cos\phi\sin\phi}{V^2} - h'_x\frac{g(n-1)\cos\phi}{V} = N_A + N_T$$

$$I_{xy}\frac{g^2(n-1)^2\cos^2\phi}{V^2} - I_{zx}\frac{g^2(n-1)^2\cos\phi\sin\phi}{V^2} - h'_x\frac{g(n-1)\cos\phi}{V}$$
$$= \begin{bmatrix} C_{n_0} + C_{n_\beta}\beta + C_{n_q}\frac{qb}{2V} + C_{n_r}\frac{rb}{2V} \\ + C_{n_{i_{LoCE}}} i_{LoCE} + C_{n_{\delta_{LoCE}}}\delta_{LoCE} \\ + C_{n_{i_{DiCE}}} i_{DiCE} + C_{n_{\delta_{DiCE}}}\delta_{DiCE} \\ + C_{n_{i_{LaCE}}} i_{LaCE} + C_{n_{\delta_{LaCE}}}\delta_{LaCE} \end{bmatrix}\bar{q}Sb$$
$$+ \sum_{i=1}^{n} T_i\sin\alpha\left(\cos\phi_{T_i}\sin\psi_{T_i}z_T - \sin\phi_{T_i}y_T\right)$$
$$+ \sum_{i=1}^{n} T_i\cos\alpha\left(\cos\phi_{T_i}\sin\psi_{T_i}x_T - \cos\phi_{T_i}\cos\psi_{T_i}y_T\right) - \sum_{i=1}^{n}\Delta N_{D_i}$$

$$I_{xy}\frac{g^2(n-1)^2\cos^2\phi}{V^2} - I_{zx}\frac{g^2(n-1)^2\cos\phi\sin\phi}{V^2} - h'_x\frac{g(n-1)\cos\phi}{V}$$
$$= \begin{bmatrix} C_{n_0} + C_{n_\beta}\beta + C_{n_q}\frac{gb(n-1)\cos\phi}{2V^2} - C_{n_r}\frac{gb(n-1)\sin\phi}{2V^2} \\ + C_{n_{i_{LoCE}}} i_{LoCE} + C_{n_{\delta_{LoCE}}}\delta_{LoCE} \\ + C_{n_{i_{DiCE}}} i_{DiCE} + C_{n_{\delta_{DiCE}}}\delta_{DiCE} \\ + C_{n_{i_{LaCE}}} i_{LaCE} + C_{n_{\delta_{LaCE}}}\delta_{LaCE} \end{bmatrix}\bar{q}Sb$$
$$+ \sum_{i=1}^{n} T_i\cos\alpha\left(-\cos\phi_{T_i}\sin\psi_{T_i}z_T + \sin\phi_{T_i}y_T\right)$$
$$+ \sum_{i=1}^{n} T_i\sin\alpha\left(\cos\phi_{T_i}\sin\psi_{T_i}x_T - \cos\phi_{T_i}\cos\psi_{T_i}y_T\right) - \sum_{i=1}^{n}\Delta N_{D_i} \tag{A.11.46c}$$

A.11.5 Steady State Rolling Performance

In the following the 6-DOF static EOM are derived for the LaCE deflection required, to satisfy a prescribed roll helix angle$pb/2V$ with a value of, e.g., 0.07. The EOM have to be solved for the trimmed condition, and in particular the aileron deflection required to enforce the prescribed roll rate, p. The flight condition of interest is horizontal flight ($\theta = \gamma = 0$). The aircraft performs a steady roll manoeuver and the situation of particular interest is, when the aircraft rolls through $\phi = 0$. However, the variable $\phi \neq 0$ is permitted to allow to trim the aircraft in y-direction (consider only small angles of ϕ) (Fig. A.11.4).

Since a quasi-steady flight condition is considered in stability axes, the following conditions apply:

$$v \geq \leq 0 \tag{A.11.47a}$$

$$w = 0 \tag{A.11.47b}$$

$$\dot{u} = \dot{v} = \dot{w} = 0 \tag{A.11.48}$$

$$q = r = 0 \tag{A.11.49}$$

$$\dot{p} = \dot{q} = \dot{r} = 0 \tag{A.11.50}$$

$$\theta = \gamma = \psi = 0 \tag{A.11.51}$$

$$\dot{\theta} = \dot{\psi} = 0 \tag{A.11.52}$$

$$Y_T = L_T = N_T = 0 \tag{A.11.53}$$

Equations (A.9.55a–c) to (A.9.59a–c) written in stability axes become

$$X_A + X_T = 0 \tag{A.11.54a}$$

$$Y_A + mg \sin \phi = 0 \tag{A.11.54b}$$

$$Z_A + Z_T + mg \cos \phi = mpv \tag{A.11.54c}$$

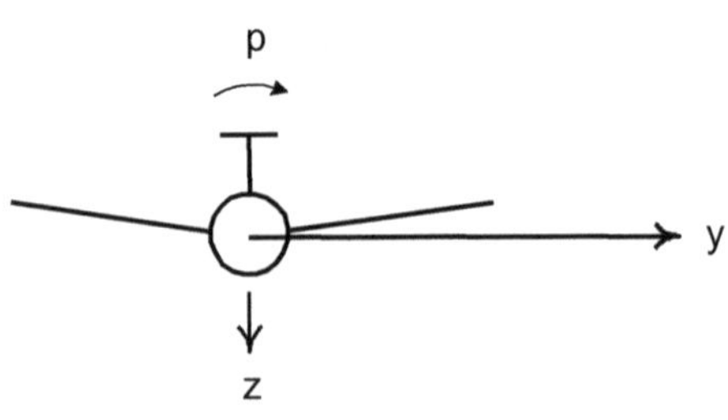

Fig. A.11.4 Roll performance at $\phi = 0$

$$L_A = 0 \tag{A.11.55a}$$

$$M_A + M_T = I_{zx}p^2 - ph'_z \tag{A.11.55b}$$

$$N_A = -I_{xy}p^2 + ph'_y \tag{A.11.55c}$$

$$p = \dot{\phi} \tag{A.11.56a}$$

$$q = 0 \tag{A.11.56b}$$

$$r = 0 \tag{A.11.56c}$$

Equations (A.11.47a–c) to (A.11.56a–c) can be assembled to the *Non-Linear Trim EOM for Steady State Rolling Flight*. A thorough check of the units in Eqs. (A.11.57a–c) and (A.11.58a–c) has been performed.

$$0 = -D + X_T$$

$$0 = -\begin{bmatrix} C_{D_0} + C_{D_\alpha}\alpha + C_{D_p}\frac{p\bar{c}}{2V} \\ + C_{D_{i_{LoCE}}} i_{LoCE} + C_{D_{\delta_{LoCE}}}\delta_{LoCE} \\ + C_{D_{i_{DiCE}}} i_{DiCE} + C_{D_{\delta_{DiCE}}}\delta_{DiCE} \\ + C_{D_{i_{LaCE}}} i_{LaCE} + C_{D_{\delta_{LaCE}}}\delta_{LaCE} \end{bmatrix}\bar{q}S + \sum_{i=1}^{n} T_i\left(\cos\phi_{T_i}\cos\psi_{T_i}\cos\alpha + \sin\phi_{T_i}\sin\alpha\right) \tag{A.11.57a}$$

$$mg\sin\phi = Y_A$$

$$mg\sin\phi = \begin{bmatrix} C_{Y_0} + C_{Y_p}\frac{pb}{2V} \\ + C_{Y_{i_{LoCE}}} i_{LoCE} + C_{Y_{\delta_{LoCE}}}\delta_{LoCE} \\ + C_{Y_{i_{DiCE}}} i_{DiCE} + C_{Y_{\delta_{DiCE}}}\delta_{DiCE} \\ + C_{Y_{i_{LaCE}}} i_{LaCE} + C_{Y_{\delta_{LaCE}}}\delta_{LaCE} \end{bmatrix}\bar{q}S \tag{A.11.57b}$$

$$-mg\cos\phi + mpv = -L_A + Z_T$$

$$-mg\cos\phi + mpV\sin\beta = -\begin{bmatrix} C_{L_0} + C_{L_\alpha}\alpha + C_{L_p}\frac{p\bar{c}}{2V} \\ + C_{L_{i_{LoCE}}} i_{LoCE} + C_{L_{\delta_{LoCE}}}\delta_{LoCE} \\ + C_{L_{i_{DiCE}}} i_{DiCE} + C_{L_{\delta_{DiCE}}}\delta_{DiCE} \\ + C_{L_{i_{LaCE}}} i_{LaCE} + C_{L_{\delta_{LaCE}}}\delta_{LaCE} \end{bmatrix}\bar{q}S + \sum_{i=1}^{n} T_i\left(-\cos\phi_{T_i}\cos\psi_{T_i}\sin\alpha + \sin\phi_{T_i}\cos\alpha\right) \tag{A.11.57c}$$

$$0 = L_A$$

$$0 = \begin{bmatrix} C_{l_0} + C_{l_p} \frac{pb}{2V} \\ + C_{l_{i_{LoCE}}} i_{LoCE} + C_{l_{\delta_{LoCE}}} \delta_{LoCE} \\ + C_{l_{i_{DiCE}}} i_{DiCE} + C_{l_{\delta_{DiCE}}} \delta_{DiCE} \\ + C_{l_{i_{LaCE}}} i_{LaCE} + C_{l_{\delta_{LaCE}}} \delta_{LaCE} \end{bmatrix} \bar{q}Sb \tag{A.11.58a}$$

$$I_{zx}p^2 - ph'_z = M_A + M_T$$

$$I_{zx}p^2 - ph'_z = \begin{bmatrix} C_{m_0} + C_{m_\alpha}\alpha + C_{m_p} \frac{p\bar{c}}{2V} \\ + C_{m_{i_{LoCE}}} i_{LoCE} + C_{m_{\delta_{LoCE}}} \delta_{LoCE} \\ + C_{m_{i_{DiCE}}} i_{DiCE} + C_{m_{\delta_{DiCE}}} \delta_{DiCE} \\ + C_{m_{i_{LaCE}}} i_{LaCE} + C_{m_{\delta_{LaCE}}} \delta_{LaCE} \end{bmatrix} \bar{q}S\bar{c} + \sum_{i=1}^{n} T_i\left(\cos\phi_{T_i}\ \cos\psi_{T_i} z_T - \sin\phi_{T_i} x_T\right) \tag{A.11.58b}$$

$$-I_{xy}p^2 + ph'_y = N_A$$

$$-I_{xy}p^2 + ph'_y = \begin{bmatrix} C_{n_0} + C_{n_p} \frac{pb}{2V} \\ + C_{n_{i_{LoCE}}} i_{LoCE} + C_{n_{\delta_{LoCE}}} \delta_{LoCE} \\ + C_{n_{i_{DiCE}}} i_{DiCE} + C_{n_{\delta_{DiCE}}} \delta_{DiCE} \\ + C_{n_{i_{LaCE}}} i_{LaCE} + C_{n_{\delta_{LaCE}}} \delta_{LaCE} \end{bmatrix} \bar{q}Sb \tag{A.11.58c}$$

The following estimates the time to bank, by considering a single-degree of freedom approximation given with Eq. (A.11.56a) as

$$L_A = I_x\dot{p}$$

$$\left[C_{l_0} + C_{l_p}\frac{pb}{2V} + \sum\left(C_{l_{i_{LaCE}}} i_{LaCE} + C_{l_{\delta_{LaCE}}} \delta_{LaCE}\right)\right] qSb = I_x\dot{p} \tag{A.11.59}$$

Solved for the roll acceleration, $\dot{p}$, we receive

$$\dot{p} = \frac{qSb}{I_x}\left[C_{l_0} + C_{l_p}\frac{pb}{2V} + \sum\left(C_{l_{i_{LaCE}}} i_{LaCE} + C_{l_{\delta_{LaCE}}} \delta_{LaCE}\right)\right] \tag{A.11.60}$$

The coupling effects with the LoCE and DiCE are neglected and the following definitions apply:

$$C_{l_0} \text{ is defined at } \alpha = 0, \text{not at } C_L = 0$$
$$C_{l_p} = \frac{\partial C_l}{\partial\left(\frac{pb}{2V}\right)}$$
$$C_{l_{i_{LaCE}}} = \frac{\partial C_l}{\partial i_{LaCE}}$$
$$C_{l_{\delta_{LaCE}}} = \frac{\partial C_l}{\partial \delta_{LaCE}}$$

Equation (A.11.60) illustrates the dependence of roll performance on the roll control derivative coefficient, the roll damping term, and moment of inertia. The term C_{l_0} should normally be zero. Assuming the initial conditions

$$p(0) = 0$$
$$\phi(0) = 0$$

an analytical expression for the roll rate can be obtained when considering the most common conditions in (A.11.60), which represents a linear first order equation to which an analytical solution is available:

$$\dot{p} = \frac{qSb}{I_x}\left[C_{l_0} + C_{l_p}\frac{pb}{2V} + C_{l_{\delta_{LaCE}}}\delta_{LaCE}\right] \tag{A.11.61}$$

The solution to this differential equation is

$$p = \dot{\phi}(t) = -\frac{2V}{b}\frac{C_{l_{\delta_{LaCE}}}\delta_{LaCE}}{C_{l_p}}\left(1 - e^{L_p t}\right) \tag{A.11.62}$$

where the roll angular acceleration per unit roll rate is given with

$$L_p = \frac{qSb^2}{2VI_x}C_{l_p} \tag{A.11.63}$$

and where $-L_p^{-1}$ is the roll time constant. Note from (A.11.62), that the maximum steady state roll rate for the particular magnitude of LaCE step input is given with

$$p = -\frac{2V}{b}\frac{C_{l_{\delta_{LaCE}}}\delta_{LaCE}}{C_{l_p}} \tag{A.11.64a}$$

$$\frac{pb}{2V} = -\frac{C_{l_{\delta_{LaCE}}}\delta_{LaCE}}{C_{l_p}} \tag{A.11.64b}$$

where (A.11.64b) is the roll helix angle. At this stage it is possible, to compare the results obtained with the above 1-DOF simulation and the 6-DOF simulation set up before. Finally, the bank angle is obtained by integrating the roll rate given with (A.11.64a):

$$\int p dt = \phi(t) = -\frac{2V}{b}\frac{C_{l_{\delta_{LaCE}}}\delta_{LaCE}}{C_{l_p}}\left[t + \frac{1}{L_p}\left(1 - e^{L_p t}\right)\right] \qquad \text{(A.11.65)}$$

The bank angle response given with (A.11.65) consists of a first term varying linearly with time, and a second term varying exponentially with time. The second term vanishes for infinite time, resulting in an overall linear bank angle response relationship (constant roll rate manoeuver) with time.

A.11.6 Quasi-Steady, Straight Take-Off Rotation Manoeuver

The instant during the take-off rotation manoeuver considered in the present context is the moment, when nose-wheel lift-off is commanded, see Fig. A.11.5.

This special DCFC is defined by the following:

1. The forward speed of the aircraft is the lift-off speed, $V = V_{lift-off}$.
2. The LoCE has commanded nose wheel lift-off.
3. The pitching moment generated by the LoCE just balances the vehicle with the nose gear fully extended (no weight on the nose gear, no contact of the nose gear with the runway).
4. The rotational speed is still zero, $q = 0$, but the angular acceleration, $\ddot{\theta}$, is maximum.
5. The load factor is still one, $n = 1$.
6. The aircraft is in a forward acceleration process, $\dot{V} \neq 0$.
7. Aircraft with a tricycle landing gear and taildraggers are considered.
8. The take-off run along a horizontal runway is considered only, $\gamma = 0$.
9. The c.g. is assumed to lie laterally between the main gear contact points; thus the aircraft weight is equally distributed between these two contact points.
10. A take-off is considered with $\phi = 0$ but $\beta \geq \leq 0$.
11. The angular velocity vector of the aircraft relative to inertial space is $\vec{\omega} = [0 \quad 0 \quad 0]^T$.
12. The orientation angle rates of change are $\dot{\phi} = \dot{\theta} = \dot{\psi} = 0$.

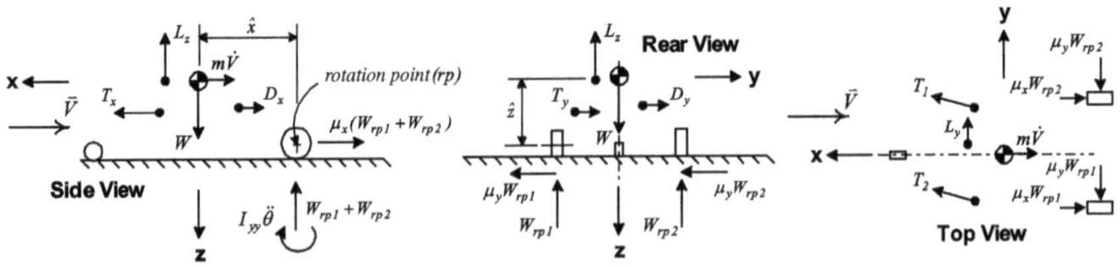

Fig. A.11.5 Take-off rotation 'snap-shot'

The primary interest lies in determining the control power required to generate a predefined angular acceleration $\ddot{\theta}$ about the main gear axel (no moments are transmitted through the axel) during the beginning of the take-off rotation manoeuver where $\dot{\theta} = 0$. Figure A.11.5 illustrates the instant during the take-off rotation manoeuver considered.

With $\phi = 0$ and the c.g. placed on the centre-line of the aircraft, we assume each pair of reaction forces (normal- and friction force) of the main gear to be of equal size

$$\left|\vec{W}_{rp1}\right| = \left|\vec{W}_{rp2}\right| \tag{A.11.66a}$$

$$\left|\mu_x \vec{W}_{rp1}\right| = \left|\mu_x \vec{W}_{rp2}\right| \tag{A.11.66b}$$

$$\left|\mu_y \vec{W}_{rp1}\right| = \left|\mu_y \vec{W}_{rp2}\right| \tag{A.11.66c}$$

The forces described with (A.11.66c) aim to arrest any lateral movement of the aircraft as induced by aerodynamic- or thrust asymmetry conditions. The ground reaction forces in x- and y-direction are friction forces see (A.11.66b) and (A.11.66c), whereby the contact forces in z-direction are normal forces at the rotation point between the ground and the main gear, see (A.11.66a). Overall, the forces described with (A.11.66a–c) are a function of aircraft weight, W_{rp1} and W_{rp2}, themselve a function of aircraft mass, aerodynamic forces, and thrust forces.

The tire-to-runway friction coefficient μ has the following values:

$$\mu_x < 1.0 \tag{A.11.67a}$$

$$\mu_y < 1.0 \text{ (in case sidedrift is taken into account)} \tag{A.11.67b}$$

$$\mu_x \ll \mu_y \tag{A.11.67c}$$

The forces in y-direction pose a specific problem. The tire-to-runway friction coefficient μ_y is not existant for the static problem where no sideward drift exists. For this condition, the y-equation becomes trivial and the algorithm reduces to a 5-DOF problem, since the lateral force equilibrium is assumed to be satisfied at all times.

$$\sum Y_{rp_i} = \sum Y_{external_i} = L_y + T_y \tag{A.11.68}$$

However, when modelling sideward drift, the take-off flight case becomes a 6-DOF problem with additional modelling complexity, since detailed information of the landing gear dynamics is required. Consequently, it has been decided to consider the take-off rotation flight condition without sideward drift.[11]

Equations (A.9.55a–c) to (A.9.59a–c) become the *Linear 5-DOF Take-Off Rotation Manoeuver* written in stability axes (a thorough check of the units in Eqs. (A.11.69a, b) and (A.11.70a–c) has been performed).

$$m\dot{V} = X_A - X_{rp} + X_T$$

$$m\dot{V} = -D - \mu_x W_{rp} + X_T$$

$$m\dot{V} + \mu_x W_{rp} = -\begin{bmatrix} C_{D_0} + C_{D_\alpha}\alpha \\ + C_{D_{i_{LoCE}}} i_{LoCE} + C_{D_{\delta_{LoCE}}} \delta_{LoCE} \\ + C_{D_{i_{DiCE}}} i_{DiCE} + C_{D_{\delta_{DiCE}}} \delta_{DiCE} \\ + C_{D_{i_{LaCE}}} i_{LaCE} + C_{D_{\delta_{LaCE}}} \delta_{LaCE} \end{bmatrix} \bar{q}S \qquad \text{(A.11.69a)}$$

$$+ \sum_{i=1}^{n} T_i\left(\cos\phi_{T_i}\cos\psi_{T_i}\cos\alpha + \sin\phi_{T_i}\sin\alpha\right)$$

$$-mg = -L - W_{rp1} - W_{rp2} + Z_T$$

$$mg - W_{rp} = L - Z_T$$

$$mg - W_{rp} = \begin{bmatrix} C_{L_0} + C_{L_\alpha}\alpha \\ + C_{L_{i_{LoCE}}} i_{LoCE} + C_{L_{\delta_{LoCE}}} \delta_{LoCE} \\ + C_{L_{i_{DiCE}}} i_{DiCE} + C_{L_{\delta_{DiCE}}} \delta_{DiCE} \\ + C_{L_{i_{LaCE}}} i_{LaCE} + C_{L_{\delta_{LaCE}}} \delta_{LaCE} \end{bmatrix} \bar{q}S \qquad \text{(A.11.69b)}$$

$$- \sum_{i=1}^{n} T_i\left(-\cos\phi_{T_i}\cos\psi_{T_i}\sin\alpha + \sin\phi_{T_i}\cos\alpha\right)$$

$$-I_{xy}\dot{q} = L_A$$

$$I_{xy}\ddot{\theta} = -\begin{bmatrix} C_{l_0} \\ + C_{l_{i_{LoCE}}} i_{LoCE} + C_{l_{\delta_{LoCE}}} \delta_{LoCE} \\ + C_{l_{i_{DiCE}}} i_{DiCE} + C_{l_{\delta_{DiCE}}} \delta_{DiCE} \\ + C_{l_{i_{LaCE}}} i_{LaCE} + C_{l_{\delta_{LaCE}}} \delta_{LaCE} \end{bmatrix} \bar{q}Sb \qquad \text{(A.11.70a)}$$

[11]The problem of generating reasonably accurate analytical ground handling results is a challenging task even during *detail* design.

$$I_y\dot{q} = M_A + M_T - M_{rp}$$

$$I_y\ddot{\theta} + W_{rp}(\mu_x\hat{z} + \hat{x}) = \begin{bmatrix} C_{m_0} + C_{m_\alpha}\alpha + C_{m_q}\frac{q\bar{c}}{2V} \\ + C_{m_{i_{LoCE}}} i_{LoCE} + C_{m_{\delta_{LoCE}}} \delta_{LoCE} \\ + C_{m_{i_{DiCE}}} i_{DiCE} + C_{m_{\delta_{DiCE}}} \delta_{DiCE} \\ + C_{m_{i_{LaCE}}} i_{LaCE} + C_{m_{\delta_{LaCE}}} \delta_{LaCE} \end{bmatrix} \bar{q}S\bar{c} \tag{A.11.70b}$$

$$+ \sum_{i=1}^{n} T_i\left(\cos\phi_{T_i}\ \cos\psi_{T_i} z_T - \ \sin\phi_{T_i} x_T\right)$$

$$I_{xy}\dot{q} = -N_A$$

$$I_{xy}\dot{q} = -\begin{bmatrix} C_{n_0} + C_{n_\beta}\beta \\ + C_{n_{i_{LoCE}}} i_{LoCE} + C_{n_{\delta_{LoCE}}} \delta_{LoCE} \\ + C_{n_{i_{DiCE}}} i_{DiCE} + C_{n_{\delta_{DiCE}}} \delta_{DiCE} \\ + C_{n_{i_{LaCE}}} i_{LaCE} + C_{n_{\delta_{LaCE}}} \delta_{LaCE} \end{bmatrix} \bar{q}Sb \tag{A.11.70c}$$

References

1. Etkin, B. and Reid, L.D., *"Dynamics of Flight – Stability and Control,"* Third Edition, John Wiley & Sons, Inc., 1996.
2. Brüning, G., Hafer, X., and Sachs, G., *"Flugleistungen,"* Third Edition, Springer, 1993.

Author Index

B. Chudoba, *Stability and Control of Conventional and Unconventional Aerospace Vehicle Configurations*, Springer Aerospace Technology,
https://doi.org/10.1007/978-3-030-16856-8

Subject Index

B. Chudoba, *Stability and Control of Conventional and Unconventional Aerospace Vehicle Configurations*, Springer Aerospace Technology,
https://doi.org/10.1007/978-3-030-16856-8

B

C

Q

R

MIX
Papier aus verantwortungsvollen Quellen
Paper from responsible sources
FSC® C105338

If you have any concerns about our products,
you can contact us on
ProductSafety@springernature.com

In case Publisher is established outside the EU,
the EU authorized representative is:
Springer Nature Customer Service Center GmbH
Europaplatz 3, 69115 Heidelberg, Germany

Printed by Libri Plureos GmbH
in Hamburg, Germany